Natural Resources, Tourism and Community Livelihoods in Southern Africa

This book examines the connections between natural resources, tourism and community livelihood practices in Southern Africa, highlighting the successes and constraints experienced over the last 50 years.

Questioning how natural resources, tourism and community livelihoods relations can positively contribute towards development efforts, this book adopts an interdisciplinary approach to understand socio-ecological systems that characterize the dynamics for sustainable development. It explores the history of conservation and natural resource management in Southern Africa and traces the development and growth of nature-based tourism. Boasting a wide range of tourism landscapes, including national parks, wetlands, forests and oceans, the book draws on case studies from a variety of Southern African countries, including Botswana, Namibia and South Africa, and considers the political challenges for implementing policies and practices. Furthermore, it analyses broader issues such as the impact of climate change, human–wildlife co-existence and resulting conflicts, poor access to funding and poverty in local communities. The book argues that the links between conservation and livelihoods can be best understood by considering the different approaches to reconciling the demands of conservation and livelihoods that have evolved over the past decades.

Containing contributions from natural and social sciences the book provides guidance for practitioners and policymakers to continue to shape policies and practices that are in line with the key tenets of sustainable development. It will also be of great interest to students and scholars researching Southern Africa, sustainable tourism and conservation.

Moren T. Stone is Senior Lecturer in the Department of Environmental Science, University of Botswana.

Monkgogi Lenao is Senior Lecturer in the Department of Tourism and Hospitality, University of Botswana.

Naomi Moswete is Senior Lecturer in the Department of Environmental Science, University of Botswana.

Routledge Studies in Conservation and the Environment

This series includes a wide range of inter-disciplinary approaches to conservation and the environment, integrating perspectives from both social and natural sciences. Topics include, but are not limited to, development, environmental policy and politics, ecosystem change, natural resources (including land, water, oceans and forests), security, wildlife, protected areas, tourism, human-wildlife conflict, agriculture, economics, law and climate change.

Natural Resources, Tourism and Community Livelihoods in Southern Africa
Challenges of Sustainable Development
Edited by Moren T. Stone, Monkgogi Lenao and Naomi Moswete

For more information about this series, please visit: www.routledge.com/series/RSICE

Natural Resources, Tourism and Community Livelihoods in Southern Africa

Challenges of Sustainable Development

Edited by
Moren T. Stone, Monkgogi Lenao and Naomi Moswete

Routledge
Taylor & Francis Group
LONDON AND NEW YORK

First published 2020
by Routledge
2 Park Square, Milton Park, Abingdon, Oxon OX14 4RN

and by Routledge
52 Vanderbilt Avenue, New York, NY 10017

Routledge is an imprint of the Taylor & Francis Group, an informa business

First issued in paperback 2021

British Library Cataloguing-in-Publication Data
A catalogue record for this book is available from the British Library

Library of Congress Cataloging-in-Publication Data
A catalog record for this book has been requested

ISBN: 978-0-367-25412-4 (hbk)
ISBN: 978-1-03-208517-3 (pbk)
ISBN: 978-0-429-28942-2 (ebk)

Typeset in Bembo
by Apex CoVantage, LLC

Contents

List of illustrations viii
List of contributors x
Acknowledgements xiii
Abbreviations xiv

1 Contextualizing and conceptualizing relationships in natural resources, tourism and community livelihoods: an introduction overview 1
NAOMI N. MOSWETE, MONKGOGI LENAO AND MOREN T. STONE

PART I
Natural resources management and conservation in Southern Africa: history and trajectories 9

2 Historical evolution of conservation and tourism in Southern Africa: the case of Botswana 11
BONGANI GLORIOUS GUMBO

3 Nature tourism, wildlife resources and community-based conservation: the case study of Malawi 26
JAMES MALITONI CHILEMBWE

4 The context and future of tourism in Africa's national parks: could privatization within protected areas be the panacea? 38
CLEOPAS NJEREKAI

5 A review of community social upliftment practices by tourism multinational companies in Botswana 52
WAME L HAMBIRA

6 **Navigating community conservancies and institutional complexities in Namibia** 64
PAUL HEBINCK, RICHARD DIMBA KIAKA AND RODGERS LUBILO

7 **Fourteen years of tourism and climate change research in Southern Africa: lessons on sustainability under conditions of global change** 78
GIJSBERT HOOGENDOORN AND JENNIFER M. FITCHETT

PART II
Natural resource-based tourism development and growth in Southern Africa: policy issues, challenges and practices 91

8 **The analysis of conflict and coexistence of traditional and contemporary land uses** 93
PATRICIA K. MOGOMOTSI, MELVILLE SAAYMAN AND ANDREA SAAYMAN

9 **Local people's perspectives on wildlife conservation, ecotourism and community livelihoods: a case study of Lusaka National Park, Zambia** 108
VINCENT R. NYIRENDA, CASTRO MILIMO AND NGAWO NAMUKONDE

10 **Tourism and poverty alleviation in the Global South: emerging corporate social responsibility in the Namibian nature-based tourism industry** 123
MARY-ELLEN KIMARO AND JARKKO SAARINEN

11 **Revisiting devolution in community-based natural resources management in Zimbabwe: towards inclusive governance approaches** 143
JONES MUDIMU MUZIRAMBI, REGIS MUSAVENGANE AND KEVIN MEARNS

PART III
Conservation and tourism development debates in Southern Africa: origins, narratives and progress 159

12 **The economics and governance of the wildlife economy in drylands in Southern Africa** 161
BRIAN A. CHILD

13 Role of forest resources in local community livelihoods: implications for conservation of Chobe Forest Reserve, Botswana 176
JOYCE LEPETU AND HESEKIA GAREKAE

14 An assessment of supply-side factors and ecotourism in Mauritius 190
SEETANAH BOOPEN AND SANNASSEE R.V

15 Investigating sustainable development goals, livelihoods and tourism development in a rural protected area: the iSimangaliso Wetland Park, South Africa 205
FELICITE A FAIRER-WESSELS

PART IV
Southern Africa in the world: cases and lessons 219

16 Natural resources, sustainable tourism development and community livelihoods relationships: a comparison between Botswana and the USA 221
MOREN T. STONE, GYAN P. NYAUPANE, DALLEN J. TIMOTHY AND LESEGO S. STONE

17 Economic assessment of tourism-based livelihoods for sustainable development: a case of handicrafts in Southern and Eastern Africa 235
GAGOITSEOPE MMOPELWA AND LESEGO MACKENZIE

18 Comparative evaluation of visitors' perceptions of park characteristics in Southern Africa 254
BRIJESH THAPA

PART V
Conclusion 267

19 Conclusion: natural resources, tourism and community livelihoods relationships: contemplating the future 269
MONKGOGI LENAO, NAOMI MOSWETE AND MOREN T. STONE

Index 277

Illustrations

Figures

4.1	The Solimar International Integrated Development model	46
6.1	Situational map ǂKhoadi ǁHôas Conservancy	66
6.2	Situational map Wuparo Conservancy	70
8.1	The locations of the four study sites	94
8.2	CBPP fences in the Okavango Delta	97
8.3	Study areas and respondents' ploughing fields (*Masimo*)	98
8.4	Reasons for not getting compensation	101
8.5	Reasons for not reporting livestock predation	104
9.1	Lusaka National Park and its surrounding areas, Zambia	110
9.2	Relationship between level of education attained, gender, marital status and occupation	115
10.1	Africa's corporate social responsibility pyramid	127
10.2	An integrative framework for anti-poverty tourism research	129
13.1	Map of Chobe District depicting the study areas	179
16.1	CECT villages	225
16.2	Appreciative inquiry process	228
17.1	Relationship between vulnerability context, livelihood capitals, handicraft sustainability and human wellbeing	238
18.1	Three national parks in Southern Africa	256

Tables

1.1	Positive and negative socio-economic and environmental impacts of tourism	2
3.1	Characteristics of respondents	31
3.2	Showing summary of emerged themes	32
4.1	The key tourism and conservation activities pursued by Africa's national parks	39
8.1	Details of the key informants	95
8.2	Crop-raiding experiences and reporting	99
8.3	Reasons for not reporting crop raiding	100

9.1 Major predictors of perceptions among Shantumbu people 114
10.1 Mapping the CSR discourses 128
10.2 Conceptual framework for poverty alleviation through CSR towards responsible tourism 131
10.3 Poverty alleviation through CSR towards responsible tourism 135
10.4 Company A dominant CSR practices 136
10.5 Company B dominant CSR practices 137
10.6 Percentages of those employed at different levels and their educational attainment 138
13.1 Background characteristics of the respondents 180
13.2 Mean annual value (in pula currency) of forest products per year per household 182
14.1 Key tourism indicators 192
16.1 Population by State and County surrounding GSENM 224
16.2 Chobe Enclave Conservation Trust (CECT) study area population 226
16.3 Highly exploited wild plant species 230
16.4 Wild plants species less frequently used 231
17.1 Annual exchange rates between US$ and Eastern and Southern African countries (2016) 240
17.2 Returns to labour for basket and non-craft enterprises in Botswana (2016 prices) 246
17.3 Summary of financial and economic viability of ten craft enterprises in Ngamiland, Botswana (2016 prices) 247
18.1 Quality of park characteristics at two national parks in Zambia 259
18.2 Quality of park characteristics at three national parks 260
18.3 Type of wildlife observed at three national parks 261

Contributors

Brian Child received a BSc (Hons) in Agricultural Economics from the University of Zimbabwe and a DPhil from Oxford University, where he was a Rhodes Scholar.

James Malitoni Chilembwe is reading towards a PhD in Tourism Management at Glasgow Caledonian University (GCU), Scotland, UK.

Felicite A Fairer-Wessels is a senior lecturer at the Department of Marketing Management, Division of Tourism Management, University of Pretoria, South Africa.

Jennifer M. Fitchett is a senior lecturer at the School of Geography, Archaeology and Environmental Studies, University of the Witwatersrand, South Africa.

Hesekia Garekae holds an MPhil in Natural Resource Management from Okavango Research Institute, University of Botswana, with a bias towards forests, livelihoods and governance. He is currently a Ph.D. candidate with the Department of Environmental Science, Rhodes University, South Africa.

Bongani Glorious Gumbo is a senior lecturer in the Department of History, University of Botswana.

Wame Lucretia Hambira is Senior Research Scholar with the Okavango Research Institute of the University of Botswana.

Paul Hebinck is Associate Professor in the Sociology of Development and Change group at Wageningen University, The Netherlands, and Adjunct Professor at the University of Fort Hare, South Africa.

Jona Heita holds an MA Degree in Culture and Environment in Africa from the University of Cologne, Germany. He is currently pursuing a PhD in Geography.

Gijsbert Hoogendoorn is Associate Professor at the Department of Geography, Environmental Management and Energy Studies, University of Johannesburg, South Africa.

Richard Dimba Kiaka is a post-doctoral researcher with the LINGS project (Local Institutions in Global Societies) at the Institute of Social and Cultural Anthropology at the University of Hamburg, Germany.

Mary-Ellen Kimaro is Lecturer of Tourism Studies in the Department of Geography, History and Environmental Studies at the University of Namibia.

Monkgogi Lenao is a senior lecturer in the Department of Tourism and Hospitality at the University of Botswana.

Joyce Lepetu is a senior lecturer at Botswana University of Agriculture and Natural Resources (BUAN). She has a PhD in Forest Sciences (2007), University of Florida, USA.

Rodgers Lubilo is a PhD holder in Development Sociology from Wageningen University, Netherlands. He is currently heading the North Luangwa Ecosystem Project, Zambia, for the Frankfurt Zoological Society.

Lesego Mackenzie is a lecturer in the Department of Environmental Science at the University of Botswana.

Kevin Mearns works at University of South Africa, Department of Environmental Sciences, College of Agriculture and Environmental Sciences, Florida Campus, South Africa.

Castro Milimo is a BSc holder in Wildlife Management, Department of Zoology and Aquatic Sciences, School of Natural Resources, The Copperbelt University, Kitwe, Zambia.

Gagoitseope Mmopelwa is Associate Professor of Environment and Development at the University of Botswana.

Patricia Kefilwe Mogomotsi is Senior Research Fellow and Coordinator of Training at the Okavango Research Institute, Botswana.

Naomi Moswete (PhD) is Senior Lecturer of Human Geography and Tourism Science at the University of Botswana.

Regis Musavengane has a PhD in Geography and Environmental Studies from the Witwatersrand University, Geography, Archeology and Environmental Studies School. He currently holds a Post-Doctorate Research Fellow at School of Tourism and Hospitality in the University of Johannesburg, South Africa.

Jones Mudimu Muzirambi is a PhD Alumni of the University of South Africa, as well as an Educator/Researcher in Mpumalanga Department of Education, Republic of South Africa.

Ngawo Namukonde is a lecturer in the Department of Zoology and Aquatic Sciences, School of Natural Resources, Copperbelt University, Zambia.

Cleopas Njerekai is a final year PhD student with the Midlands State University in Zimbabwe and is also a lecturer in the Department of Tourism at the same university.

Gyan P. Nyaupane is Professor and Graduate Program Director at Arizona State University (ASU), Tempe, Arizona, USA.

Vincent R. Nyirenda is a senior lecturer under the Department of Zoology and Aquatic Sciences, School of Natural Resources, Copperbelt University, Zambia.

Jarkko Saarinen is Professor of Geography at the University of Oulu, Finland and School of Tourism and Hospitality, and University of Johannesburg, Auckland Park, Johannesburg, South Africa.

Andrea Saayman is Professor of Economics and a researcher in the research unit: Tourism Research in Economics, Environs and Society (TREES) at North-West University, South Africa.

Melville Saayman is Professor in Tourism Management and Director of the Research Unit: Tourism Research in Economics, Environs and Society (TREES) at North-West University, South Africa.

R.V. Sannassee is a professor in the Department of Finance at the University of Mauritius, Mauritius.

Boopen Seetanah is Associate Professor in Economics and Finance and the Faculty Research Advisor at the Faculty of Law and Management of the University of Mauritius, Mauritius.

Lesego S. Stone is a research scholar at the Okavango Research Institute in the University of Botswana.

Moren T. Stone is Senior Lecturer of Environmental Science and Tourism at the University of Botswana.

Brijesh Thapa is a US Fulbright Senior Specialist and Professor in the Department of Tourism, Recreation and Sport Management at the University of Florida, USA.

Dallen J. Timothy is Professor of Community Resources and Development at Arizona State University and Senior Sustainability Scientist at the Julie Ann Wrigley Global Institute of Sustainability, USA.

Acknowledgements

The Editors of the book would like to express their deep gratefulness to the book chapters' contributors who submitted wide-ranging, rich and insightful works to make this volume possible. Their received chapters demonstrated in many ways how diverse the issues are in view of natural resources, tourism and community livelihoods relationship dynamics in Southern Africa, hence questioning the achievement of sustainable development in the region. We acknowledge their patience throughout the project processes whereby they persistently worked on the reviews on time to improve their chapters. We also recognise the efforts of those whose chapters could not be accepted. Acknowledgements are also conveyed to the Routledge's Taylor & Francis advisory team that guided us throughout – from the commencement of the book proposal to its completion – the likes of Hannah Ferguson and Amy Johnston. We are also indebted to thank the reviewers who reviewed our book proposal and individual chapters.

Finally, we thank our families, friends and colleagues for the support and words of inspiration that energised us to make sure that we execute this project within the set timelines.

Abbreviations

AA	Appropriate Authority
AHRIM	Association des Hôteliers et Restaurateurs de l'île Maurice
AI	Appreciative Inquiry
AMCEN	African Ministerial Conference on Environment
APT	Anti-Poverty Tourism
APU	Anti-Poaching Unit
BCC	Botswana Christian Council
BRDC	Binga Rural District Council
BSE	Botswana Stock Exchange
BWP	Botswana Pula
CAMPFIRE	Communal Areas Management Programme for Indigenous Resources
CBC	Community-Based Conservation
CBNRM	Community Based Natural Resources Management
CBOs	Community Based Organisations
CBPP	Contagious Bovine Pleuro-Pneumonia
CBR	Benefit-Cost Ratio
CBT	Community-Based Tourism
CECT	Chobe Enclave Conservation Trust
CFI	Corporate Finance Institute
CFR	Chobe Forest Reserve
CHA	Controlled Hunting Areas
CHAs	Controlled Hunting Areas
CI	Conservation International
CITES	Convention on International Trade in Endangered Species
CKGR	Central Kalahari Game Reserve
CNP	Chobe National Park
CPIs	Consumer Price Indices
CPR	Common-Pool Resource
CSI	Corporate Social Investment
CSR	Corporate Social Responsibility
DAFF	Department of Agriculture, Forestry and Fisheries
DAHP	Department of Animal Health and Production

DFRR	Department of Forestry and Range Resources
DWNP	Department of Wildlife and National Parks
EIA	Environmental Impact Assessment
EU	European Union
FGD	Focused Group Discussion
GCNS	Global Compact Network Spain
GDP	Gross Development Product
GRI	Global Reporting Initiative
GSENM	Grand Staircase-Escalante National Monument
HATAB	Hotel and Tourism Association of Botswana
HDS	Hotel Development Strategy
ICA	Intensive Conservation Areas
ICT	Information, Communication Technology
IPCC	Inter-governmental Panel on Climate Change
IPZ	Intensive Protection Zone
IRDNC	Integrated Rural Development and Nature Conservation
IUCN	International Union for the Conservation of Nature
KAP	Knowledge, Attitudes and Practices
KAZA TFCA	Kavango-Zambezi Transfrontier Conservation Area
KNP	Kafue National Park
Ksh	Kenya Shilling
KZN	KwaZulu Natal
LINGS	Local Institutions in Globalized Societies
LNP	Lusaka National Park
MET	Ministry of Environment and Tourism
MID	Mauritius Ile Durable
MIKE	Monitoring the Illegal Killing of Elephants
MNCs	Multinational National Companies
N$	Namibian Dollar
NACSO	Namibian Association of CBNRM Support Organisations
NGO	Non-Governmental Organisation
NPV	Net Present Value
NTDP	National Tourism Development Plan
NTFPs	Non-Timber Forest Products
ODMP	Okavango Delta Management Plan
PAC	Problem Animal Control
PPG	Planning Policy Guidance
PPP	Public-Private-Partnership
RDC	Rural Development Council
RDCs	Rural District Councils
SANParks	South African National Parks
SDG	Sustainable Development Goal
SIDS	Small Island Development States
SLF	Sustainable Livelihood Framework
SME	Small and Medium Enterprise

SNLP	South Luangwa National Park
SPSS	Statistical Package for Social Sciences
SUCCESS	Sustainable Coastal Communities and Ecosystems
TA	Traditional Authority
TFCAs	Trans-frontier Conservation Areas
TOCADI	Trust for the Okavango Cultural and Development Initiatives
TZS	Tanzanian Shilling
UNCTAD	United Nations Conference on Trade and Development
UNDP	United Nations Development Programme
UNDP	United Nations Development Programme
UNEP	United Nations Environment Program
UNESCO	United Nations Educational, Scientific and Cultural Organization
UN-SDGs	United Nations Sustainable Development Goals
UNWTO	United Nations World Tourism Organisation
US$	United States Dollar
USAID	United States Agency for International Development
USA	United States of America
VDC	Village Development Committee
WECD	World Commission on Environment and Development
WMAs	Wildlife Management Areas
WWF	World Wildlife Fund
ZAR	South African Rand

1 Contextualizing and conceptualizing relationships in natural resources, tourism and community livelihoods

An introduction overview

Naomi N. Moswete, Monkgogi Lenao and Moren T. Stone

Southern Africa is endowed with natural resources that abound in diverse areas of the region. The wealth of natural resources found traversing the Southern African landscape include but are not limited to the Le Morne Brabant world heritage site in Mauritius, Lake Chilwa Wetland Biosphere Reserve of Malawi, Okavango Delta, Botswana and Victoria Falls in Zimbabwe/Zambia.

Natural resources in and outside gazetted areas and /or protected regions of the world have become an important component of the tourism sector, local communities and conservation in Southern Africa and the world. Due to the attached value and importance of natural resources (forests, wildlife) they ought to be preserved and conserved for posterity. Hitherto, protected areas as attractions for nature tourism have become vital livelihood options for communities located in the vicinity of wildlife reserves and other secure sites of universal importance, which include the iSimangaliso Wetland Park of South Africa, the Chobe National Park in Botswana and many others. Additionally, examples of distinctive natural resources from outside Africa are also discussed in this edited book, including the Grand Staircase-Escalante National Monument in the United States. In today's lingo, tourism and natural resources are interlinked as people travel the world to experience well-kept wilderness and non-traversed secrets of Southern Africa. It is against this backdrop that managing protected areas (parks and reserves) and other associated resources presents opportunities and challenges for both national and international governments alike. Opportunities in this case include protecting the natural, cultural and recreational values of national parks (conservation) and opening access to visitors and tourists to enjoy and appreciate (outdoor recreation) (Manning, 2009). In spite of vast opportunities alluded to, park authorities face management challenges which lead to reduced quality of experience due in part to trampling, littering or pollution of water sources (Ewert, Dieser, & Voight, 1999; Manning, 1999).

Introducing tourism and sustainability

According to the United Nations World Tourism organisation (UNWTO, 2007), tourism involves travel away from areas of abodes for an overnight stay for reasons that include leisure, outdoor recreation, education or relaxation. Destinations chosen by tourists include nature reserves, wildlife refuge sites, and heritage sites. Recently, tourism to natural areas has experienced an upsurge of tourists and visitors alike, thereby implying the need for new modalities of management of resources to alleviate its negative impacts on people and the environment (Moswete, Nkape, & Tseme, 2017; Richards & Hall, 2000).

Thus, the tourism industry is dependent on the natural resource base for an environment for its continued existence. Without a well-kept and clean physical environment, the tourism sector cannot be sustained. As stated by Goeldner and Ritchie (2012, p. 23), "there is no question that tourism delivers benefits, but tourism is not perfect." As well echoed, tourism industries often create negative impacts on the environment, society, culture, and sometimes even on the economy (Richards & Hall, 2000). For example, trampling is a common negative impact of tourism as it destroys ground cover, widens walkways, causes vegetation loss, trail deepening and loss of habitats, especially for burrowing animals (Manning, 2009). Creation of illegal roads and tracks by tourists' vehicles in some environmentally sensitive areas, such as the Chobe National Park (Moswete et al., 2017), Camp Xakanaka or Okavango Delta (Mbaiwa, 2003) has had negative impacts on the fragile vegetation communities of the area, reducing the picturesque quality of some parts of the delta in Botswana (Mbaiwa, 2003). There are socio-economic and environmental costs and benefits associated with sustainable tourism (see Table 1.1).

Thus, sustainable tourism has become a new method through which natural and other tourism resources could be protected (UNEP/WTO, 2005). Sustainable tourism is devoted as tourism that takes full account of its current and future economic, social and environmental impacts, addressing the needs of visitors, the industry, the environment and host communities (Holden,

Table 1.1 Positive and negative socio-economic and environmental impacts of tourism

Socio-economic benefits	*Environmental benefits*
+ Promotes a global community	+ Justifies environmental protection
+ Promotes international understanding & peace	+ Reinforces preservation of rare & endangered wildlife
+ Increases incomes	+ Protection of critical water resources
+ Spreads development	- Degrades the natural environment (landscape)
+ Provides employment opportunities	- Damage of vegetation in parks & forest areas
- Commercialises culture, traditions & the arts	- Trampling of scenic & sensitive areas by tourists
- Develops excess demands for resources	- Water pollution from resorts on the coastal areas
- Degrades the cultural environment	
- Creates misunderstanding	

Source: Compiled from (Goeldner & Ritchie, 2012; Holden, 2008; Mbaiwa, 2003)

2008; UNWTO, 2007). Therefore, the guiding principles of sustainable tourism include but are not limited to the following:

- Tourism activities and development should respect the scale, nature and character of the place in which they are sited.
- In any location, harmony must be sought between the needs of the visitor, the destination/place and the host community.
- The relationship between tourism and the environment must be managed so that the environment is sustainable in the long run.
- It must respect social and cultural traditions in the host country.
- It must be non-exploitative of the local people and ensure, as far as possible, that benefits flow to local people.

(Carter, Garrod, & Low, 2015; Holden, 2008)

Conservation, tourism and local communities: challenges for development

In Southern Africa, vast areas of land have been set aside for conservation of wildlife and other natural resources. This shows a positive move and commitment to conservation and protection of resources. Conflict over resource use arises when several groups of people compete for limited resources (Nelson, 2010). For instance, government may see a game reserve or national park as a wildlife habitat where wildlife is to be protected, whilst the local people residing in or adjacent to a protected area regard wildlife and the preserved areas as their own land where they should be allowed to hunt and gather food freely (Bolaane, 2004; Nelson, 2010). In Botswana the San of Khwai residents were relocated from Moremi Game Reserve and resettled at the northern gate of the reserve when the reserve was created in 1963 (Bolaane, 2004; Mbaiwa, 2003). The action denied these communities free access to their traditional land and resources, consequently affecting their traditional hunting and gathering lifestyles. Such actions have caused conflicts and disputes between natural resources (e.g. wildlife), resource managers and tourism developers (Nelson, 2010).

Conceptual framework

This book is guided by the principles of sustainable development and/or sustainable tourism development in assessing connections between natural resources, tourism and community livelihoods.

The starting point is that the local community should benefit through participation in nature-based tourism enterprises. These initiatives have taken different formats across the sub-region. Topical amongst which are multiple community conservation and tourism enterprise arrangements where two or more communities are expected to share in the management and benefits arising from a single initiative. The different socio-economic and ecological benefits accruing from these types of arrangements have been widely documented.

Among the key conclusions arrived at is the realisation that these partnerships have contributed towards improving both livelihoods and human–wildlife co-existence. On the other hand, nature-based tourism in Southern Africa has experienced a number of limitations and challenges over the past decades. These include political instabilities in certain countries, poor tourism infrastructure, human–wildlife conflicts, lack of institutional and human capacity, poor access to funding as well as top-down governance approaches to natural resources management. These have weakened nature-based tourism development in many natural resource-rich Southern African countries. Collective research interest in trying to link the livelihoods of people living near natural resources to the conservation of these resources has been developing over the last few years. Through these prolific discussions it has been discovered, among other things, that local communities often rely on products and services from nearby natural areas to meet their livelihood needs. Therefore, their use constitutes a demand on the biodiversity/resources of these areas while their conservation objectives, coupled with those of the state and outside groups, constitute another. The resulting conflict, compatibility or complementarity between the demands created by livelihoods activities and conservation objectives has been the focal point of much discussion and effort over the last decades. Be that as it may, efforts aimed at systematically defining and assessing the dynamics of these linkages have not been given enough attention. To this end, the actual nature of these linkages remains largely understudied and, therefore, unknown.

This edited book focuses on natural resources, tourism and community livelihoods practices pulled from some countries in the Southern African region. The book highlights successes and constraints experienced in this part of the region. The book adopts an interdisciplinary approach as it attempts to understand socio-ecological systems that characterize the dynamics of the relationships between natural resources, tourism and community livelihoods. Chapter contributors to the book are researchers in multiple disciplines in the natural and social sciences with a diversity of views; a necessary condition for the design of best practices and policies. Through this diversity in scholarship, the book is suitable for practitioners, planners, policy developers and interested stakeholders to learn, adapt and continue to review and align policies and practices that are in line with the key tenets of sustainable development. The chapters reflect what these relationships are and what has been accomplished in various geographical areas within the Southern African natural resources and tourism landscapes. The concept of relationships between conservation and livelihoods can perhaps be best understood by considering the different approaches to reconciling the demands of conservation and livelihoods that have evolved over the past years.

Chapters in this edited book consist of theories and case studies drawn from different Southern African regions, demonstrating practical cases related to natural resource conservation, tourism, local communities and sustainability. The chapters cover a variety of issues that affect natural resources use and sustainability, conservation, tourism and local communities.

Outline of the book

This book is divided into four parts: Part 1 is the introduction, which contextualizes and conceptualizes relationships in natural resources, tourism, and community livelihoods. It is in this chapter that tourism is introduced, and resource use and conservation tourism is discussed. Part 2 has six (6) chapters which discuss natural resources management and conservation in Southern Africa, including history and trajectories. For example, Gumbo talks about the historical evolution of conservation in Southern Africa, while Chilembwe, Njerekai and Hebinck discuss wildlife resources, community conservation and CBNRM with examples from Botswana, Nambia and Malawi. Hambira is concerned with the ways in which multinational tourism companies engage in local community social upliftment activities through corporate social responsibility in natural resources conservation in and around protected areas. Hoogendoorn and Fitchett bring about a new twist in which they reveal the negative impact of climate on nature-based tourism and the need to conduct more research on the relationship of tourism and climate change.

Part 3 focuses on issues related to natural resource-based tourism development and growth in Southern Africa covering policy issues, challenges and practices. There are four chapters in this part. Mogomotsi, Saayman and Saayman write about conflict and the coexistence of traditional and contemporary land uses using a case study of Botswana's Okavango Delta region. They argue that the Okavango Delta has experienced negative natural resource dynamics, such as increasing competition and conflicts over natural resources, biodiversity loss and some cases of natural resource depletion. Nyirenda, Milimo and Namukonde adopt the sustainable tourism development concept, with emphasis on its application to protected area management. The chapter discusses local people's perspectives on wildlife conservation, ecotourism and community livelihoods in Zambia. Similarly, Kimaro and Saarinen explore issues of tourism and poverty alleviation through corporate social responsibility by different tourism-related companies in Namibia. And Muzirambi, Musavengane and Mearns explore issues related to devolution of CBNRM and governance of natural resource management and communities.

Part 4 of this book consists of five (5) chapters. At most it covers topics that include conservation and tourism development debates in Southern Africa. Brian Child adopts the sustainable governance approach to write on the economics and governance of wildlife economy in drylands in Southern Africa. This chapter explains the institutional history of private conservation and CBNRM, interrogating the cost and benefits of the CBNRM approach in community conservation. Child recommends a new paradigm shift in which communities become shareholders and not stakeholders so that they can acquire full rights to own land and wildlife. Lepetu and Garekae's chapter focuses on the role of forest resources and conservation using a case study of the Chobe Forest Reserve in Botswana. This chapter emphasises the importance of ecotourism and how it can be used to manage resources. Seetanah and Sannassee's contribution is on a similar topic, with a specific focus placed on institutional arrangements and the manner in which they can either aid or impede the development of ecotourism in the sub-region. The final contribution in this part comes from Fairer-Wessels, where she discusses the relationship between sustainable development goals, livelihoods and tourism development in a rural protected area in iSimangaliso Wetland Park, South Africa. Here she decries rural poor's apparent inability to leverage on the opportunities provided by the resource base in order to lift themselves out of poverty.

Part 5 attempts to situate Southern Africa in the world regarding the tourism, conservation and livelihoods nexus. It presents cases from Southern Africa as well as lessons that may be learnt in comparison to the other regions of the world. This part has four (4) chapters. Stone et al.'s chapter focuses on natural resources, sustainable tourism development and community livelihoods relationships, making a comparison between Botswana and the United States. In the end, Stone et al. found that in both communities conflicts were reported, such as increased human – wildlife at CECT, vandalism of canyons and off-track driving at GSENM. Mmopelwa and Mackenzie's chapter focuses on the economic assessment of tourism-based livelihoods for sustainable development. Drawing from handicraft-related literature spurning Botswana, Namibia and Tanzania, they argue that communities found near tourism-based resources must reap significant benefits that have contributed to improved livelihoods. Finally, Thapa's chapter focuses on a comparison evaluation of visitors' perceptions of park characteristics in Southern Africa. In this chapter, he investigates how protected areas in the Southern African region may utilise their resource endowments (comparative advantage) to improve on their competitiveness, especially through infrastructural and other tourism-related investments.

Finally, Chapter 6 concludes the book and brings along synthesis and conclusions which are derived from all chapters or outcomes of the book. The overall conclusion of the chapter is that the many case studies presented by this book indicate that the relationships between natural resources, tourism and community livelihoods are dynamic, complex and context specific. However, it has emerged that, if natural resources (either in protected areas or buffer zones) are to remain feasible in the future, local communities must be given a greater role not only in their management, but also their livelihood issues must be sufficiently addressed. Due to the propensity of nature-based tourism opportunities in Southern Africa, several countries see nature tourism as a significant foundation of revenues, while local communities view it as a prospect to improve their livelihood conditions.

This chapter has set the scene for an edited volume that aims to contribute to the growing knowledge on natural resources, tourism and community livelihoods connections by examining how different natural resources have emerged as tourism resources, how tourism contributes to conservation and generating to livelihoods improvement or lack thereof and how these resources are governed.

References

Bolaane, M. (2004). The impact of game policy on the river San/Bushmen of Botswana. *Social Policy and Administration, 38*(4), 399–417.

Carter, C., Garrod, B., & Low, T. (Eds.). (2015). *Encyclopedia of sustainable tourism.* Wallingford: CABI.

Ewert, A. W., Dieser, R. B., & Voight, A. (1999). Conflict and the recreational experience. In E. Jackson & T. L. Burton (Eds.), *Leisure studies: Prospects for the twenty first century* (pp. 335–342). Pennsylvania: Venture Publishing Inc.

Goeldner, C., & Ritchie, J. R. B. (2012). *Tourism: Principles, practices, philosophies* (12th ed.). New York: John Wiley & Sons.

Holden, A. (2008). *Environment and tourism* (2nd ed.). London: Routledge.

Manning, R. E. (2009). *Parks and people: Managing outdoor recreation at Acacia National Park.* Hanover: University of Vermont Press.

Mbaiwa, J. (2003). The socio-economic and environmental impacts of tourism in the Okavango Delta, Northwestern Botswana. *Journal of Arid Environments, 54*(2), 447–468.

Moswete, N., Nkape, K., & Tseme, M. (2017). Wildlife tourism safaris, vehicle decongestion routes and impact mitigation at the Chobe National Park, Botswana. In L. Ismar & R. Green (Eds.), *Wildlife tourism, environmental learning and ethical encounters* (pp. 71–88). Australia: Springer.

Nelson, F. (Ed.). (2010). *Community rights, conservation & contested land: The politics of natural resource governance in Africa.* London: Earthscan.

Richards, G., & Hall, D. (2000). *Tourism and sustainable community development.* London: Routledge.

United Nations Environmental Programme (UNEP/WTO). (2005). *Making Tourism More Sustainable: A Guide for Policy Managers.* UNEP/Division of Technology, Industry and Economics. Paris and Madrid: WTO.

United Nations World Tourism Organisation (UNWTO). (2007). *UNWTO ST- EP programme: An initiative of the world tourism organisation in conjunction with the UNWTO ST- EP foundation.* Madrid, Spain.

Part I

Natural resources management and conservation in Southern Africa

History and trajectories

2 Historical evolution of conservation and tourism in Southern Africa

The case of Botswana

Bongani Glorious Gumbo

Introduction

Community conservation has been defined as political and economic ideology that works largely to eliminate government's power to influence the affairs of private businesses; neo-liberalism is faulted for advancing market forces above local interests or social justice (Leistyna, 2007, p. 97). Market forces have historically been associated with controlling the levers of economic power, thus emasculating the power of local communities. The chapter begins with an exploration of the concept and practice of conservation during pre-colonial times, interrogating socio-cultural practices in which subsistence hunting and conservation were compatible. It then historicises the coming of Europeans to Botswana in the 19th century and the genesis of fortress conservation or protected areas, arguing that parks were a colonial landscaping with its roots in the West. The establishment of the then Chobe Game Reserve in 1962 and that of the Moremi Game Reserve in 1963 represented the actualisation of the philosophy of fortress conservation, keeping away local communities who were viewed as a threat to wildlife (Bolaane, 2004). As it unfolds, the chapter considers in detail, the factors that informed the establishment of these game reserves. With independence in 1966, it shall be seen that the post-colonial government introduced, albeit painstakingly slowly, transformative strategies, namely, participatory approaches that sought to integrate communities in the management of wildlife resources as well as in the tourism sector. Numerous as they are, definitions of the term "conservation" converge on the notion of "the maintenance of genetic, species, and ecosystem diversity in the natural abundance in which they occur" (Hammes, 2007, p. 178). Research on conservation of wildlife and biodiversity, and tourism development in Southern Africa, especially in Botswana, has largely been dominated by narratives on thoughts, values and practices of modern Western conservation trajectories (Collingwood, 1997). These studies have, until fairly recently, lacked a historical dimension to conservation as practised by local communities prior to colonial establishment in Botswana, creating an impression that wildlife conservation only began with the arrival of Europeans in the region in the 19th century (Matlhare, 1997). Similarly, debates on the tourism sector have tended to emphasise contemporary developments such as policy matters, socio-economic effects and emergent global trends, ignoring the early history of the establishment of the sector and how it has evolved over time (Mbaiwa & Darkoh, 2008; Moswete & Mavondo, 2003; Saarinen & Manwa, 2008). This chapter is therefore a historical analysis of the complex evolution of wildlife conservation, the thinking that shaped approaches to conservation at different times from pre-colonial, across the colonial and up to the post-colonial times in Botswana. The chapter also traces the historical developments of the tourism industry over the same period.

Conceptual framework

Even though this chapter is primarily historical, its thesis is anchored on the sustainable development framework. Hitchcock (1990, p. 161), for instance, states that "tourism has been recommended as a strategy for sustainable development by numerous governments and international development agencies." A growing body of literature has attempted to define the concept of sustainable development and develop appropriate policy frameworks. Bearing a multidisciplinary "makeup" of socio-economic and environmental dimensions, the concept of sustainable development emphasises the need for appropriate management of and access to resources for both the present and future generations (Hilty, Sefert & Treibert, 2005, p. vii). The most widely accepted definition of sustainable development was proffered by the World Commission on Environment and Development (WCED) in 1987, as a "pattern of development that meets the needs of the present without compromising the ability of future generations to meet their own needs" WCED, 1987, p. 15). This definition is also shared by Kojo and Moyo (2008) who view it as responsive to social aspects, arguing that development ought to be cognisant of current and future generations. Invoking a political economy framework, the duo argue that genuine sustainable development can be achieved through embracing a strong ethical case for social justice and equitable power relations that recognise obligations towards livelihoods, the environment and economic development. They are explicit in emphasising the need for strong involvement of local communities. In their words, "local communities should be empowered to manage their resources with the help of governments" (Kojo & Moyo, 2008, p. 220). They also emphasise intergenerational and intra-generational access to resources and equity, arguing for overcoming poverty while at the same time promoting economic growth and ensuring the survival of natural resources without depleting them. They frown at neoliberals' tendency of stressing markets and profits more than social justice, even when dealing with structural problems. Thus, this chapter posits that, in the case of Botswana, as elsewhere in developing countries, at the core of the success of sustainable wildlife conservation and tourism ought to be the recognition of participatory processes of local communities in management and utilisation of the resources and leveraging on historical realities and indigenous knowledge systems. Put differently, sustainable development practitioners should consider the "natural resources conservation-economic development-livelihoods" nexus, which brings together mutually inclusive linkages between these key components of sustainability within wildlife conservation and tourism development in Botswana for today and posterity.

Methodology

The chapter draws from both primary and secondary sources. It utilises, as primary sources, archival data which was gleaned from the Botswana National Archives. These include official records comprising minutes of meetings, communication between colonial officials, colonial officials and local authorities, memoranda, and letters and official documents relating to post-colonial developments. It is largely these archival sources that informed the historical and chronological arrangement of the evolutionary stages of conservation and tourism developments. These sources also revealed how colonial attitudes were sometimes prejudicial against local communities, who were viewed not as makers of history but mere adjuncts to the historical processes, hence excluding them in discussions on the establishment of games reserves. Secondary sources include published and unpublished materials such as books, research theses, policy documents, newspapers and internet sources, most of which deal with the recent past and contemporary matters.

Conservation debates

Almost invariably throughout the conversations on conservation in Africa, the narrative has been that of European constructions and objectives on the subject, with only very limited reference to conservation systems as practised by indigenous communities in pre-colonial times. Such Western views, often bordering on stereotypes, were influenced by Victorian ideas that viewed Africa as a continent with "primitive cultures [that] have sometimes seriously disturbed their local environment" (Hinz, 2003, p. 19). Africans were perceived as people who saw game as either food or predator worthy of destruction on sight. As one colonial administrator wrote on East Africa: "To some leaders of the new nations, game preservation is a white man's foible; they see lion only as a killer of their cattle, elephant only as a destroyer of crops, and buck only as a meal" (Botswana National Archives, hereafter referred to as BNA S. 568/13/2, 1960).

But recent literature has questioned such stereotypes, arguing that in Southern Africa the so-called "primitive cultures achieved a far more stable environmental adaptation than presently assumed by industrial civilizations" (Hinz, 2003, p. 20). Indicting the West, an expert on Botswana history, Thomas Tlou (1985), stated that the advent of European hunters and traders with guns into the Chobe and Ngamiland Districts of Botswana in the 1850s led to the killing of large numbers of wild animals and exporting game products to Europe. A British hunter, Frederick Courteney Selous, reportedly killed about two hundred buffaloes. There are many other examples of trophy-hunting escapades in the Zambezi region (the woodlands and floodplains of the Zambezi and Chobe Rivers), captured by scholars that include Maria Fisch (1999), who extensively documented German activities in the then Caprivi Strip in German South West Africa (Namibia). These hunters recruited African assistants, making them accomplices in the slaughter of game animals (Gumbo, 2002). This chapter debunks the myth that it was mainly the local communities that were responsible for the extermination of wildlife, thereby necessitating fortress conservation by colonial governments. Most African communities have historically hunted for "the pot," for subsistence as part of sources of animal protein in their livelihoods as well as skins for clothing (Butynski & Von Richter, 1975; Morton & Hitchcock, 2014). Subsistence hunting was carried out using rudimentary equipment such as bows and arrows, spears and sometimes traps, arguably killing fewer animals for subsistence utility than did trophy hunting by Europeans in the 19th century. The hunters later used the *kgobela*, a muzzleloader, obtained from European traders (Tlou, 1985).

Conservation landscape in the pre-colonial period

As elsewhere in the Southern African region, the idea and practice of conservation did not arrive with European colonialism. This is illustrated herein with a vignette of local conservation as practised in pre-colonial Botswana. While hunting was part of daily life for these communities, conservation was in fact embedded in the practice of hunting. Traditional conservation took many forms. For example, traditional leaders/chiefs (*dikgosi*) played an important role in the conservation of wild animals. Game animals were communal property under the custodianship of the chief, who established age regiments and authorised hunting expeditions that took place during specific seasons (Campbell, 1997). Informed by indigenous knowledge systems and lived experiences, the hunting season was usually declared when most targeted female animals had weaned their young. As a measure of control and thus conservation, hunting was governed by rules, and high premiums were placed on prized species such as leopords. They were hunted sparingly only by authorised age regiments, and the skins were given to the chief to be "tanned" into special costumes for his dress during royal ceremonies. It was also customary for hunters to present certain portions of hunted meat to the chief. Punitive measures were meted out to those who did not comply (BNA S. 416/6, 1951).

Wildlife conservation in traditional societies was also integrated in complex social cultural institutions and practices such as totems. A totem is a symbolic relationship between people and natural phenomena such as specific animals or plants, entailing mystical and religious beliefs. Thus, "totemism is manifested in various forms and types in different contexts and is most often found among populations whose traditional economies relied on hunting and gathering mixed farming with hunting and gathering, or emphasized the raising of cattle" (Haekel, n/d). The totem is thought to interact with a given kin group or an individual and to serve as their emblem or symbol (Totemism, 2015). Almost all Batswana clans identified with a specific totem or (*sereto*). By associating with a specific animal, the clan treated it with reverence, respecting it and refraining from harming, hunting or eating the animal (Schapera, 1955; Campbell, 1995). As Spinage (1991, p. 8) observes, "One of the commonest obligations concerning a totem animal was that it could not be killed, or even touched. This protection was conferred on a range of animals." Among the Tswana, for instance, the Bangwato's totem was a duiker/*phuti*, the Batawana (a breakaway group of the Bangwato) had some not eating the duiker and others a lion; the Bakwena revered a crocodile and fish, the Barolong a kudu/*tholo*, the Balete a buffalo, while the Bakalanga and the Ndebele in North-Eastern Botswana had multiple totems involving zebra/*dube*, elephant/*ndlovu*, buffalo/*nyathi* (Gumbo, 2010). Despite its imperfections, totemism as a conservation strategy saved the different species of wild animals in a socio-economic environment that depended in part on hunting for livelihood, as those who identified with specific species would not kill the totem animals (Ntiamoa-Baidu, 2008).

Conservation in the colonial era

During the colonial period, Western conservation initiatives in Africa can be traced to 18th- and 19th-century European utopian ideas of an idyllic "Edenic search" where Europeans could visit to experience the pristine environment teaming with "unique wildlife species" in rural Africa (Ghimire & Pimbert, 1997; Grove, 1995). With only very few wild animals surviving in Europe, Africa was the place that provided an opportunity for Europeans to escape from the hurly burly of an industrialised and urbanised society (Bolaane, 2004; Murombedzi, 2003). Informed by narratives on the increasing destruction of wildlife species through hunting and widespread deforestation in Africa, "a decline from a prior state of pristine wilderness," conservation activists were disturbed by such reports and lobbied for fortress conservation, which entailed game reserves and national parks (Jones, 2006, p. 484). These would demarcate specific conservation areas where man and wildlife would not meet, in the belief that national parks and game reserves would be safe havens for wildlife. This was the genesis of game parks for both the preservation of the different species for posterity and human viewing. As indicated earlier, the Zambezi region, including the Chobe and Ngamiland Districts, had borne the brunt of the escapades in trophy hunting, largely by South African and German hunting parties operating from the then Caprivi Strip, Namibia.

So important for posterity, hence conservation, was the idea of wildlife preservation to the British Monarchy that King George VI, on the occasion of his visit to the then Bechuanaland Protectorate (present day Botswana) in 1938, remarked, "the wildlife of today is not ours to dispose of as we please. We must account for it to those who come after us' (BNA S 568/13/1). Benchmarking on the Kruger National Park in South Africa, Sir Charles Rey, the British Resident Commissioner in the Protectorate, was greatly impressed with the tree canopy in the riverine vegetation and the abundance of wild animals in the acacia and *mophane* thickets in the Chobe District. Rey was so impressed with this natural beauty such that he proposed the "development of a game reserve for tourist traffic. I feel strongly that in the interests of the territory . . . we should make every effort to overcome the difficulties which stand in the way of developing this scheme" (BNA RC 14/3, 1929–1935). He believed that the territory should be developed economically for the benefit of the indigenous populations as well as European settlers (Parsons & Crowder, 1988). Accusations had been levelled against previous colonial administrators over the slow pace of development in the protectorate, save for the collection of taxes from the indigenous people (Makgala, 2004). For Rey, however, a game reserve in the Chobe region thus dovetailed well with the general vision of the territory's economic development in the form of a tourism industry that would also benefit, albeit fortuitously, the inhabitants of the protectorate. It was this dream of Charles Rey in the early 1930s that continued to rank at the top on the agenda of the successive administrations, which culminated in the creation of the Chobe Game Reserve in 1960.

In order to efficiently coordinate the activities associated with game conservation, the colonial administration established the Department of Wildlife in 1956, which underwent various evolutionary stages. When it was first conceived, it was known as the Elephant Control Unit within the Public Works Department (Gumbo, 2002). Its main task was to control menacing animals such as elephants from causing harm to communities. It was later elevated to the status of a Game Department. In 1958, the department changed its name to the Department of Wildlife, National Parks and Tourism. In 1985, the Department of Wildlife and National Parks (DWNP) was separated from Tourism. Major Bromfield was the first game warden for the Chobe National Park, tasked with practical functioning of the Game Park. Bromfield had, himself, been influenced by James Stevenson-Hamilton, who was credited with the establishment of the Kruger National Park in the then Transvaal, South Africa (Carruthers, 1995). The DWNP was the institutional body responsible for the management of wildlife habitat and biodiversity in national parks. It was also responsible for enforcing laws relating to wildlife protection.

In 1963, seeing the value of wildlife and, therefore, the need for its protection, the Batawana of Ngamiland also established the Moremi Game Reserve, making it the first game reserve to be established by an African ethnic group (Bolaane, 2005). Subsequently, the colonial administration developed a Game Policy to resonate with the principal mandate – the protection of wildlife. It stipulated visiting hours for tourists in the Game Reserve, initially from 8 a.m. to noon and from 2 p.m. to 5 p.m. daily (BNA S 568/13/2). Entry into the Game Reserve was subject to a permit issued by a Game Reserve official at the payment of entry fees. While permitted, hunting was restricted to areas outside the boundaries of the Game Reserve (BNA S 145/3).

Although this chapter is not a political ecology of conservation, it would be remiss to not highlight some of the effects of the establishment of the Game Reserve as a conservation strategy. This is important because the net effect of the impacts was largely negative on the local communities, thereby compromising the latter's cooperation with the colonial conservation initiatives. Colonial authorities neither consulted the communities when demarcating the park boundaries nor did they involve them in the formulation of the Game Policy. Communities lost large tracts of lands which were annexed to the Chobe Game Reserve, thus limiting expansion for settlement as the population grew and also restricting the sizes of crop paddocks for their livestock, including restrictions on collecting firewood for fuel energy from what had now become a protected area (Taylor, 2000). More insidiously, the colonial administration evicted San/Basarwa communities from Mababe and Sankuyo on the southern borders of the reserve, depriving them of their historical hunting grounds. The Game Policy also affected fishing, a traditional subsistence activity for these riparian communities, by prohibiting the activity on the Chobe River around Kasane, the argument being that Kasane was part of the Chobe Game Reserve. Politically, traditional leaders lost the power to authorise hunting, as permission for hunting now rested with the colonial authorities who would use their discretion to issue or withhold the permit (Gumbo, 2002). Other growing pains emanated from a new disturbing phenomenon of human–wildlife conflict involving crop destruction by browsing animals and predation of livestock by carnivorous animals, who were also a threat to human life.

Post-colonial conservation trajectories

Botswana attained political independence in 1966. Building on the foundations laid by the colonial administration, the post-colonial government enacted several "comprehensive" wildlife conservation laws, a recognition of the fragility of the wildlife resource and a demonstration of the unwavering commitment to protect game under the auspices of the DWNP. At the same time, in 1967, the government elevated the status of the Chobe Game Reserve to that of a national park. The first post-colonial conservation law was the Fauna Conservation (Amendment) Act of 1967, followed by the Unified Hunting Regulations of 1979, amended by the Fauna Conservation (Amendment) Act of 1979. These were succeeded by the Wildlife Conservation and National Parks Act No. 28 of 1992. The essence of the successive laws demonstrated the urgency with which the government desired to conserve wildlife. Game poaching, both from within and outside the country, was heightening, especially during the period of political instability occasioned by the Southern African liberation wars involving neighbouring countries such as Namibia, Zimbabwe, Angola and South Africa, spilling over into Botswana (Parsons, 2008). It was important to act decisively in the circumstances, and it was hoped that the enforcement of the various pieces of legislation would have the desired effect in protecting wildlife.

It would appear that the objectives of conservation notwithstanding, the new laws brought in too many changes too quickly for the local communities to adjust to, given the long history of their relationship with wildlife. According to Spinage (1991), post-independent Botswana adopted tighter controls than did the colonial administration. Subsequently, there were protests against the new laws at *kgotla* (the village assembly) meetings in the Chobe District in the 1970s. The Fauna Conservation (Amendment) Act of 1979 irked these communities by zoning some areas into Controlled Hunting Areas (CHA) for citizen hunting. Their concern emanated from the fact that the CHAs lumped together local subsistence communities with affluent urbanites from the country's cities, yet for the locals, hunting was a livelihood activity that had historically sustained them, while it was leisure for urban dwellers turned sportsmen. Above all, however, it was the human–wildlife conflict that accentuated poverty levels through the destruction of fields, crops and predation on livestock, causing resentment in supporting conservation by local communities.

In the 1990s, in an attempt to assuage communities' enduring anger resulting from the game menace, the government hastened to introduce the Community-Based Natural Resource Management (CBNRM) programmes. This was done with the goal to resuscitate community trust through participation in the management and utilisation of wildlife. A significant amount of literature exists on the CBNRM programmes in Botswana (Tamuhla, 1997; Twyman, 2000; Mbaiwa, 2004; Kgathi & Ngwenya, 2005). The CBNRM aimed to "alleviate poverty and advance conservation by strengthening rural economies and empowering local communities to manage resources for their long-term social, economic and ecological benefits" (Lepetu, Makopondo, and Darkoh (2008, p. 114). Sponsored by the United States Agency for International Development (USAID), the first project, the Chobe Enclave Conservation Trust (CECT) was launched in the western villages of the Chobe District in 1990. Among other objectives, CECT aimed to empower local communities in decision-making, engendering the value of wildlife as an economically viable natural resource (Tamuhla, 1997). The communities were thus induced into participating in wildlife conservation in return for annual hunting quotas. The quotas could be turned into cash by selling them to the highest bidder among safari hunting companies (Lepetu et al., 2008). The income was invested in the form of brick-moulding projects in Kachikau, a co-operative shop in Mabele and campsites for potential tourist visits in Kavimba (Tamuhla, 1997). Arguably, the new conservation strategy was yielding the desired results in terms of benefits for the community and conservation imperatives. A sense of shared ownership was developing within members of the local communities.

The hunting ban

The relative modicum of "calm" brought about by the CBNRM programmes notwithstanding, the debilitating human–wildlife conflict did not disappear. The proximity of the park to the villages increased contact between the local communities and wild animals, creating an acrimonious relationship between communities and the government. Fast forward, and in the midst of the unfolding conflicts, the government announced a blanket ban on hunting in 2014, due ostensibly to a countrywide decline in wildlife species (Mbaiwa, 2018).

The ban effectively prohibited both subsistence and safari hunting while promoting photographic tourism, eliciting local and international condemnation. The elephant numbers have since soared, threatening the carrying capacity as well as human life and farmers' property, once more bringing back the anti-conservation attitudes among the local communities (Reid, 2019). According to Mbaiwa (2018), an outspoken critic of the ban on hunting, the action was irrational as it was not premised on scientific studies. He avers that these were politics of control of the tourism industry, a war between trophy hunting and photographic tourism, with the latter enjoying the support of the then-sitting President Ian Khama (2008–2018), a self-proclaimed conservationist opposed to hunting. Khama's political interference in the tourism industry is not new. He is accused of appointing his brother, Tshekedi Khama, to the Ministry of Tourism in order to safeguard personal interests in the industry. He is also accused of masterminding the eviction of the San/Basarwa from the Central Kalahari Game Reserve (CKGR) in order to establish a hotel with Wilderness Holding, a company in which he has a 5% share (Ditlhase, 2012). The ban led to huge losses in terms of meat from hunting, income generated through hunting safaris and jobs by those employed in the CBNRM projects associated with safari hunting, thereby accentuating cases of poaching and thus undermining the conservation agenda (Mbaiwa, 2018). With the hunting ban in place, the human–wildlife conflict has, once again, created a silhouette in conservation debates, creating a sense of uncertainty, animosity and portending poverty to the communities. It is not enough to attempt to protect the animals without concomitant utilisation of the resource by the communities. Against a strong wave of anti-ivory trade by some international wildlife conservation organisations, such as Conservation on International Trade in Endangered Species of Wild Fauna and Flora (CITES), Monitoring the Illegal Killing of Elephants (MIKE) and Elephants Without Borders, leaders of 'wildlife economy' countries in Southern Africa (Botswana, Namibia, Zambia and Zimbabwe) recently converged in Kasane, Botswana, and later Victoria Falls in Zimbabwe to defend the need to sustainably reduce elephant numbers, arguing for legal trade in ivory. Speaking on behalf of other Southern African governments, Zimbabwean President Emmerson Mnangagwa argued that they be "allowed to trade on the basis that legal sales will eliminate illegal markets and that the money raised will fund wildlife protection . . . and improve lives of communities around them" (Bafana, 2019, p. 5). This view is also shared by VanderPost (2007, p. 227), who asserts that "it is almost impossible for any African country to support large-scale and long-term conservation without generating significant wildlife value for both national and local interests."

The unchecked growing numbers of elephants is an immense threat to the woodlands and forests, potentially hastening climate change in a country that is already exposed to fast-encroaching desertification with rising temperatures. Faced with the vagaries of weather, such as prolonged droughts, changing rainfall patterns and now destruction of forests by elephants, Botswana ought to reconsider the alternative of culling elephants before the northern wildlife habitat is completely destroyed (Child, 2019; Crous, 2019). This may not only assist in mitigating climate change but also reduce incidents of human–wildlife conflicts in which many people have been killed by elephants while going about their livelihood chores (Dust, 2018).

Having looked at the theme of conservation, the section that follows interrogates the historical evolution of the tourism industry, also in the context of the sustainable development thesis.

Historical development of tourism in Botswana

Arguably, very little has been written on the early history of the tourism industry. This section of the chapter attempts to fill the lacuna. As indicated earlier, the colonial administration created the Chobe Game Reserve for the protection of game animals, and, much more importantly, to lure tourists into the country for the viewing of the different animal species roaming the park. The success of the newly completed Chobe Game Reserve in attracting tourist visits would depend, inter alia, on other support systems and infrastructure of various kinds. One of the most essential of these was a hotel, which would provide accommodation to tourists visiting the game sanctuary. The colonial administration contracted an intrepid couple, Colonel Charles F. Trevor and his wife Ethnee, from the then Southern Rhodesia (now Zimbabwe) (Gumbo, 2002). The hotel, built largely of rondavels, was named the Chobe River Hotel because it was overlooking the banks of the river – a tourist fascination. The hotel is what is today known as the Chobe Safari Lodge, having changed hands over the years. By 1960, the hotel was open to the public, but this excluded Africans. Exclusion of Africans in entertainment places was a racial norm in the region, with apartheid in South Africa and racial discrimination in Southern Rhodesia. Thus, it was not surprising that Whites in the Bechuanaland Protectorate regarded themselves superior to Blacks/Africans.

Tourism development in post-colonial Botswana started very late because of liberation wars which were fought between freedom fighters and colonial forces, as already argued. Owing to the geographical position in relation to the countries that were directly involved in these wars, Botswana was caught in the crossfire between these warring parties, forcing it to focus more on the security of the country while neglecting tourism (Gumbo, 2014). In the meantime, the private sector, led by South African capital, dominated the Chobe River front, covering the area between the Chobe Safari Lodge and the Chobe rapids to the east. In Botswana, tourism is based on both consumptive wildlife utilisation such as safari hunting and non-consumptive utilisation including viewing the wilderness and wildlife.

Touted as a conduit of development of the rural economies, as well as enabling developing countries to participate in the global economy (Telfer & Sharpley, 2008), Botswana's tourism is the second largest foreign exchange earner after diamonds, with prospects of employment creation in rural areas (Gumbo, 2014). In 2017, the tourism sector contributed 11.5% to the Gross

Domestic Product (GDP), a graphic illustration of the importance of the sector to the country's economic growth. The tourist industry is located mainly in the northern wetland region of Botswana and in the Ngamiland and Chobe Districts. The fauna and wilderness landscapes that constitute the main tourist resource stretch along a geographical "tourist belt" that includes riverine forests in southern Zambia, the world heritage site ofVictoria Falls and Hwange National Park in Zimbabwe, extending to the Chobe and Moremi National Park in Botswana and Etosha National Park in Namibia, makes tourism a transnational activity.

When peace returned to the region in the 1990s, Botswana began to partner with the private sector in developing tourism. The first regulatory instrument on tourism was the Tourism Policy of 1990 (Government of Botswana (GoB), 1990). It was shortly followed, in 1992, by the Tourism Act of 1992 (GoB, 1992) and the Botswana Tourism Master Plan of 2000 (btdp, 2000). Thus, from 1990 onwards, the government regarded tourism as an important industry, warranting a fully fledged department that would plan, coordinate and regulate tourism activities. The objectives of the Tourism Policy suggested a radical departure in terms of, at least, addressing rural poverty through tourism (GoB, 1990). Government undertook to create employment opportunities, especially in those areas adjacent to national parks and games reserves, and, in so doing, reduce rural-urban migrations. It was also hoped that the government would increase its revenue and foreign currency earnings as well as use tourism to market the country internationally (Government of Botswana, 1990). The private sector involvement manifested in the formation of the Hotel and Tourism Association of Botswana, (HATAB), which was the link between tourism operators and government. By 1991, hotels in Maun and Kasane and lodges in Okavango accounted for 14.5% of the total number of hotels, lodges and safari camps in the country, implying an expansion of tourism facilities and growth of the industry (Republic of Botswana, 1991).

With the democratisation of South Africa in 1994, the Southern African region became the "fastest growing tourist destination in Africa" (Republic of Botswana, 1997, p. 306). In 1997, the tourism expenditure in Botswana amounted to BWP1.1 billion, although out of this amount, BWP605 million was retained outside Botswana in leakages through commissions paid to agents and importation of tourists' food (Bentinck, 2002). The tourism industry also increased job opportunities. In 1998, it was estimated that approximately 9,000 people were employed in tourism-related occupations (Bentinck, 2002). In 2012, the tourism industry's direct contribution to the GDP was 3.0%, whereas its total contribution to it was 7.7%. Its direct contribution to employment was 4.7%, while its total contribution to it was 9.1% (Turner, 2013, p. 1). Cooperation between neighbouring states comprising Angola, Botswana, Namibia, Zambia and Zimbabwe led to the formation of the Kavango-Zambezi Transfrontier Conservation Area (KAZA TFCA) with a goal to jointly managing cross-border conservation. Despite these accolades, however, the industry was criticised for various malpractices, including, among others, poor working conditions, especially for unskilled local staff, racism and denying workers the right to unionise so as to bargain for better working conditions (Gumbo, 2010).

Conclusion

This chapter has examined the evolution of wildlife conservation and the historical development of tourism in Botswana. It has considered the different landscapes of the concept and practice of conservation over time, spanning between the pre-colonial and post-independence periods. While contributing to protecting wildlife species and their habitat, the parks and game reserves competed with human land-use needs. Damage to property by game animals accentuated and aggravated the "people-park conflicts." However, wildlife authorities slowly learned to find ways to manage wildlife in order to appropriately mitigate and balance the competing needs of conservation and local livelihoods. These efforts marked a move towards a more people-centred conservation. The chapter has also considered the historical evolution of tourism. Notably, the industry is credited with a significant contribution to the growth of the country's economy. There are, however, growing concerns over malpractices such as poor working conditions and racism in the largely foreign-controlled sector.

References

Amanor, K. S., & Moyo, S. (2008). *Land and sustainable development in Africa*. London: Zed Books.

Bafana, B. (2019, July 14). Let us trade. *The Patriot on Sunday*, p. 5.

Bentinck, K. (2002, April). *Developing a national eco-tourism strategy for Botswana tourism; A catalyst for sustainable development in Africa, Seminar Proceedings, Abuja, Nigeria* (pp. 26–27). Madrid: WTO.

Bolaane, M. (2004). *Wildlife conservation and local management: The establishment of Moremi Park, Okavango, Botswana in the 1950s–1960s*. Unpublished PhD thesis, University of Oxford.

Bolaane, M. (2005). Chiefs, hunters and adventurers: The foundation of the Okavango/Moremi National Park, Botswana. *Journal of Historical Geography*, *31*(2), 241–259.

Botswana National Archives. (1935). BNA RC 14/3, *Official papers- 1929–1935, C.F. Rey. Tours in the Bechuanaland Protectorate*, Vol. 1.

Botswana National Archives. (1951). BNA S. 416/6, *History of the bechuanaland protectorate tribes: The hambukushu*, by Thomas Larson.

Botswana National Archives. (1960). BNA S 568/13/2, *Game reserve – northern, establishment of, letter No. 64 from Philip K. Crowe, the chobe game reserve.*

Botswana Tourism Development Program [BTDP]. (2000). *Botswana tourism master plan*. Gaborone: Government Printer.

Butynski, T. M., & Von Richter, W. (1975). Wildlife management in Botswana. *Wildlife Society Bulletin*, *3*(1), 19–24.

Campbell, A. (1995). Utilization of Wildlife in Botswana From Earliest Times to AD 1900. In K. Leggett, (Ed.), *The Present Status of Wildlife and its Future in Botswana, The Proceedings of a Symposium/Workshop organized by the Kalahari Conservation Society and the Chobe Wildlife Trust*. Gaborone: The Kalahari Conservation Society, 53–60.

Campbell, C. (1997). A History of Wildlife in Botswana to 1966. In F. Monggae (Ed.), *Conservation and management of wildlife in Botswana – strategies for twenty first century: Conference hosted by the department of wildlife and national parks in collaboration with the kalahari conservation society* (pp. 7–31). Gaborone: The Kalahari Conservation Society.

Carruthers, J. (1995). *The kruger national park: A social and political history*. Pietermaritzburg: University of Natal Press.

Central Statistic Office (CSO). (1992). *Tourism statistics*. Gaborone: Botswana.

Child, B. (2019, May 3). A crisis of too many elephants, not too few, Part 2. *Mmegi online, retrieved from* https://www.google.com/search?q=A+crisis+of+too+many+elephants%2C

Collingwood, R. (1997). The role of donor support in wildlife conservation and management. In F. Monggae (Ed.), *Proceedings of a national conference on conservation and management of wildlife in Botswana: Strategies for the twenty first Century, 13th–17th October 1997* (pp. 79–84). Gaborone: The Kalahari Conservation Society.

Crous, R. (2019, March 29). The future of big trees and elephants in the Okavango. *Mmegi online*.

Ditlhase, Y. (2012, November 2). Khama Inc: All the President's family, friends and close colleagues. *Mail & Guardian*.

Dust, T. (2018, June 7). Blame elephant killings on Government. *The Patriot on Sunday*, online.

Fisch, M. (1999). *The Caprivi Strip During German colonial Period, 1890 to 1914*. Windhoek: Out of Africa.

Ghimire, K. B., & Pimbert, M. P. (Eds.). (1997). *Social change and conservation: Environmental politics and impacts of national parks and protected areas*. London: Earthscan Publications.

Government of Botswana (GoB). (1990). Tourism policy. *Government paper No. 2 of 1990*. Gaborone: Government Printers.

Government of Botswana (GoB). (1992). *Tourism act*. Gaborone: Government Printer.

Grove, R. H. (1995). *Green imperialism: Colonial expansion, tropical island Edens and the origins of environmentalism, 1600–1860*. Cambridge: Cambridge University Press.

Gumbo, B. (2002). *The political economy of development in the chobe: Peasants, fishermen and tourists, 1960–1995*. Unpublished M.A. Thesis, Department of History, University of Botswana.

Gumbo, B. (2010). *Economic and social change in the communities of the wetlands of chobe and ngamiland, with special reference to the period since 1960*. Unpublished PhD Thesis, University of Cape Town.

Gumbo, B. (2014). Southern frican liberation wars: The halting development of tourism in Botswana, 1960–1990s. *South African Historical Journal, 66*(3), 572–587.

C:\Users\Admin\AppData\Local\Temp\Temp1_15044-3109 Stone to CE.zip\02 to CE\15044-3109-Ref Mismatch Report.docx - LStERROR_4Hammes, R. (2007). The ecologically noble savage debate. *Annual Review of Anthropology, 36*, 177–190.

Haresnape, G. (1974). *The great hunters*. Cape Town: Purnell.

Hilty, L. M., Seifert, E. K, & Treibert, R. (2005). *Information Systems for Sustainable Development*. London: Idea Group Publishing.

Hinz, M. O. (2003). *Without chiefs there would be no game: Customary law and nature conservation*. Windhoek: Out of Africa.

Hitchcock, R. K. (1990). Tourism and sustainable development among remote area populations in Botswana. In L. Pfoteinhauer (Ed.), *Tourism in Botswana: Proceedings of a symposium held in Gaborone, Botswana, 15–19 October, 1990* (pp. 161–175). Gaborone: The Botswana Society.

Jones, S. (2006). A political ecology of wildlife conservation in Africa. *Review of African Political Economy, 33*(109), 483–495.

Kaekel, J. (n/d). *Totemism*. Retrieved from www.britannica.com/to pic/totemism-religion.

Kgathi, D. L., & Ngwenya, B. N. (2005). Community based natural resource management and social sustainability in Ngamiland. *Botswana Notes and Records, 37*, 61–79.

Leistyna, P. (2007). Neoliberal non-sense. *Counterpoints, 299*, 97–123.

Lepetu, J., Makopondo, R. O. B., & Darkoh, M. B. K. (2008). Community-based natural resource management and tourism partnership in Botswana: Which way forward? *Botswana Notes and Records, 39*, 113–124.

Makgala, C. J. (2004). Taxation in the tribal areas of the Bechuanaland protectorate, 1899–1957. *Journal of African History, 45*(2), 279–303.

Matlhare, J. M. (1997). Changes in wildlife policies and regulations. In F. Monggae (Ed.), *Proceedings of a national conference on conservation and management of wildlife in Botswana: Strategies for the twenty first century 13th–17th October 1997* (pp. 185–190). Gaborone: Kalahari Conservation Society.

Mbaiwa, J. E. (2004). The success and sustainability of community-based natural resource management in the Okavango delta, Botswana. *South African Geographical Journal, 86*(1), 44–53.

Mbaiwa, J. E. (2018). Effects of the safari hunting tourism ban on rural livelihoods and wildlife conservation in northern Botswana. *South African Geographical Journal, 100*(1), 41–61.

Mbaiwa, J. E., & Darkoh M. B. K. (2008). The socio-economic and environmental effects of the implementation of the tourism policy of 1990 in the Okavango delta, Botswana. *Botswana Notes and Records, 39*, 138–155.

Ministry of Finance and Development Planning [MFDP]. (1977). *National Development Plan NDP* (p. 8). Gaborone: Ministry of Finance and Development Planning.

Morton, F., & Hitchcock, R. (2014). Tswana hunting: Continuities and changes in the Transvaal and Kalahari after 1600. *South African Historical Journal, 66*(3), 418–439.

Moswete, N., & Mavondo, F. T. (2003). Problems facing the tourism industry in Botswana. *Botswana Notes and Records, 35*, 69–77.

Murombedzi, J. (2003). Pre-colonial and colonial conservation practices in Southern Africa and their legacy today. In Webster Whande, Thembela Kepe, & Marshall Murphree (Eds.), *Local communities, equity and conservation in Southern Africa: A synthesis of lessons learnt and recommendations from a Southern African technical workshop* (pp. 23–24). Cape Town: Programme for Land and Agrarian Studies (PLAAS).

Ntiamoa-Baidu, Y. (2008). *Indigenous beliefs and biodiversity conservation: The effectiveness of sacred groves, taboos and totems in Ghana for habitat and species conservation*. Retrieved from https://eds.a.ebscohost.com/eds/pdfviewer/pdfviewers?vid=1&sid=691b3657-9c91-4914-9957-b77e1b6cc936%40sessionmgr4006 [accessed on 2 July 2019].

Parsons, N. (2008). The Pipeline: Botswana's reception of refugees, 1956–68. *Social Dynamics, 34*(1), 17–32.

Parsons, N., & Crowder, M. (Eds.). (1988). *Sir Charles Rey: Monarch of all I survey: Bechuanaland diaries 1929–1937*. Gaborone: The Botswana Society.

Republic of Botswana (1991). *Tourism Statistics*. Gaborone: Central Statistics Office.

Republic of Botswana (1997). *National Development Plan NDP 8*. Gaborone: Ministry of Finance and Development Planning.

Reid, I. (2019, 24 March). The elephant debate – from the mouths of Botswana farmers. *Sunday Standard*, p. 12.

Saarinen, J., & Manwa, H. (2008). Tourism as a socio-cultural encouter: Host-guest relations in tourism development in Botswana. *Botswana Notes & Records, 39*, 43–53.

Schapera, I. (1955). *A handbook of Tswana law and custom: Compiled for the Bechuanaland protectorate administration*. London: International African Institution.

Spinage, C. (1991). *History and evolution of the fauna conservation laws of Botswana*. Gaborone: The Botswana Society.

Tamuhla, A. G. (1997). *Factors influencing community participation in sustainable wildlife utilization: A case of the chobe conservation trust, Botswana*. M.Sc. Thesis, University of Botswana.

Tamuhla, A. G. (2001). *Factors influencing community participation in sustainable wildlife utilization: A case of the chobe conservation trust, Botswana*. M.Sc. Thesis, University of Botswana.

Taylor, M. (2000). *Life, land and power: Contesting development in northern Botswana*. PhD Thesis, University of Edinburgh.

Telfer, D. J., & Sharpley, R. (2008). *Tourism and development in the developing world*. London & New York: Routledge.

Tlou, T. (1985). *A history of Ngamiland, 1750–1906: The formation of an African state*. Gaborone: Macmillan.

Totemism. (2015). http://www.sociology-article.blogspot.com/2015/03/totemism.html [accessed on 28 June 2019].

Turner, R. (2013). *Travel and tourism economic impact 2013: Botswana*. London: World Travel & Tourism.

Twyman, C. (2000). Community participation? Community-based natural resource management in Botswana. *The Geographical Journal, 166*(4), 323–335.

VanderPost, C. (2007). Geographic prospects for large-scale African mammal wildlife conservation. *GeoJournal, 69*(4), 223–237.

World Commission on Environment and Development – WCED. (1987). *Our common future*. London: Oxford University Press.

3 Nature tourism, wildlife resources and community-based conservation

The case study of Malawi

James Malitoni Chilembwe

Introduction

Community-Based Conservation (CBC) enhances local participation and promotes economic, social and cultural well-being of a defined society (Butcher, 2007; Butcher & Smith, 2010). It is a type of tourism that is linked to sustainable development due to its conservation and development goals. Due to the nature of its activities, community-based conservation is similar to ecotourism because of local involvement in control, participation and ownership over the tourism activities in their communities.

International agencies highly promote community-based tourism (CBT) as a means to reduce poverty among communities, particularly those in developing countries (Chaudhary & Lama, 2014; Spenceley, 2008). The promotion of CBT is also aimed at ensuring community participation in decision-making and sharing the benefits of development (Bello, Lovelock, & Carr, 2016; Setokoe, Ramukumba, & Ferreira, 2019; Sebele, 2010; Thetsane, 2019). It further reduces conflicts among communities, tourism operators and resources and promotes peace in the society (Hlengwa & Maruta, 2019; Mohanty, Rout, & Samal, 2019). Moreover, places that practice community-based tourism can reduce community problems while increasing their economic status in an area in terms of infrastructure development (UNWTO, 2018). Generally, some activities of community-based tourism involve traditional beliefs which demand respect. In addition, communities involved in CBT often have no other sources of income besides natural resources, wildlife, lakes, mountains and trees. CBC is essential to tourism planning through initiating discussions about conservation measures and development plans which may improve livelihoods.

CBC may either impact communities positively or negatively. As Butcher (2007) explains, tourism influences a community's economic, social and cultural wellbeing. Through tourism, economies of countries can improve. These would include generating regions, transit and host destinations. The tourism industry can contribute a substantial portion of national incomes. A good example is the generation of foreign exchange, as Basariya and Ahmed's (2019) study indicates that in 2009 India's tourism industry generated about US$100 billion. In Malawi the tourism industry has been articulated in the growth and development poverty reduction strategy I, II & III, which plans to increase promotion and local communities' involvement so that everyone achieves direct benefits. The locals who stay along the Lake Malawi shore and surrounding natural attractions, where they meet visitors directly, enjoy the benefits more. These locals are also provided with sporting venues, entertainment facilities, restaurants and a better range of food and beverages.

On the other hand, there are also limitations of community-based conservation on the communities and development (Bello, Lovelock, & Carr, 2017; Butcher, 2007) that should not be ignored. For example, in most areas where tourism is active, locals are faced with a high cost of living as well as different types of pollution related to noise and water. Some areas experience overcrowding from visitors, sometimes resulting in conflicts between the locals and a negative attitude towards tourism. Moreover, the negative perception of the communities towards tourism initiates challenges to tourism's sustainability, conservation and development (López, Virto, Manzano, & Miranda, 2018; Mathews & Sreejesh, 2017). Consequently, communities engage in malpractices that may destroy or disturb the natural environments that attract tourists. It also becomes worse if tourists join forces with the communities in destroying the same nature.

This chapter is based on empirical research. It focuses on nature tourism, wildlife resources and community-based tourism. The first section briefly introduces and discusses the concepts of conservation as well as community-based tourism and wildlife. The second section critically discusses nature-based tourism and climate change, followed by the third section on community-based and sustainable tourism in Malawi. A methodological approach for this chapter is provided, followed by results, discussion and finally the conclusion.

Community-based tourism, wildlife resources and conservation

The success of conservation and community-based tourism in many areas depends on attitudes and behaviours of communities living in villages around the protected areas (Chilembwe & Mweiwa, 2019; Myung, 2018; Snyman, 2014; Shale & Rantšo, 2019). Managing and understanding community needs and their attitudes on socio-economic issues can bring more efficient, equitable and sustainable community-based conservation. As a result, there is a need for managing community relationships and determining long-term support for the protected areas and their communities. For example, an ecotourism joint venture between the private sector and communities can be used as one way of managing community-based tourism and conservation. A study by Snyman (2014) based on 1,400 community interviews in Botswana, Malawi, Namibia, South Africa, Zambia and Zimbabwe reveals conservation implications on human–wildlife conflict negatively impact attitudes and calls for policy focus areas. Other studies on community-based environmental resources conservation confirm that there are challenges as well as failures. For example, Shale and

Rantšo (2019) point out that the lack-of-enforcement approach by many governments is due to state-centred conservation programmes. Moreover, lack of clear resource policies aggravates the problem (Weicheselgartner & Kasperson, 2010). As a result, there are no real benefits to communities, as well as failures, to provide start-up capital in remote areas where projects are being implemented. They, therefore, proposed a people- or community-centred approach to natural resource conservation to minimise resource depletion. Bello et al. (2017) add that communities fail to participate in protected area-based tourism because of unfair distribution of benefits, lack of coordination, and human–wildlife conflicts. In many developing countries, community participation is dominated by a few people who are labelled as educated and are vocal, but also wealthier, in the community. The majority are mere followers, as they seem to have very little ownership of community projects. This gap brings in lack of coordination within community-based tourism projects.

Nature-based tourism

Tourism influences the economic viability of tourist destinations through community activities, tourist behaviours and the entire tourism system (Hall et al., 2015). Nature-based tourism destinations, such as Kilimanjaro and Mulanje Mountains in Tanzania and Malawi, respectively, are dependent on favourable environmental conditions. However, due to human activities, Mount Kilimanjaro is being negatively affected (Kilungu, Leemans, Munishi, Nicholls, & Amelung, 2019). In Malawi, Nyika National Park is not spared from environmental degradation (Nyirenda & Chilembwe, 2015). Communities around Lake Chilwa in Liwonde, Mulanje Mountain, and Lake Malawi National Park have equally experienced environmental degradation. The disappearance of cider trees in Mulanje Mountain and the yearly drying up of water reserves in Lake Chilwa have presented local communities with problems. Community sources of employment and income are to a larger extent related to nature-based tourism, more especially water fishing resources. Similarly, the "Mwalawamphini" (tattoo rock) at Cape Maclear in Lake Malawi National Park has faced massive degradation due to the behaviours of both tourists and local communities. The study on environmental impacts of tourism on Chose Hills at Nyika National Park discovered that tourists, as well as local community activities, contribute to degradation (Nyirenda & Chilembwe, 2015). They noted that activities such as cutting down trees, poaching and playing with tattoo rocks more often can deplete the natural environment and cause displacement of wildlife from their habitats.

Mountain environmental degradation occurs when there is rock outcropping and removal of top soils, which are caused by not only nature but also human activities, as they frequently visit the resources (Scott, Jones, & Konopek, 2007). For example, evidence of this is observable at Nyika National Park where the rock is outcropping and informal trails appear, especially at the Chose Hills (Nyirenda & Chilembwe, 2015).

Community-based tourism and sustainable tourism in Malawi

Chongoni Art Rock in Dedza, Malawi, is surrounded by the community. In the past, there were no consultations between the Department of Forestry, Department of Antiquities and local communities regarding the management of this site. Each institution was managing its own interests within the forest reserve. The local communities viewed the tourist forest reserve as a government property. As a result, they did not care to help in the conservation of the natural environment. They engaged in massive cutting down of trees, setting bush fires, poaching wildlife and lighting fire in the rocks' shelters (Chilembwe, 2014). However, this malpractice changed when communities were fully involved through community-based projects such as beekeeping, mushroom farming, tour-guiding activities and community cultural village activities. They embarked on awareness campaigns to sensitize the local communities on the need to conserve the forests, rock paintings and wildlife. In the process, local communities were encouraged to become part owners of the natural resources. Positively, there is reduced tension between natural resources managers (government officials), non-governmental organisations and local communities. Meanwhile, the local communities are benefiting economically as:

- they are engaged in businesses like beekeeping and mushroom farming;
- self-employment as tour guides, porters and patrolmen;
- recruitment is prioritised in the local community (i.e. tour guides)

In Blantyre's Michiri Tourist Conservation Area, the traditional doctors are allowed to enter the protected area to extract roots for medicine. They interact with the natural environment, especially trees, which can lead to problems of deforestation and also scarcity of displacement of small animals in the park. However, as a matter of sustainability, traditional doctors are asked to plant ten trees every year to maintain the environment. The challenge is the lack of monitoring mechanisms to see to it that a minimum number of trees per year are planted (Chilembwe & Mponda, 2016). Other community tourism-based organisations in Malawi are Donija Cultural Village, Chigwere Cultural Village and Mpale Cultural Village, which are strategically located in various parts of Malawi. While examples are not limited to these, however, it should be noted that not all community tourism-based organisations receive government support in Malawi. Therefore, there is a need to be supported in terms of training, skills acquisition, small-scale income-generating activities and the provision of start-up capital so that they are concentrated on community-based activities rather than engaging in malpractices that can destroy nature and lead to environmental degradation or depletion, such as illegal mining, wildlife poaching and cutting down trees for charcoal-making to generate income.

Out of Malawi's total size of 118,480 km^2, land covers 94,270 km^2 while the rest is water. National parks and wildlife reserves cover 11 percent, while forests and protected hill slopes take 10 percent of the total land cover (Novelli, Scarth, 2007).

The main research objective of the study was to explore community-based tourism challenges (activities) in the context of nature tourism and wildlife resources environments (national parks).

In the interest of this study, a community is limited to people living in national parks. In order to achieve the objective, the following questions were asked:

- What malpractices are there leading to natural environmental degradation change?
- How has the natural environmental degradation affected the community-based tourism?
- Are there any measures to mitigate or minimise the negative impacts of environmental degradation in community-based tourism?

Research methods

This is a qualitative exploratory study. Data was collected through face-to-face in-depth interviews with eight park officials. The two senior tourism regulators were also interviewed in order to triangulate information obtained from park officials. An interview guide was developed comprised of five key questions and pilot tested on two park managers. The interviews were conducted in December 2018 for the period of three weeks. All interviewees were asked the same type of questions except follow-up and probing questions, which varied depending on responses. The interviews were conducted in the English language, but responses were a mixture of English and vernacular language. Interviews were audio recorded except in a few cases, when advised not to record, though notes were captured. The total average interview times were 1 hour, 13 minutes. The interviews were transcribed, and those recorded in the local language (Chichewa) were also translated into English. The study relied heavily on the key informants and observations during periodic visits to the sites due to geographical location. The local communities (non-park dwellers) were excluded from interviews for two reasons: first, at Nyika National Park, the communities are 35 kilometres away from the main park and deemed to have little knowledge about the community-based activities inside the park. Those who live inside the park were only relatives and family members of the park officials, and it was necessary to only interview officials who, apart from being employees and key informants, were also representatives of the local community inside and outside the park by the nature of their daily work. Second, at Lake Malawi National Park, the local communities stay inside the park and are also part of the park because there are no dangerous animals. However, the local people are restricted by the boundaries within the park when carrying out their activities. As a result, they were also excluded. Observations were conducted on the two national parks under study and were an on-going process to observe how natural resources and natural attractions have changed over time.

The study used a very small sample size of ten key informants, of which eight were residents within the wildlife reserves (national parks) and two were from the regulatory department. In addition the study involved government, employees, families and relatives confined to the parks as already justified in the methodological approach. A snowball sampling procedure was utilised. According to Lewis-Beck, Bryman and Futing Liao (2004), snowball sampling can be applied to allow for the studies to take place where it is deemed difficult to find respondents due to lack of or a limited number of participants. In the case of this study, each interview relied on referrals from one key informant respondent to another until there were no more referrals. The data were thematically analysed.

Each interview was assigned a number ranging from one to ten. Others factors considered were: gender, work experience and nature of work, as shown in Table 3.1 below.

Table 3.1 Characteristics of respondents

Interview	*Gender*	*Years of Experience*	*Nature of Work*
1	Male	13	Park management
2	Male	6	Regulator
3	Male	17	Tour guide
4	Female	8	Education and training
5	Male	11	Conservation and education
6	Female	6	Education and Training
7	Male	18	Regulator
8	Male	7	Tour guide
9	Male	5	Education and training
10	Female	9	Park management

Source: Author – Fieldwork (2018)

Results and discussions

The total number of the respondents was ten (10), of which seven (7) were males while three (3) were females. It comprised two (2) park managers, three (3) environmental educators, two (2) tour guides, two (2) regulators and one (1) responsible for conservation and education. The minimum work experience was 5 years while the highest was 18 years.

This study presents three themes. First, the malpractices that led to environmental degradation in community-based tourism. It is followed by the consequences of natural environmental degradation to the community-based tourism destination. Third, the measures to minimise the negative impacts of natural depletion in community-based tourism, as summarised in Table 3.2 below.

Table 3.2 Showing summary of emerged themes

Malpractices	*Consequences*	*Measures*
- wildlife poaching - deforestation - setting up bush fires - setting up traps - stone stepping - visitors' interaction with natural resources - local community conflict with natural resources	- reduction in wildlife population - human–wildlife conflicts - erosion - loss of vegetation cover - reduced tour guiding activities - loss of destination demand - lack of tourism-based activities - lack of coordination and trust - unfair distribution of wealth	- community-based awareness - training community on natural environmental issues and protection - awareness on the use of protected areas' resources - benefits and information sharing of resources - empowering the communities on decision-making - encouraging community-based tourism participation

Source: Author – Fieldwork (2018)

The first theme identified the malpractices leading to natural environmental degradation in community-based tourism. The malpractices, such as wildlife poaching, deforestation, setting up bush fires, setting up traps, stone stepping, visitors' interaction with natural resources, and local community conflicts with natural resources were found to be the main causes of why nature tourism is facing problems. Wildlife resources and community-based tourism face challenges for sustainability and development of tourism. The local communities are engaged in unwelcomed practices; for example, poaching and deforestation because of poverty, and they look for human survival, as the excerpts of respondents reveal below:

> You see, community based-tourism does not provide enough means for survival. So sometimes local communities go for deforestation and poaching because they want food for human survival. That is why they cut down the trees to make charcoal, which earns them income, then they sell. However, they also poach because some of these local people like eating bush meat.
>
> [Interview 1]

> The local communities are facing challenges because they cut down trees which could have given them other alternatives of employment, like a guiding tourist who would like to visit the forests. Trees benefit both the locals and business of tourism; without trees, then tourism suffers because of the nature of our tourist activities, and wildlife will not exist due to lack of animal habitat[s].
>
> [Interview 10]

Buhalis (1999) provided several examples of limitations to tourism development, for example, service failure and being unable to attract markets or tourists. In the context of this study, poaching and deforestation were among them. Indeed, when human food survival is prioritised, it becomes difficult to think of natural resources and wildlife resources conservation. The practice can ultimately lead to other environmental problems, for example, the scarcity of rainfall and water flooding due to natural disturbances. This can bring in problems to community-based tourism during their human living and including other creatures in the environment. Further, it can lead to lack of touristic activities to a destination for community-based tourism, as the destination becomes no longer attractive for domestic and international tourism. It, therefore, calls for environmental tourism education among the communities surrounding the natural resources, as proposed by Chilembwe (2013).

The findings further show that the challenges of nature tourism, wildlife resources and community-based tourism for the sustainability and development of tourism in Malawi were also contributed by the touristic activities, as the interview excerpt below depicts:

> Tourists bring seeds of destruction to our destinations when they visit. Besides providing us employment and contributing to the country's economic development, they also destroy, disrupt and spoil the natural real environment that attracted them. You can see, the hills have been altered, this is not [the] original appearance.
>
> [Interviews 3, 8]

Touristic activities, such as the removal of topsoil on the top of hills and mountains and tampering with trees to understand the formation or nature of a particular tree, for example, culled trees and tattoo trees, and mountains and Mouth Mountains, can destroy attractions. Moreover, both domestic and international tourists' attitudes can contribute to the environmental destruction of the area or tourism sustainability (Mathews & Sreejesh, 2017; López et al., 2018). However, the impacts of natural resources' depletion will likely affect everyone if there is a lack of coordinated efforts from all of the stakeholders to mitigate the effects (Michailidou, Vlachokostas, & Moussiopoulos, 2016; Weir, 2017). Lack of community involvement to conserve natural resources and lack of proper enforcement of tourism laws by regulators due to corruption is likely to amplify environmental problems in tourism.

Several factors were identified on the second theme about the consequences of natural environmental degradation to the community-based tourism destination reduction in the wildlife population. These were humanhuman–wildlife conflicts, erosion of top soils, loss of vegetation cover, reduced tour-guiding activities, loss of destination demand, lack of tourism-based activities, lack of coordination and trust and unfair distribution of wealth. These findings are in line with Bello et al. (2017) and Snyman (2014) who found that communities fail to participate in community-based tourism because of unfair distribution of benefits, lack of coordination and human–wildlife conflicts. On a similar note, Nyirenda and Chilembwe (2015) and Kilungu et al. (2019) discovered that human conflict with the environment causes the erosion and removal of top soils in the protected areas. The excerpt from the respondents provides evidence that conflicts exist between local people and the government:

> I feel the local people have their challenges; so does the government, but there is no joint effort to find solutions to end problems of deforestations.
>
> [Interviews 3, 7]

Sebele (2010) asserts that lack of community tourism benefits is likely to cause conflicts among the stakeholders. Therefore, local communities should always be engaged when sharing benefits surrounding the wildlife areas and natural tourism environment in order to preserve the resources. Tourism operators should help the local people to establish community tourism ventures to spread the economic and social benefits within the communities (Hlengwa & Maruta, 2019).

The third and final theme was about the measures put in place to minimise the negative impacts of natural depletion in community-based tourism. A number of measures were identified, for example, increasing community-based awareness, training the community on natural environmental issues and protection, bringing awareness to the use of protected areas' resources, acknowledging the benefits and sharing information about resources, empowering communities on decision-making and encouraging community-based tourism participation. This is also evidenced by some respondents who said:

> Our core values are simply that without a nurtured and well-looked-after environment, we cannot sustain the natural resource[s] and wildlife tourism. That is why some of us are conservationists, in order to advocate for the protection of our natural environment.
>
> [Interviews 2, 7]

> There is a need for [a] coordinated effort to deal with the present challenges. I say this often, and I would love to go to the government and find out about their afforestation projects. We are nothing but conservationists, and we will be happy to buy any views that align with our beliefs and values.
>
> [Interviews 5, 6, 9]

Accordingly, Shale and Rantšo (2019) assert that lack of law enforcement contributes to natural environmental problems in the communities. Government should take the lead not only on enforcement of rules but also in the coordination process, as Bello et al. (2017) observed. In community-based tourism activities, the locals can be provided with environmental training awareness (Chilembwe, 2013, 2014), as well as provided with knowledge on how to formulate and run a community-based tourism organisation, as one way of preventing them from destroying wildlife resources and the environment.

Conclusion

Nature tourism and wildlife resources are critical to the development of a country, and there is a need to apply effective conservation measures to sustain tourism (Chilembwe, 2014). However, the malpractices, such as wildlife poaching, deforestation, setting up bush fires, setting up traps, stone stepping, visitors' interaction with natural resources and local community conflict with natural resources, all need to be considered. Although the changes in environmental conditions are caused by the local communities surrounding the natural resources, the study has also indicated that, to some extent, visitors have played a role in contributing to destruction due to their touristic activities. Many tourist destinations have suffered due to deforestation, poaching of animals and other human-induced activities from which Malawi is not immune. Consequently, environmental degradation occurring at a community-based tourism destination includes a reduction in the wildlife population, human–wildlife conflicts, erosion of top soils, loss of vegetation cover, reduced tour-guiding activities, loss of destination demand, lack of tourism-based activities, lack of coordination and trust, unfair distribution of wealth (Butcher, 2007; Bello et al., 2017) and lack of law enforcement (Shale & Rantšo, 2019). Therefore, there is a need for a coordinated effort by all stakeholders to mitigate the practices. The practicing of sustainable tourism at "home and away" is one of the best approaches to be adopted by both community tourism-based organisations (residents) and tourists in general in order to preserve nature tourism and wildlife resources (Barr, Shaw, Coles, & Prillwitz, 2010). Finally, there is a need for continuous awareness and tourism environmental education if tourism sustainability and development is to be accomplished.

References

Barr, S., Shaw, G., Coles, T., & Prillwitz, J. (2010). A holiday is a holiday: Practicing sustainable, home and away. *Journal of Transport Geography*, *18*(3), 474–481.

Basariya, S. R., & Ahmed, R. R. (2019). The influence of 'adventure tourism activities' in promoting tourism business in mountain stations. *African Journal of Hospitality, Tourism and Leisure*, *8*(2). Retrieved from www.ajhtl.com/uploads/7/1/6/3/7163688/article_5_vol_8__2__2019.pdf [accessed on 2nd June 2019].

Bello, F. G., Lovelock, B., & Carr, N. (2016). Enhancing community participation in tourism planning associated with protected areas in developing countries: Lessons from Malawi. *Tourism and Hospitality Research*, *0*(0), 1–12. doi:10.1177/1467358416647763

Bello, F. G., Lovelock, B., & Carr, N. (2017). Constraints of community participation in protected area-based tourism planning: The case of Malawi. *Journal of Ecotourism*, *16*(2), 131–151.

Buhalis, D. (1999). Limits of tourism development in peripheral destinations: Problems and challenges. *Tourism Management*, *20*, 183–185. doi:10.1016/S0261-5177(98)00082-X

Butcher, J. (2007). *Ecotourism, NGOs and development.* London: Routledge.

Butcher, J., & Smith, P. (2010). 'Making a difference': Volunteer tourism and development. *Tourism Recreation Research, 35*(1), 27–36.

Chaudhary, M., & Lama, R. (2014). Community-based tourism development in Sikkim of India – study of Darap and Pastanga villages. *Transnational Corporation Review, 6*(3), 228–237.

Chilembwe, J. M. (2013). Sustainable tourism and environmental education in Malawi. In H. F. C. Mwale, S. Chiotcha, & W. Phalira (Eds.), *Rural livelihoods, environmental sustainability and climate change in Malawi: Annotated bibliography* (No. 368, p. 137). Zomba: University of Malawi, LEAD Southern and Eastern Africa.

Chilembwe, J. M. (2014). Examination of socio-cultural impacts of tourism in Chembe village in Mangochi District, Malawi. *International Journal of Business Quantitative Economic & Applied Management Research, 1*(1), 60–90.

Chilembwe, J. M., & Mponda, I. K. (2016). Tourism sustainable governance practices in Malawi as a tourist destination: Challenges and opportunities for tourism development. *Tourism Spectrum, 2*(1), 1–10.

Chilembwe, J. M., & Mweiwa, V. R. (2019). Responsible travel and tourism adventure: Evidence from Malawi as a tourist destination. In A. Sharma (Ed.), *Sustainable tourism development: Futuristic approaches* (pp. 31–53). FL: Apple Academic Press.

Hall, C. M., Amelung, B., Cohen, S., Eijgelaar, E., Gössling, S., Leemans, R., . . . Scott, D. (2015). On climate change skepticism and denial in tourism. *Journal of Sustainable Tourism, 23*(1), 4–25.

Hlengwa, D. C., & Maruta, A. T. (2019). Community based tourism ventures opt for communities around the save valley conservancy in Zimbabwe. *African Journal of Hospitality, Tourism and Leisure, 8*(2). Retrieved from www.ajhtl.com/uploads/7/1/6/3/7163688/article_10_vol_8_2__2019.pdf [accessed on 3rd June 2019].

Kilungu, H., Leemans, R., Munishi, P. K.T., Nicholls, S., & Amelung, B. (2019). Forty years of climate change and land-cover change and its effects on tourism resources in Kilimanjaro national park. *Tourism Planning & Development, 16*(2), 235–253. doi:10.1080/21568316.2019.1569121

Lewis-Beck, M. S., Bryman, A., & Futing Liao, T. (2004). *The sage encyclopaedia of social science research methods.* Thousand Oaks, CA: Sage Publications. doi:10.4135/9781412950589

López, M. F. B., Virto, N. R., Manzano, J. A., & Miranda, J. G. (2018). Residents' attitude as a determinant of tourism sustainability: The case of Trujillo. *Journal of Hospitality and Tourism Management, 35*, 36–45.

Mathews, P.V., & Sreejesh, S. (2017). Impact of responsible tourism on destination sustainability and quality of life of the community in tourism destinations. *Journal of Hospitality and Tourism Management, 31*, 83–89.

Michailidou, A. V., Vlachokostas, C., & Moussiopoulos, N. (2016). The interaction between climate change and the tourism sector: Multiple-criteria decision analysis to assess mitigation and adaptation options in tourism areas. *Tourism Management, 55*, 1–12.

Mohanty, P. P., Rout, H. B., & Samal, A. (2019). Community based sustainable tourism development – a tool for fostering and promoting peace: A case study of Odisha, India. *African Journal of Hospitality, Tourism and Leisure, 8*(2). Retrieved from www.ajhtl.com/uploads/7/1/6/3/7163688/article_20_vol_8_1__2019.pdf [accessed on 3rd June 2019].

Myung, E. (2018). Environmental knowledge, attitudes, and willingness to pay for environmentally friendly meeting – an exploratory study. *Journal of Hospitality and Tourism Management, 36*, 85–91.

Novelli, M., & Scarth, A. (2007). Tourism in protected areas: Integrating conservation and community development in Liwonde national park (Malawi). *Tourism and Hospitality Planning & Development, 4*(1), 47–73. doi:10.1080/14790530701289697

Nyirenda, B. M. C., & Chilembwe, J. M. (2015). Environmental impacts of tourism: Chose Hill at Nyika national park. *Impact: International Journal of Research in Business Management, 3*(9), 17–24.

Scott, D., Jones, B., & Konopek, J. (2007). Implications of climate change and environmental change for nature-based tourism in the Canadian rocky mountains: A case study of Weterton lakes national park. *Tourism Management, 28*(2), 570–579.

Sebele, L. S. (2010). Community-based tourism ventures, benefits and challenges: Khama rhino sanctuary trust, central district, Botswana. *Tourism Management, 31*, 135–145.

Setokoe, T. J., Ramukumba, T., & Ferreira, I. W. (2019). Community participation in the development of rural areas: A leaders' perspective of tourism. *African Journal of Hospitality, Tourism and Leisure, 8*(2). Retrieved from www.ajhtl.com/uploads/7/1/6/3/7163688/article_30_vol_8_1__2019.pdf [accessed on 3rd June 2019].

Shale, L., & Rantšo, T. (2019). A community-based model of environmental resource conservation and livelihoods: Non-governmental organisation supported Tšenekeng botanical gardens in Semonkong, Lesotho. *African Journal of Hospitality, Tourism and Leisure, 8*(1). Retrieved from www.ajhtl.com/uploads/7/1/6/3/7163688/article_22_vol_8_1__2019.pdf [accessed on 2nd June 2019].

Snyman, S. (2014). Assessment of the main factors impacting community members' attitudes towards tourism and protected areas in six southern African countries. *Koedoe, 56*(2). Art. #1139. doi.org/10.4102/koedoe.v5612.1139.

Spenceley, A. (2008). Local impacts of community-based tourism in Southern Africa. In A. Spenceley (Ed.), *Responsible tourism: Critical issues for conservation and development*, 286–303. London: Earthscan.

Thetsane, R. M. (2019). Local community participation in tourism development: The case of Katse villages in Lesotho. *Athens Journal of Tourism, 6*(2), 123–140. doi:10.30958/ajt.6-2-4

UNWTO. (2018). *Tourism for development- volume II: Good practices*. Madrid, Spain: World Tourism Organisation. Retrieved from www.e-unwto.org/doi/pdf/10.18111/9789284419746 [accessed on 2nd June, 2019].

Weicheselgartner, J., & Kasperson, R. (2010). Barriers in the science-policy-practice interface: Toward a knowledge- action- system in global environmental change research. *Global Environment Change, 20*, 266–277.

Weir, B. (2017). Climate change and tourism – are we forgetting lessons from the past? *Journal of Hospitality and Tourism Management, 32*, 108–114.

4 The context and future of tourism in Africa's national parks

Could privatization within protected areas be the panacea?

Cleopas Njerekai

Introduction

According to the International Union for the Conservation of Nature (IUCN, 2016), national parks are clearly defined and publicly managed geographical areas established to achieve the long-term conservation of nature and the ecosystems and cultural values within them. They are supposed to be intensive zones of protection for flora and fauna, and these parks vary in type, shape and size (Europarc Federation, 2010). Today, some of these parks cut across national boundaries, giving rise to vast conservation areas commonly referred to as Trans-frontier Conservation Areas (TFCAs) or 'Peace Parks'. In order to give the readership an idea of the size of such amalgams, South Africa, Zimbabwe and Mozambique's Greater Limpopo Trans-frontier Park equates to the land area of England in terms of size. The Kruger National Park alone is the same size as Wales or Israel. (Hanks, 2003).

The African continent is home to some of the most extensive (largest by area) and magnificent national parks in the world, such as Kafue National Park (Zambia), Etosha National Park (Nambiba), Central Kalahari National Park (Botswana) and Luvushi Manda National Park (Malawi). These parks are renowned for their natural beauty and diverse wildlife and have become the essence of tourism in most key African tourist destinations such as Tanzania, Kenya, South Africa, Zimbabwe, the Seychelles and many more. The parks cater to discerning travelers akin to visiting unique and largely un-spoilt locations away from mainstream tourism destinations. These parks are also of immense value to the African economies in which they are housed because, besides biodiversity conservation, they contribute significantly to national income, investment, employment, poverty alleviation and overall community development (Lindsey, 2008). As an example, Zimbabwe national parks occupy almost 12.5% of the country's total land area and contribute immensely to the country's total investment (Zimparks Handbook, 2015). These above-cited park roles are being fulfilled through a number of tourism and conservation activities, summarised in Table 4.1.

Despite the value of the national parks to the continent's tourism and national economic development, a plethora of issues and challenges are bedeviling their

Table 4.1 The key tourism and conservation activities pursued by Africa's national parks

Tourism-related activities	*Conservation activities*
Camping, fishing, boat cruising, rafting, game viewing, birdwatching, accommodation, canopy tourism, zip lining, mountain climbing, cycling, scuba diving, snorkeling and small to medium enterprise development	Protection of wildlife, breeding and the creation of organism banks, research and development, education and training, formulation of legislation, sourcing and provision of funding for the rehabilitation and conservation of the parks

operations. These key issues and challenges were aptly highlighted at the 1st Pan-African Conference on "Sustainable Tourism Management in African National Parks," which this writer attended, and was organized by the United Nations World Tourism Organisation (UNWTO). The conference was organized in order to share experiences and map the way forward in relation to conservation and the development of tourism in these areas. Although the conference was last held in Arusha, Tanzania, in October 2012, these challenges have remained persistent, and the search for more permanent solutions should continue. More than 30 out of the 54 countries on the continent were represented, and the presenters at the conference were highly qualified professionals and practitioners in the management of national parks across the continent. This chapter seeks to bring these key issues and challenges to the fore and then present and critique one business model, which received overwhelming support by the presenters at the conference, to enhance tourism, conservation and community development in the continent's national parks. The idea is to instigate further debate on the model and possibly invite even more and better models to manage the tourism conservation conundrum in the continent's national parks. The academic problem addressed by this chapter relates to the lack of an over-arching model to address the key issues and challenges facing the development of tourism in the continent's national parks. This chapter, therefore, seeks to:

(a) examine the context in which tourism is occurring in Africa's national parks as presented at the above conference, and then
(b) present and discuss proposed models intended to enhance tourism, conservation and community development in the continent's national parks.

Brief literature review and knowledge gaps addressed by the chapter

There is abundant literature on the key issues and challenges affecting tourism in Africa's national parks. However, most of these studies have targeted individual parks to the effect that a more holistic and continental picture of these challenges is missing. Given the size of these parks and their associated issues of accessibility and funding, studies spanning several countries to the magnitude of that represented in this chapter are almost inconceivable. In addition, most literature on the continent's national parks has been skewed towards conservation (Mahumuza & Balkwill, 2013). Research that has been undertaken on tourism in these parks has also focused more on the positive and negative impacts of tourism development in these areas and not necessarily on the future of tourism in the parks (Finnessay, 2012). Furthermore, although there are various tourism and conservation models that have been put in place to ensure sustainable tourism development in the continent's national parks, wildlife populations in the continent's national parks is continuing to decline at an alarming rate (Heltne, 2019). Comparative studies of these models in terms of their ability to address the key issues and challenges facing tourism development and wildlife conservation in these parks are scarce. In fact, individual successful or unsuccessful models dominate the literature. This knowledge gap needs to be filled so that the best model or models can be adopted to enhance tourism, conservation and community development in these parks. This is the main objective of this chapter.

Methodology

Both primary and secondary data was gathered to produce this chapter. The primary data was gathered from the proceedings of the 1st Pan-African Conference on "Sustainable Tourism Management in African National Parks" held in Arusha, Tanzania, in October 2012, which this researcher attended. This conference was organized by the UNWTO, and its main objective was to better position "park tourism" as one of the core components of overall national park management in Africa and to acquire an overview of the current challenges in terms of the demand and supply chain. The conference also intended to share experiences and map the way forward in relation to conservation and the development of tourism in these areas. Since then, no follow-up conference on the sustainable development of these parks has been held in Africa to date. Thirty-one out of the 54 countries on the continent were represented at the conference, and highly qualified professionals and practitioners in the management of national parks across the continent presented at this conference. The presentation titles are appended to the chapter references. To gather information on the key issues and challenges faced in the development of tourism in the continent's national parks, the researcher took detailed notes of these as they were presented and then later had informal discussions with the presenters where need for clarification arose. Three models to address the raised key issues and challenges emerged from this conference. The researcher again took detailed notes of the models during the presentations and then later on collected hard and soft copies of the PowerPoint presentations to fortify the notes. The researcher then developed and administered a scale to rank the extent to which each model addressed the key issues and challenges raised at the conference. The scale had the key issues and challenges on the vertical scale and the proposed models on the horizontal. The respondents were only expected to tick the model which they thought most effectively addressed each challenge. The 30 presenters at the conference were then taken as the key informants and hence a non-probability sampling technique of the key informant type was used to gather the data. Therefore, based on their opinions, the presenters ranked the models. In view of the discussed targeted respondents and the sampling and data collection methods, a qualitative design was adopted to gather data for the chapter. Journal articles related to the management of tourism in national parks provided the secondary data.

Findings

Key issues and challenges in the development of tourism in Africa's national parks

This conference revealed that tourism in Africa's national parks was occurring in an operating environment where civil unrest, human rights' abuse, crime and other conflicts were rampant. The result has been an image crisis, and, for some destinations such as Burundi, Gabon and Uganda, it has been and still is difficult to attract and retain clientele. If this trend continues, a continental image crisis is looming. However, on a more positive note, the conference revealed that most of the clientele to the continent's national parks were now predominantly comprised of visitors from fellow African countries and hence fears of over-reliance on volatile foreign markets were allayed. As an example, in South Africa, the domestic visitation figure to these parks was 77.30%. The conference also revealed that most of the continent's national parks were located in expansive, dry and fragile environments which were difficult to monitor and control. Shortages of resources were rampant in these areas, leading to numerous human–wildlife conflicts across the continent. Of particular note were the shortages of water for wildlife. Most of the national parks experiencing such shortages had resorted to sinking boreholes and installing water pumps to provide for this resource at a cost. Since most of these parks were dry most of the year, veld fires, once started, spread rapidly, destroying flora and fauna and hence the very habitats that sustain these areas. Once the flora has been destroyed, the soils become acidic and certain types of species start disappearing, for example, wildebeests, teak forests and many others. Another slow but imperceptible threat to tourism in Africa's national parks was environmental pollution. This was being compounded by locals and visitors through a culture of irresponsible waste disposal. Tourism to the continent's national parks is also seasonal. As an example, Mozambique's Gorongosa National Park faced this problem, and hence the park closed annually from December to March due to flooding of the Zambezi River. For the rest of the parks, the wet season was reportedly causing accessibility challenges. The conference therefore pointed out that there was a need to research the possibilities of ensuring that tourism to the continent's parks peaked throughout. At Gorongosa National Park, environmental degradation was also being caused by high levels of poverty in the surrounding communities, which forced them to rely heavily on park resources for their sustenance. In addition, slash and burn agriculture, illegal logging, wildlife poaching and pollution from illegal alluvial gold mining were destroying ecosystems in the park. Seychelles cited activities which were inherently destructive to the environment, such as scuba diving, yachting, and snorkeling coupled with visitors and locals with a culture of irresponsible environmental behaviour. Almost all of the presenters also sighted the problem of climate change and global warming as causing water points to recede or dry up, hence creating a serious challenge to the survival of flora and fauna in these national parks. Seychelles cited coral bleaching, coastal erosion and a rise in sea level as the key indicators, while for Tanzania, two (2) dams had already dried up. Today, the number of dams drying up on the continent cannot certainly be decreasing. If this problem persists, a looming ecological disaster awaits the continent's national parks and protected areas.

The presentations also revealed that the continent's national parks were the custodians of some of the world's most endangered species of flora and fauna. Because of the high value attached to some of the fauna, for example, the Big Five animals, tourism in these areas was occurring amid rampant poaching perpetrated by locals, foreign nationals and even tourists themselves, some of whom were coming in as wolves in sheep's clothing. This challenge was cited by almost all presenters at the conference either directly or indirectly. The problem was cited for Gorongoza National Park (Mozambique), Victoria Falls National Park (Zimbabwe), Seychelles, South Africa and many more. The existence of poaching syndicates, poorly paid anti-poaching personnel and the resultant corruption was compounding this problem.

The conference affirmed that tourism to Africa's national parks was occurring amidst a plethora of stakeholders, such as animal conservationists, local communities, governments, NGOs and international conventions, for example, CITES, tourism organizations and many more whose roles were not well coordinated and in most cases contradictory. This diversity is confirmed by various authors such as Vigor and Healy (2002). This was creating conflicts in the management of these resources. As an example, animal activists may argue that artificial game parks and other wildlife-based tourist activities should be banned, as they are excellent examples of animal cruelty. The management of Africa's protected areas is therefore in a quandary. Of particular concern were the elephant populations in some parts of the continent, such as Zimbabwe, which CITES has declared endangered on a world scale and therefore are not to be killed. This declaration remains in place despite these animals continuing to wreak havoc and threaten the existence of whole ecosystems where carrying capacities have been exceeded. At the Stanly and Livingstone Private Game Park, the elephants, especially the bull elephants, were uprooting trees as a show of power, while their kin, ring-bucked trees and left them to wither and die. This had led to serious policy considerations, with international conventions and communities declaring certain species as endangered while the locals co-existing with the species witnessed massive destruction of the environment, increased human–wildlife conflict and far-reaching alterations to whole ecosystems. The elephants are the greatest destroyers of habitats because they have very few predators.

The discovery of precious minerals in some of the continent's national parks was also causing land-use conflicts within these areas. Given the bankruptcy that most African governments are facing, wildlife conservation is always playing second fiddle to mining of these mineral deposits. These discoveries are also strengthening the belief that these national parks were initially established to conceal precious minerals from exploitation by the locals.

The recent ban on trophy hunting by some countries such as Botswana (Mbaiwa, 2018) and the ban on the transportation of trophy-hunted products by some countries such as the USA were also affecting tourism in the continent's national parks. These measures have been taken to curb poaching and a full cost-benefit analysis of this development needs to be undertaken for each individual country.

Due to limited direct flights and poorly developed internal air transport systems in many countries in Africa, many national parks on the continent were, therefore, equally inaccessible. This situation did and continues to deprive these tourist attractions of potential financial benefits. Resources to effectively popularise and market the parks' product offerings were thus limited. South Africa emerged as the country that was putting a lot of effort in this direction. The country indicated that the South African National Parks were annually allocating up to R1 million to the marketing and promotion of their parks.

The presenters to the conference also revealed that technology had brought mixed fortunes to tourism development in the continent's national parks. As an example, while technology had improved tourism and conservation activities in many ways, the same technology was also enhancing poaching through the adoption of high-communication technologies and spying equipment which were rendering anti-poaching measures ineffective. The creation of zoos and wildlife trafficking in the name of conservation was also lowering visitation levels to these parks.

In general, there was consensus at the conference that nature-based tourism, particularly in African national parks, promised to remain a large and growing global industry, partially dependent upon the attributes of the natural environment and biodiversity (wildlife and scenery) which represents one of Africa's strategic competitive advantages as a destination.

Where these were being managed and leveraged responsibly for tourism, these parks were making significant contributions to many African economies in the form of job creation, foreign exchange generation and the stimulation of SMEs. Hence, tourism was also providing additional resources for biodiversity conservation and sustainable use of public and private land.

Despite the high potential for sustainable tourism development in the continent's national parks highlighted in this chapter, many of these parks suffer from numerous constraints and limitations, and if these constraints and limitations are not addressed, the future of tourism in the continent's national parks and protected areas is bleak.

Proposed partnerships and business models to enhance tourism, conservation and community development in the continent's national parks

Three key models targeting tourism development in the continent's national parks and intended to reveal the discussed key issues and challenges in this chapter were revealed at the conference proceedings. These were as follows:

Model 1: maintaining the traditional "people out" park management model

Members to the conference were in consensus that this "people out" template that the world was currently using to manage wildlife resources was fatally flawed and doomed to fail in an increasingly crowding world. This model's probability of failure was and still is significantly high, especially on the African continent where human demands for land resources are already exceedingly high and human populations are ever increasing. There is need for a new park management model. One based on co-existence between wildlife and humans and not separation. This accurate symbiotic interpretation and understanding of nature is long overdue. Panel discussions at this conference further revealed that this model could only survive in politically peaceful and economically stable environments in which every stakeholder, including the animals themselves and conservation conscious NGOs, works for the good of each other. Such environments are increasingly becoming pipedreams for most countries on the continent (Yale environment 360, 2010).

Model 2: The Solimar International Integrated Development model

Solimar International is a sustainable tourism development and marketing firm. The firm helps destinations and organisations to develop and market sustainable tourism experiences that support conservation, preserve cultural heritages and enhance the livelihoods of the locals. Uganda is one country where this model was applied to its protected areas. The country's protected areas were facing challenges relating to a variety of problems.

Among these were the increased pressure on protected areas due to rapidly increasing populations, inadequate park revenues, stakeholder disintegration, limited market awareness and access to protected areas by guests. In view of these challenges, the government partnered with Solimar International to address them. Five basic steps were taken in the implementation of the Solimar Integrated Development model. The first step was to set up a tourism development and conservation framework using the public-private-partnership (PPP) approach. The second step involved enhancing tourism development in the protected areas. This involved an inventory and assessment of the existing tourism opportunities within the protected areas. This was followed by tourism infrastructure development and the development of new tourism products such as predator and mongoose tracking by guests.

The third step involved the enhancement of business services in the destination. In this case, an audit of the existing tourism business and other ancillary services was undertaken to determine gaps and opportunities. Thereafter, a business development program, including training of personnel, provision of one-on-one technical support and the development of market linkages was then undertaken. As an example, the Peals of Uganda market linkage development enhanced 40 local tourism products and worked with 120 national tourism businesses. Other tourism business plans were also developed and implemented within the national parks.

The fourth step involved the building of partnerships in support of the destination in which the interests of tourism and conservation development partners were taken on board. The participation of non-traditional partners (e.g. media) was also leveraged. Catalytic actions for tourism development and conservation were also agreed upon by the stakeholders, including the Pearls of Uganda and Friends of Queen Elizabeth National Park.

The final step was to promote the destination through high-quality descriptive and visual content of the country's enhanced facilities and attractions. Web-based and physical promotional materials were developed, and marketing training was undertaken. All of this resulted in the development of the Uganda Wildlife Authority Web Portal and voluminous sales and marketing materials for the parks. Within one year, page views for the website increased from 18,000 to 114,000. Targeted marketing and outreach were then used to increase visitation, length of stay and to boost revenues. Figure 4.1 summarises these steps.

The net impact of this project included:

- 60% average increase in revenue for targeted community enterprises;
- a 27% increase in household incomes in the communities around national parks;
- a 35% reduction in people's abject poverty in the surrounding communities;
- a 20% increase in people living over the poverty datum line;
- a 16% increase in park visitation, and,
- Over US$4 million leveraged in partnership support for conservation and community activities.

One great lesson from this presentation was that, for tourism/conservation and the local communities to co-exist well, wildlife must have a positive place in the livelihoods and mindsets of the surrounding communities.

Model 3: wildlife privatization within protected areas

The Stanly and Livingstone private game park in Victoria Falls, Zimbabwe, proposed this as the panacea to the continent's tourism-conservation and

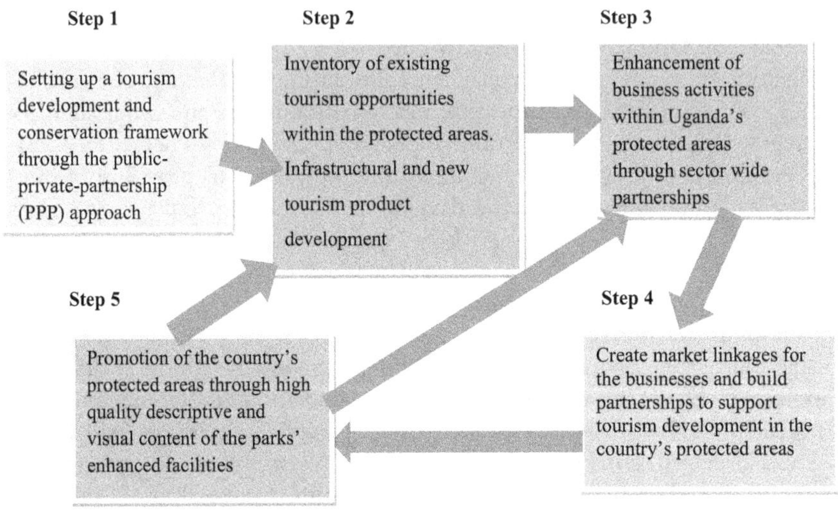

Figure 4.1 A summary of the steps involved in applying the Solimar International Integrated Development Model to Uganda's protected areas

Source: The author (2019)

community development conundrum. The park is approximately 2,500 hectares and 10km from Victoria Falls. The reserve is an Intensive Protection Zone (IPZ) for black rhinos, the Big Five and other flora and fauna. Its model is anchored on the understanding that most governments richly endowed with these wildlife resources on the continent lacked the capacity to conserve and protect them and that these national parks were too expansive to monitor and control. In worst case scenarios, one game ranger can be expected to patrol 10 km^2 of game area daily and on foot! In terms of conservation, the Stanley and Livingstone Private Game Park has managed to establish its own Anti-Poaching Unit (APU) to augment government efforts to protect endangered wildlife and other flagship species prone to poaching. It has also resorted to the employment of Game Scouts and fully armed Support Unit police officers and soldiers to protect the animals. Helicopter patrols are also undertaken to scare away poachers. All animals in the game park are under 24-hour surveillance through CCTV. The park has a full-time conservationist to ensure that eco-friendly practices resonate throughout the private game park. In 2012, US$200,000 worth of conservation equipment for the park was procured. Due to these conservation efforts, animal populations, including the hotly endangered ones such as the rhinos, have greatly improved while they are rapidly dwindling in the rest of the park authority- and government-owned national parks.

In terms of community engagement and development, the park has introduced a day centre and rhino escapade to educate local communities on wildlife issues. All of its employees are locals, and the park has a two-tier pricing system to encourage locals to participate in tourism. The park also protects community livestock from wildlife attacks through the lodge's APU. The village tours that the park embarks on encourage the buying of local crafts by tourists, resulting in locals benefitting financially and in terms of instilling pride in their local heritage and traditions. The park also adopted the Mzipah Primary School in the rural areas of Victoria Falls and donated mosquito nets, textbooks and stationery to the school and the local communities. Ablution facilities, classrooms and teachers' accommodation were also upgraded.

Two boreholes were sunk and water reticulation set up for the villages around the park, and the roads within the local communities are being maintained. The Victoria Falls public swimming pool has been resuscitated and locals are enjoying swimming. The park plans to rehabilitate the dog kennels at the local police station and to adopt and pay for one underprivileged university student at the Camp Nakavango student centre which it runs.

However, and in general, in all of the proposed models, the researcher notes with concern that the local communities in and around the national parks and other protected areas are only receivers and not initiators of tourism development in these areas. Secondly, there is no "one-size-fits-all" model to the tourism, conservation and community development conundrum facing the continent's national parks. A concoction or multi-pronged approaches could be the best way forward from here. However, one thing for certain is that any solution or model which excludes, disregards and undermines the demands and wishes of the people who live adjacent to these national parks is certainly doomed to fail.

Discussion

This conference demonstrated that the traditional "parks" concept, with its "people out" mantra, was rapidly losing popularity as a tourism and conservation strategy, and perhaps in accordance with Peace (2010)'s observation, there is a need to replace this approach with a new approach or approaches that are centred, not on separation, but on mutual benefit and coexistence between wildlife and humans. As enshrined in the Arusha Declaration after the 2012 conference, such approaches should directly involve local communities and ensure that they get concrete benefits in terms of ownership, benefit sharing and their general well-being. The Solimar International Development model and the privatization of game within protected areas received overwhelming support from 99% of the respondents. This support is buttressed by several other individual success stories on the continent where such approaches or tourism development models have been adopted. The privatization of game within national parks was strongly advocated for in terms of the model's ability to address issues relating to park extensiveness and the resultant challenges of park patrols, monitoring and control.

Conference resolutions and further recommendations

Given the models discussed in this chapter and noting that the above key issues and challenges were seriously threatening the future of tourism and conservation in Africa's national parks and protected areas, the following were the conference's resolutions:

Funding

There was a need for the continent's national parks to be weaned from government funding and be fully commercialized. In its 2012–2022 Responsible Tourism strategy, South Africa indicated that its government had moved away from providing day-to-day operational funding and was steadily moving into once-off project-based grant funding. To this effect only 13.5% of its funding for national parks now came from the government.

Planning

South Africa also pointed out that it was infeasible and counterproductive to "make all parks mean everything to everybody". Meaningful differentiation of the continent's national parks based on their strengths, potential and contributions was therefore necessary. Calls were therefore made by the presenters for the continent to stop planning in silos and to adopt a cartel approach to this resource, just like the oil-producing companies in North Africa.

Emerging destinations and new product development

The content's national parks were challenged to come up with new tourist products. In his parting message, the Presidential Advisor to the Republic of Gabon encouraged conference delegates to visit Gabon because, "something special was happening there." Whether this was a marketing gimmick or not still needs to be verified. Examples of new and exciting products revealed included canopy and guerilla tourism in Uganda and the sale of the Coco De Mer Nuts in Seychelles. The Coco De Mer seed is the world's largest seed, weighing up to 17 kilograms!

Knowledge exchange and dissemination of best practices

The conference recommended that there be more platforms for the dissemination of best practices, exchange of knowledge and technical experiences between African countries aimed at the sustainable development of tourism and conservation in the continent's national parks and other protected areas.

To this effect, it was proposed that this conference be an annual event held on a rotational basis in order to maintain a permanent forum for the subject. The United Nations World Tourism Organisation (UNWTO), in association with its member States as well as concerned International Organizations and NGOs, also urged African countries to consider the creation of observation stations to develop scientific, environmental, economic knowledge and social awareness regarding national parks and protected areas on the continent.

Community engagement

There was also a need to directly engage and co-opt local communities in the management of the continent's national parks to ensure that they gain concrete benefits in terms of employment and income generation.

Improved stakeholder collaboration

The need to create synergies among the relevant stakeholders and especially between the public authorities and private sector in the management of national parks across the continent was also highlighted. In this case countries were encouraged to consider opportunities for closer collaboration on ways to improve policies for and management of tourism in these national parks. This included the creation of more transfrontier conservation areas and the establishment of regional centres of excellence in tourism and conservation in the continent's national parks.

Conclusion

Overall, this chapter concludes that tourism in Africa's national parks is occurring amidst a plethora of key issues and challenges which need urgent attention. If these issues are left unattended, wholesome changes to park ecosystems will certainly occur, leading to the extinction of several species of flora and fauna on the continent. In the wake of these challenges, a paradigm shift in the management of the continent's park resources could rescue the situation. In this case, it is recommended that the resolutions passed at the UNWTO's 1st Pan-African Conference on "Sustainable Tourism Management in African National Parks" held in Arusha, Tanzania, in October 2012, be urgently actioned to secure the sustainability of tourism in the continent's national parks and other protected areas.

References

Europarc Federation. (2010). *European Charter for sustainable tourism in protected areas: Guide to sustainable tourism development in protected areas.* Europarc Federation, Regensburg.

Finnessay, L. (2012). *The negative effects of tourism on national parks in the United Sates.* A Thesis Submitted to Johnson & Wales University in partial fulfillment of the requirements for the degree of Bachelor of Science under the Honors Program.

Hanks, J. (2003). Transfrontier Conservation Areas (TFCAs) in Southern Africa. *Journal of Sustainable Forestry*, *17*(1–2), 127–148.

Heltne. (2019, February). A comeback for African National Parks. *New York Times*, February issue.

Hoole, A., & Berkes, F. (2010). Breaking down fences: Recoupling social-ecological systems for biodiversity conservation in Namibia. *Geoforum*, *41*(2), 304–317.

International Union on Conservation of Nature (IUCN), (2016). *An Annual Report of 2016*, IUCN, Gland.

Lindsey, P.A. (2008). Trophy hunting in Sub Saharan Africa, economic scale and conservation significance. *Best Practices in Sustainable Hunting*, *1*, 41–47.

Mbaiwa, J. E. (2018). Effects of the safari hunting tourism ban on rural livelihoods and wildlife conservation in Northern Botswana. *South African Geographical Journal*, 100(1), 41–61.

Muhumuza, M., & Balkwill, K. (2013). Factors affecting the success of conserving biodiversity in national parks: a review of case studies from Africa. *International Journal of Biodiversity*, 2013, 1–20.

Peace, F. (2010). Why Africa's national parks are failing to save wildlife. *The Yale Environment 360*, Yale school of Forestry and Environmental studies, USA.

Vigar, G., & Healy, P. (2002). Developing environmentally respectful policy programs: Five key principles, *Journal of Environmental Planning and Management*, 45(4), 517–532.

Vigor and Healy. (2002). Developing environmentally respectful policy programs: Five key principles, *Journal of Environmental Planning and Management*, *45*(4), 517–532.

Zimbabwe Parks and Wildlife Management Authority. (2015). *Hunting in Zimbabwe*. Zimparks Handbook.

Conference presenters and presentation titles

Business perspective on tourism products and marketing in protected areas, Serena Hotels in East Africa- *Mr. David Sem, Country Sales Manager, Serena Hotels, Tanzania.*

Community Involvement in conservation activities, Ngorongoro Conservation Area, Tanzania – *Dr. Justus Muumba, Ngorongoro, Conservation Area Authority, Tanzania.*

Community Involvement in Tourism in iSimangaliso Wetland Park and World Heritage Site – *Mr. Andrew Zaloumis, CEO, Kwa-Zulu Natal, South Africa.*

Community Tourism Programme, African Conservation Centre, Nairobi, Kenya.

Deserts and Ecotourism, Case of Algeria – *Mrs. Chérifa Bensedik, Professor at the National School of Tourism, Algeria.*

Developing and Marketing Tourism Products: A Case of Stanley and Livingstone Lodge, Zimbabwe – *Mr. Vincent Makamure, General Manager, Stanley and Livingstone Logde, Zimbabwe and Cleopas Njerekai, Midlands State University, Zimbabwe.*

The Economic Contribution of Tourism in Namibia's Protected Areas – *Mr. Colgar Sikopo, Director of Parks and Wildlife Management Director of Tourism, Ministry of Environment and Tourism, Namibia.*

Gorongosa National Park, Mozambique – *Mr. Mateus Mutemba, Gorongosa Park Warden, Mozambique.*

Legislation for management and development of Tourism in natural parks and protected areas and coordination among other related institutions in South Africa – *H.E. Ms. Tokozile Xasa, Deputy Minister of Tourism, South Africa.*

Local Government Involvement in Tourism and Conservation, Uganda – *Mr. Simon Jones, USAID, Uganda.*

Maputo Elephant Reserve: setting up and implementing robust processes for concessioning and investment promotion – *Mr. Armando Uleva Guenha, Project manager and technical advisor at Ministry of Tourism, Mozambique.*

Marine Tourism, Seychelles – *Mr. Senaratne Nimhan Mariel, Director of Environmental Engineering and Wetlands Section, Ministry of Environment and Energy, Seychelles.*

National Overview of Tourism in Protected Areas in Tanzania – *Mr. Allan Kijazi, Tanzania National Parks (TANAPA), Tanzania.*

National Approach to Community Benefits, *Zambia.*

A Planning Framework for South African National Parks: Improving Governance – *Dr. Nomvuselelo Söngelwa, Senior General Manager, Park Operations, SAN Parks, South Africa.*

Private Sector Partnership Model for Conservation-Based Tourism, Laikipia Wildlife Forum, Kenya – *Dr. Mordecai Ogada, Executive Director, Laikipia Wildlife Forum, Kenya.*

Site Level Approach to Community Involvement- Parc W and *Mrs. Rosalie Balima – Tourism Director-General, Ministry of Culture and Tourism, Burkina Faso, Faso and Mr. Pierre Kafando – National Park Department, Burkina Faso.*

Sustainable Tourism, Biodiversity and Protected Areas: Challenges and Opportunities in Africa *Mr. Frédéric Pierret – UNWTO Executive Director.*

Tour operator and consumer experience – *Mr. Vincent Fonvieille, General Director of La Balaguère, France.*

Tourism and Conservation In Marine Protected Areas, Tanzania – *Dr. Abdillah Chande, Marine Parks and Reserves, Tanzania.*

Tourism Business Built Around Conservation and Wildlife Watching, Asilia, South Africa – *Mr. Jeroen Harderwijk, Co-Founder and Managing Director, South Africa.*

Towards a Sustainable Tourism in Eastern Africa: Challenges and Opportunities for Nature-Based Tourism – *Dr Geoffrey Manyara, Senior Regional Tourism Adviser UNECA (SRO-EA), Ghana.*

Wildlife Watching, Case of Gabon – *Mrs. Annie Blondel, President's advisor in charge of Tourism, National Parks Agency, Gabon.*

Working with the tourism sector to develop and manage tourism in Kenya's National Parks (including tendering process for concessions) – *Mr. Benjamin W. Kavu, Deputy Director, Kenya Wildlife Service, Kenya.*

5 A review of community social upliftment practices by tourism multinational companies in Botswana

Wame L Hambira

Introduction

Developing countries continue to hold tourism in high regard because of its immense contribution to economic growth (Niang, 2014; Rogerson & Visser, 2011). Wildlife and pristine biological diversity are the main drivers of the tourism industry in many Southern African countries, resulting in nature-based tourism being adulated by national governments, including Botswana. Botswana boasts of iconic flora and fauna straddling the renowned tourist attractions of the Okavango Delta, Moremi Game Reserve, Chobe National Park, as well as Southern and Central Kalahari. These ecosystems provide unique wilderness experience and scenic beauty, including a taste of Botswana culture and heritage. The relatively pristine nature of the biotas has since attracted ever-increasing numbers of tourists from around the globe (see Botswana Tourism Development Program (BTDP), 2000; Department of Tourism (DoT), 2010). This has earned the country revenue used in solving economic and social challenges such as high unemployment rates, lack of foreign exchange, poor health and education facilities and growing populations (Kalikawe, 2001). Community livelihoods have in the process been improved directly or indirectly through poverty reduction, enhanced wellbeing, gender empowerment, inclusive quality education and reduction of income inequality. Hence participation of rural communities in tourism through programs such as Community Based Natural Resources Management (CBNRM) is an avenue for sustainable community livelihoods.

Even though the main revenue source for the Botswana economy is diamond exports, they are a non-renewable resource available in fixed stock, which implies possible exhaustion in the long run. Moreover, the global diamond industry is losing oligopoly status while conflict diamonds and artificial diamonds make diamond mining unsustainable. For these reasons, tourism remains at the top of the list, as the Botswana government explores some other means of sustaining the economy (Kalikawe, 2001). In 2017, Travel and Tourism contributed 3.8% to Botswana's Gross Domestic Product (GDP) and was estimated to grow by 5.8% in 2018 (The World Tourism & Travel Council (WTTC), 2018). However, even though tourism contributes significantly to the country's

GDP, the question still remains regarding the extent of the trickledown effect to community livelihoods. The challenges for this realisation emanate from the policy environment, which, in some instances, counteracts sustainability efforts, especially with regards to the social pillar of sustainable development. Other emerging challenges are the effects of climate change and variability on the natural capital, which forms the basis for the country's tourism industry.

It has also been argued that tourism is a private sector business (Manwa & Manwa, 2014) which utilises natural and cultural resources of the poor (Ashley & Roe, 2002). In Botswana, investment by Multinational National Companies (MNCs) who own luxury hotels, camps and mobile safaris has been encouraged by government through the "low volume high income policy". The MNCs, therefore provide luxurious accommodation and camping facilities as well as personalised high-end tours and expeditions. The fact that the tourism products that the MNCs have invested in occur in local communities' environs make it imperative for the companies to play a significant role in ensuring the communities' socio-economic wellbeing is upheld. The role of MNCs in this regard has become paramount, especially in the advent of the global sustainable development agenda which advocates for inclusive development. According to the United Nations World Tourism Organisation (UNWTO) and Global Compact Network Spain (GCNS) (2016), tourism contributes directly and indirectly to Sustainable Development Goals (SDGs). For instance, target 12b of SDG number 12 calls for the development and implementation of tools to monitor sustainable development impacts for sustainable tourism which creates jobs, promotes local culture and products (UNWTO & GCNS, 2016). Through partnerships with governments and communities, MNCs are needed in providing inclusive and quality education; access to good health and housing; conducting business operations that take cognizance of the local context, especially cultural diversity and traditions; and taking into account vulnerable groups such as youth, people with disabilities and women in provision of social upliftment opportunities, thus promoting equality of human rights (Global Reporting Initiative (GRI), 2016; Sultana, Zulkifli, & Zainal, 2018). The companies have participated in this endeavour by evoking Corporate Social Responsibility (CSR) initiatives. This is viewed as responsible tourism, especially when the enhancement of community quality of life becomes apparent (Mathew & Sreejesh, 2016).

Notwithstanding what is expected of MNCs, and with wilderness tourism being one of the fastest-growing forms of tourism (Mbaiwa, 2017), there are concerns that some of the poorest communities are found in tourism prime areas where this kind of tourism takes place. Scheyvens (2009) posits that tourism mainly serves the interests of tourists and investors and, therefore, does not automatically improve the welfare of the poorer members of the community. The purpose of this chapter is therefore to provide a review of social upliftment practices by tourism multinational companies in Botswana. Social upliftment is a term commonly used in Botswana Government development framework documents to refer to programs and activities aimed at social transformation and improved living standards for citizens. Ultimately the chapter seeks to address the questions: what social upliftment practices have been adopted by tourism MNCs in Botswana in order to contribute to sustainable livelihoods of tourism dependent communities; what are the gaps thereof? In the end, recommendations on how to address the shortcomings are provided for maximum community impact.

The politics of 'sustainable' tourism and community impact in Botswana

The luring of MNC investors into the tourism industry should not be at the expense of the communities amongst whom they operate. Political will is therefore critical in ensuring that economic benefits accrued from the tourism industry through MNCs trickle down to the concerned communities, thereby improving their quality of life. There have been some impediments in the realisation of this ideal scenario in the case of Botswana. A case in point is the repercussions of the Botswana National Tourism Policy of 1990 which promotes "high-cost, low-volume' tourism aimed at safeguarding the pristine destinations. This has resulted in prime concession areas only being accessible to up-market companies which have the financial power to invest in the lucrative tourism industry, and these companies are mainly multinationals (Mbaiwa, 2005; Mbaiwa & Hambira, 2019). They offer luxurious accommodations and non-consumptive tourist activities comprising photographic safaris, walking safaris, riding safaris, and game viewing (Ecosurv, 2012; Plantec Africa, GIS-Plan, & Fameventures, 2012; United Nations Conference on Trade and Development (UNCTAD), 2008).

While this is desirable for the much-needed foreign revenue, the control of the industry by multinationals has led to the relegation of small tourism companies, thereby compromising citizen participation in the tourism industry (Plantec Africa et al., 2012). Furthermore, even though one of the objectives of the Botswana National Tourism Policy of 1990 was to curb revenue leakages, this has proven difficult to deal with given that multinational tourism corporations are often associated with offshore banking. Hence revenue that could be used meaningfully on local communities has often leaked outside tourism destination areas. Another impediment to the meaningful positive tourism impact on local communities is the creation of "tourism bubbles" and tourism enclaves in prime tourism areas (see Mbaiwa, 2005; Mbaiwa & Hambira, 2019). Enclaves strive to inhibit unorganized encounters with local people other than employees at resorts and attractions where the relationship is very controlled and, in general, function in a way that keeps tourists separate from the host destination's authentic culture (McFarlane-Morris, 2017). Hence tourism enclaves are characterised by physical isolation of local communities pushed into peripheral areas, socio-economic segregation of the tourist space and separation of the tourist from the socio-economic fabric of the destination (Mbaiwa, Ngwenya, & Kgathi, 2008; McFarlane-Morris, 2017).

Land-use policies, plans and programs meant to enable the industry to thrive have also in some instances disadvantaged communities. For example, the demarcation of prime tourism destinations such as the Okavango Delta into concession areas (see van der Heiden, 1991) has resulted in prime tourism areas being dominated by multinational tourism companies while local communities were marginalised (Mbaiwa et al., 2008). Similarly, in some instances, lease agreements for tourism concession areas issued to some multinational companies hold a period that exceeds that which is stipulated in land management regulations. This has been seen as further alienating citizens from participating in the tourism industry in core and prime tourism areas since the long-term lease agreements ensure that citizens will not be able to penetrate the lucrative market for a long time (Ecosurv, 2012; Plantec Africa et al., 2012). Furthermore, in 2014, all prime tourism concession areas were transferred from the Ministry of Lands and Housing to the Ministry of Environment, Wildlife and Tourism to facilitate the reservation of land for tourism development through a program known as Tourism Land Bank. The program, which has since been repealed (Mokaila, 2019), required designation of land for tourism activities in prime tourism areas, thus further excluding local communities from owning such lands (Mbaiwa & Hambira, 2019). Furthermore, government introduced a hunting ban in 2014, resulting in a significant reduction of the economic value of some of the more financially marginal concession areas. This was despite the fact that safari hunting is an important land-use option for rural communities in areas not suitable for photographic tourism (Mbaiwa & Stronza, 2011). These political challenges have compromised tourism benefits to communities, thereby raising high expectations for the tourism companies to compensate the affected communities through their social upliftment programs. If left unopposed, these political challenges would ultimately impede the country's Sustainable Development agenda of which the role of the private sector cannot be overemphasised as a key partner in "the future we want."

Multinational companies and tourism-dependent communities' resilience to climate change

Despite the challenges discussed in the preceding section, the Botswana government, through its CBNRM program, has endeavoured to empower local communities to own and manage tourism ventures. The CBNRM program's main aim is to empower local communities through the use of natural resources offered by their local environments (Government of Botswana, 2000). In the process, local communities are positioned to control or own tourism activities, which motivates them to conserve the resources from which they derive their livelihoods. Communities' active involvement in CBNRM, either independently or through Joint Ventures with multinational tourism companies, has over the years accorded them financial and social benefits. (Mbaiwa & Stronza, 2011). Communities are generally poor, thus involving them in tourism activities is a good strategy to lift them out of poverty. Social upliftment programs adopted by multinational companies can therefore enable local communities to receive a significant share of economic benefits from tourism.

Notwithstanding the aforementioned and as highlighted in the introduction, one of the greatest contemporary challenges that nature-based tourism-led economies have to contend with has to do with the resilience of communities to climatic challenges. Meteorologists have recorded the highest global temperatures in recent years (Inter-governmental Panel on Climate Change (IPCC), 2014; McGrath, 2016), rendering the sustainability of nature-based tourism vulnerable. This is due to the fact that warming, as a result of climate change, leads to shifting rainfall patterns, health hazards, droughts, floods, decreasing biodiversity, wildlife extinctions and a decline in ecosystem services, thus affecting the very base of nature-based tourism (IPCC, 2013, 2014). That is, the attractiveness of destinations; length, timing and quality of tourist seasons; as well as tourist preferences are vulnerable to climate change (IPCC, 2014; United Nations Environment Program (UNEP) & African Ministerial Conference on Environment [AMCEN], n.d). While multinational tourism companies may have the capacity to deal with the impending effects of climate change on their operations, the local communities remain relatively more vulnerable. Social upliftment programs adopted by multinational companies, therefore, should also consider communities' resilience to climate change.

Theoretical foundations: sustainable tourism development framework

This study is embedded in the Sustainable Tourism Development Framework which has its foundations in the Sustainable Development discourse. The World Commission on Environment and Development (1987) describes sustainable development as development that considers societal, economic and environmental aspects of the economy along equity lines for the benefit of current and future generations. Similarly, sustainable tourism advocates for tourism development that is not at the expense of the environment and societal wellbeing. Hence it is important to determine the extent to which tourism activities benefit not only the environment and economy but the community as well. That is, communities must be enabled to generate wealth through nature-based tourism. At the same time, since tourism destinations rely on a well-functional society, healthy environment and stable economy, it is imperative that tourism businesses ensure that the three dimensions are well maintained (Tourism for SDGs Platform, n.d.). Ultimately, companies aim to generate profit, conserve the environment they operate in and create a positive social impact through, among others, their CSR programs (Tourism for SDGs Platform, n.d.). The private sector in the tourism industry is thus pivotal in local socio-economic development in terms of employment creation, skills training and development, as well as through philanthropic development projects (Snyman, 2016). The contribution of tourism to the wellbeing of communities acts as an incentive for conservation of the natural capital essential for nature-based tourism. In the end, both the community and businesses would benefit and ultimately the economy as a whole, hence sustainable development and tourism.

Multinational tourism companies in Botswana

Multinational companies operate in their home country as well as elsewhere around the world. According to the Corporate Finance Institute (CFI) (2019), not only should a multinational company maintain operations in other countries but should actually make a foreign direct investment there. MNCs are characterised by high physical and financial asset turnover, a network of branches and subsidiaries, control, continued growth through upgrades, mergers and acquisitions, sophisticated technology, especially for production and marketing, right skills capable of handling huge funds, using advanced technology, managing workers, and running a huge business entity, forceful marketing strategies, and, last but not least, good quality products (CFI, 2019). Investments made by these big corporations are therefore expected to contribute to the quality of life of the communities in which they operate. By so doing, the companies gain customer loyalty, corporate reputation and legitimacy. For this reason, this study targeted multinational tourism companies in Botswana to appreciate the kind of social upliftment practices they have in place as part of their corporate social responsibility.

Methods

Data collection and analysis

A qualitative approach using the Content Analysis method was applied in the study. Cavanagh (1997) in Hsieh and Shannon (2005) describes Content Analysis as a flexible approach for analysing text data. Coding was done on the basis of a predetermined best practice reference list adapted from United Nations Development Programme (UNDP) and Botswana Stock Exchange (BSE) (2018). The best practice reference list indicators associated with social upliftment programs that companies could adopt included partnerships with communities, investment in sustainability projects, investor dialogue, human rights, investment in employee and community social welfare, and national aspirations. The reference list therefore informed the Content Analysis conducted on company web-based data and annual reports. The two sources of data provided the companies' sustainability or CSR information which were relevant for the study. Guthrie, Cuganesan, and Ward (n.d) acknowledges that the internet has revolutionised how companies disseminate information to their stakeholders. Hence the internet was relied upon to provide information for the content analysis. The Hospitality and Tourism Association of Botswana (HATAB) website was used to search for the multinational tourism companies in Botswana. The companies were categorised as camps and lodges (4), mobile sector (2), hotel sector (4), and services sector (3). The camps and lodges sector offered luxury accommodation, personalised high-end tours, small group self-departure expeditions, game drives, private river cruises, birdwatching, scenic flights, photographic opportunities, seasonal mokoro trips, motorised boat cruises and fishing. The hotel sector focused mainly on accommodations and conferencing. The services sector's niche is on car rental, risk management and insurance, as well as construction and facilities management.

Findings

Botswana tourism MNCs social upliftment practices

The results of the social upliftment practices adopted by the companies fell into the categories of best practice reference list: partnerships with communities, investment in sustainability projects, investor dialogue, human rights, investment in employee and community social welfare, and national aspirations. The *partnerships with communities* best practice category points to practices where companies adopt social upliftment practices such as setting aside time for community projects, fair compensation to communities that supply the companies with goods, and partnerships with academic institutions (UNDP & BSE, 2018). The results revealed that companies engage in this endeavour by providing school supplies to students and sourcing traditional food and crafts from the community. The *Investment in sustainability projects* category on the other hand refers to companies' involvement in community sustainability initiatives (UNDP & BSE, 2018). The practices that came out of the content analysis in this regard include tourism and conservation excursions for students as well as community educational and environmental projects. UNDP and BSE (2018) on the other hand describe the *Investment in community social welfare* best practice category as ways in which companies invest in community and employee education, safety and health as well as upholding gender balance ethos. Some of the practices adopted by the companies to contribute to social upliftment include providing students attachments in, for instance, the food and beverage section of the company; contributing to sport development by sponsoring sporting activities; incentivising employees with benevolent funds and study loans; as well as provision of scholarships to deserving young people. The *national aspirations indictor* refers to contribution to nationally defined development targets such as government poverty eradication programs (UNDP & BSE, 2018). From the analysis, the companies supported programs such as the Presidential Housing Appeal and the Ministry of Basic Education's adopt-a-school program and provision of food rations to the less privileged. The directed Content Analysis did not reveal company social upliftment practices associated with the investor dialogue and the human rights best practice categories. With regards to *Investor dialogue*, companies are expected to engage local communities in order to accelerate investment in sustainable livelihoods, such as farming (UNDP & BSE, 2018). As for *Human rights*, big corporations should take an active role in combating inequalities and discrimination (UNDP & BSE, 2018).

Results synthesis: social upliftment practices, sustainable development and identified gaps

The Tourism SDGs Platform (n.d.) posits that the SDGs are reframing the discussion on CSR, and by playing an active role in the achievement of SDGs, companies will discover new growth opportunities and lower their risk profiles. The results from the directed content analysis have shown that the social upliftment practices by tourism MNCs contribute largely (directly or indirectly) to SDG4 (quality education). To a lesser extent, the social upliftment activities practiced by the MNCs contributed directly or indirectly to SDG 1 (no poverty), SDG2 (zero hunger), SDG8 (decent work and economic growth), SDG11 (sustainable cities and communities) and SDG15 (life on land). MNCs engaged in tourism, the fastest-growing economic sector in the world, and should be well positioned to contribute to sustainable tourism development and have a significant impact at the community level. For example, social upliftment practices could include playing a part in national poverty reduction by promoting entrepreneurship and small businesses as well as empowering disadvantaged groups such as women (SDG1); promoting agro-tourism (SDG2); adopting policies that promote diversification through tourism value chains and ultimately enhancing positive socio-economic impacts (SDG8); supporting preservation of cultural and natural heritage assets that support tourism (SDG11); not only conserving fragile biodiversity that supports tourism but also generating revenue as an alternative livelihood to local communities (SDG15) (Tourism for SDGs Platform, n.d.).

In terms of realised gaps, the fact that there were no identified social upliftment practices associated with the investor dialogue and human rights categories of the best practice reference list means that there are gaps in how companies engage local communities in order to accelerate investment in sustainable livelihoods as well as the role that the MNCs play in combating inequalities and discrimination. Furthermore, pertinent issues associated with climate change and community resilience have been largely left out. Given that climate change has become topical because of its impact on tourism and communities, companies could improve their social upliftment efforts by financing community adaptation strategies. Infusing climate investment in CSR is one avenue through which MNCs could help societies to build resilience, thereby making an impact in the lives of vulnerable communities. Ramawati, Jiang, and DeLacy (2019, p. 1) opine that "CSR of the tourism industry can help in building community adaptive capacity to climate change both for the individuals living in a community and for the community leader to lead collective action." This is imperative as many of these MNCs operate as a result of nature-based attractions which have been found to be vulnerable to climate change (see Hambira, 2011). The impact of climate change on the tourism industry will automatically put communities in a vulnerable position (Ramawati et al., 2019) since affected communities work directly and indirectly in the tourism sector. Using CSR initiatives to build community resilience requires a comprehension of the climate change-tourism industry nexus and responding to the risks thereof (Ramawati, DeLacy, & Jiang, 2015). Hence, the MNCs community educational programs could take the two initiatives into consideration. In Bali, for example, tourism businesses' social responsibility initiatives aimed at enhancing community adaptive capacity to climate change include provision of economic power to the host community; provision of access to financial support to local people through grants or charities; providing education and skills in specific areas; provision of information and knowledge regarding climate change risks; improving health care for employees and their families; emergency planning that includes the whole community outside of the hotel buildings; promotion of utilisation and promotion of traditional knowledge. In the process, sustainable tourism is attained.

Conclusion

MNCs have a responsibility to give back to society to end poverty and enhance human well-being within the limits of the earth. Even though the actions aimed at addressing societal needs are voluntary and warranted by company profits, they do benefit the companies ultimately. The tourism multinationals in Botswana have taken considerable strides in contributing to the social upliftment of communities. For instance, the companies provide school supplies, a market for community traditional food stuffs and crafts, undertake community educational and environmental projects, offer student attachments and scholarships, contribute to sport development, as well as cover basic needs such as food and shelter. However, with the increasing knowledge of impending challenges associated with climate change, there is a growing expectation for corporate actors to exercise their moral responsibility in this regard (Allen & Craig, 2016). This chapter has therefore identified this as the main missing link with respect to social upliftment practices that the companies pursue. The chapter therefore makes a clarion call to corporations not only to consider climate change as related to their environmental activities, but it should also form a major component in their social upliftment programs. This will enhance community resilience to climate change, a major challenge facing current and future generations. The study is, however, limited in that the findings are based on information available on the company websites and annual reports (where available). Future studies should therefore aim for a survey of the companies to source the required information. A future study could also be expanded to the communities to determine the extent to which they find the MNCs beneficial to their general well-being in line with the aspirations of sustainable development.

References

Allen, M. W., & Craig, C. A. (2016). Rethinking corporate social responsibility in the age of climate change: A communication perspective. *International Journal of Corporate Social Responsibility*, 1–11.

Ashley, C., & Roe, D. (2002). Making tourism work for the poor: Strategies and challenges in South Africa. *Development Southern Africa, 19*, 61–82.

Botswana Tourism Development Program (BTDP). (2000). *Botswana tourism master plan*. Gaborone, Botswana: BTDP.

Cavanagh, S. (1997). Content analysis: concepts, methods and applications. *Nurse Researcher*, 4(3), 5–16.

Corporate Finance Institute. (2019). *Multinational corporation*. Retrieved from corporatefinanceinstitute.com/resources/knowledge/strategy/multinational-corporation/.

Department of Tourism (DoT). (2010). *Botswana tourism statistics: 2006–2009*. Gaborone, Botswana: DoT.

Ecosurv. (2012). *Management plan for controlled hunting area (CT/2, volume 2 of 2: baseline information)*. Gaborone, Botswana: Ecosurv.

Füssel, H. (2008). *Who is most affected by the impacts of climate change? The magnitude of regional climate changes is only one factor of vulnerability*. Special Issue of the Project 'Climate Change and Justice. Climate policy as a component of fair globalisation and sustainable poverty reduction' in co-operation with 'welt-sichten'. Retrieved from http://area-net.org/wp-content/uploads/2016/01/ClimateChange_poverty.pdf.

Global Reporting Initiative (GRI). (2016). *GRI's contribution to sustainable development: 2016–2020*. Retrieved from www.globalreporting.org/resourcelibrary/GRI%27s%20Contribution%20to%20Sustainable%20Development%202016-2020%20(2).pdf.

Government of Botswana. (2000). *Community-based natural resource management policy*. Gaborone, Botswana: Government Printer.

Guthrie, J., Cuganesan, S., & Ward, L. (n.d.). Legitimacy theory: A story of reporting social and environmental matters within the Australian food and beverage industry. Retrieved from http://ssrn.com/abstract=1360518http://ssrn.com/abstract=1360518.

Hambira, W. L. (2011). Screening for climate change vulnerability in Botswana's tourism sector in the bid to explore suitable adaptation measures and policy implications: A case study of the Okavango Delta. *International Journal of Tourism Policy, 4*(1), 51–65.

Hsieh, H-F., & Shannon, S. E. (2005). Three approaches to qualitative content analysis. *Qualitative Health Research, 15*(9), 1277–1288.

Intergovernmental Panel on Climate Change (IPCC). (2013). Summary for policymakers. In T. F. Stocker, D. Qin, G-K. Plattner, M. Tignor, S. K. Allen, J. Boschung, . . . P. M. Midgley (Eds.), *Climate change 2013: The physical science basis*. Contribution of Working Group I to the Fifth Assessment Report of the Intergovernmental Panel on Climate Change. Cambridge: Cambridge University Press.

Intergovernmental Panel on Climate Change (IPCC). (2014). Summary for policy makers. In C. B. Field, V. R. Barros, D. J. Dokken, K. J. Mach, M. D. Mastrandrea, T. E. Bilir, . . . L. L. White (Eds.), *Climate change 2014: Impacts, adaptation and vulnerability: Part A: Global and sectoral aspects*. Working Group II contribution to the 5th assessment report of the Intergovernmental Panel on Climate change. Cambridge: Cambridge University Press.

Kalikawe, C. (2001). *Botswana: Intergrating biodiversity into the tourism sector. A country case study*. 40. UNEP – Biodiversity Planning Support Programme. Gaborone, Botswana: Department of Tourism. Retrieved from www.cbd.int/doc/nbsap/tourism/BOTSWANA(Tourism).pdf.

Manwa, H., & Manwa F. (2014). Poverty alleviation through pro-poor tourism: The role of Botswana forest reserves. *Sustainability, 6*, 5697–5713. doi:10.3390/su6095697

Mathew, P. V., & Sreejesh, S (2016). Impact of responsible tourism on destination sustainability and quality of life of community in tourism destinations. *Journal of Hospitality and Tourism Management, 31*, 83–89.

Mbaiwa, J. E. (2005). Enclave tourism and its socio-economic impacts in the OD, Botswana. *Tourism Management, 26*(2), 157–172.

Mbaiwa, J. E. (2017). Poverty or riches: who benefits from the booming tourism industry in Botswana? *Journal of Contemporary African Studies*, 35(1), 93–112.

Mbaiwa, J. E., & Hambira, W. L. (2019). Enclaves and shadow state tourism in the Okavango Delta, Botswana. *South African Geographical Journal*. doi:10.1080/03736245.2019.1601592

Mbaiwa, J. E., Ngwenya, B. N., & Kgathi, D. L. (2008). Contending with unequal and privileged access to natural resources and land in the Okavango delta. *Singapore Tropical Geographical Journal, 28*(2), 155–172.

Mbaiwa, J. E., & Stronza, A. L. (2011). Changes in resident attitudes towards tourism development and conservation in the Okavango Delta, Botswana. *Journal of Environmental Management, 92*, 1950–1959.

McFarlane-Morris, S. A. (2017). The other side of the enclave: Local perspectives on the onset of mass tourism in Jamaica (Doctoral dissertation abstract). Retrieved from https://ir.lib.uwo.ca/etd/5062/

McGrath, M. (2016). Climate change: 2015 'shattered' global temperatures record by wide margin. *BBC News – Science and Environment*. Retrieved from www.bbc.com/news/science-environment-35354579.

Mokaila, O. K. (2019). Keynote address by the Minister of Environment, Natural Resources Management and Tourism: Hospitality and Tourism Association of Botswana (HATAB) annual conference held at Kasane under the theme 'Sustainable tourism: A driving force for job creation, economic growth and development'. Aired on Duma FM on 26th April, 2019.

Niang, I., Ruppel, O. C., Abdrabo, M. A., Essel, A., Lennard, C., Padgham, J. & Urquhart, P. (2014). Africa. In V. R. Barros, C. B. Feild., D. J. Dokken, M. D. Mastrandrea, K. J. Mach, T. E. Bilir, M. Chatterjee, K. L. Ebi, Y. O. Estrada, R. C. Genova, B. Girma, E. S. Kiseel, A. N. Levy, S. MacCracken, P. R. Matrandrea, & L. L. White (Eds.), *Climate change 2014: Impacts, adaptation, and vulnerability. Part B: Regional aspects. Contribution of working group II to the fifth assessment report of the Intergovernmental Panel on Climate Change* (pp. 1199–1265). Cambridge: Cambridge University Press.

Plantec Africa, GISPlan, & Fameventures. (2012). *Mid-term review and gap analysis of the Okavango Delta Management Plan (Scoping and gap analysis report)*. Gaborone, Botswana: Botswana Tourism Organisation.

Preston-Whyte, R., Brooks, S., & Ellery, W. (2006). Deserts and Savannah regions. In Gossling, S. & C. M. Hall (Eds.), *Tourism and global environmental change: Ecological, social, economic and political interrelationships*. London: Routledge.

Ramawati, P. I., DeLacy, T., & Jiang, M. (2015). A conceptual framework to link corporate social responsibility and climate change strategies in tourism to build community adaptive capacity. In *Proceedings of the 25th annual Council for Australasian University Tourism and Hospitality Education (CAUTHE) Conference held in Southern Cross University, Gold Coast*, 2–5 February 2015 (pp. 302–3013). Lismore, Australia: School of Business and Tourism, Southern Cross University.

Ramawati, P. I., Jiang, M., & DeLacy, T. (2019). Framework for the stakeholder collaboration in harnessing corporate social responsibility implementation in tourism destination to build community adaptive capacity to climate change. *Corporate Social Responsibility and Environmental Management*, 1–11. https://doi.org/10.1002/csr.1745.

Rogerson, C. M., & G. Visser (2011). African tourism geographies: Existing paths and new directions. *Tijdschrift voor economische en sociale geografie, 102*, 251–259.

Scheyvens, R. (2009). Pro-poor tourism: Is there value beyond the rhetoric? *Tourism Recreation Research, 34*, 191–196.

Snyman, S. (2016). The role of private sector ecotourism in local socio-economic development in southern Africa. *Journal of Ecotourism, 16*(3), 247–268.

Sultana, S. Zulkifli, N., & Zainal, D. (2018). Environmental, social and governance (ESG) and investment decision in Bangladesh. *Sustainability, 10*, 1831. doi:10.3390/su10061831

Tourism for SDGs platform (n.d) Companies CSR & SDGs: How do CSR strategies relate to SDG? – key message for private sector. Retrieved from http://tourism4sdgs.org/tourism-for-sdgs/companies-csr-sdgs/.

UNEP & AMCEN. (n.d.) Africa's adaptation gap2: Bridging the gap-mobilising sources. *Technical Report*. Retrieved from http://wedocs.unep.org/bitstream/handle/20.500.11822/9092/Africas_adaptation_gap_2.pdf?sequence=1&isAllowed=y.

United Nations Conference on Trade and Development (UNCTAD) (2008). *FDI and tourism: The development dimension – East and Southern Africa*. New York and Geneva, Switzerland: United Nations.

United Nations Development Program (UNDP) and Botswana Stock Exchange (BSE) (2018). *Profiling private sector sustainability practices in Botswana: Botswana stock exchange listed domestic companies*. Gaborone, Botswana: UNDP & BSE.

UNWTO and GCNS. (2016). *The tourism sector and the sustainable development goals: Responsible tourism, a global commitment*. Madrid: UNWTO & GCNS. Retrieved from http://cf.cdn.unwto.org/sites/all/files/pdf/turismo_responsable_omt_acc.pdf.

van der Heiden, L. 1991. *Land use and development plan: Kwando and Okavango wildlife management areas*. Maun, Botswana: Government of Botswana, District Administration.

World Commission on Environment and Development (WCED). (1987). *Our common future*. Oxford, UK: Oxford University

World Economic Forum. (2013). *The travel and tourism competitive report: Reducing barriers to economic growth and job creation*. 485p. Geneva, Switzerland: World Economic Forum.

World Tourism & Travel Council (WTTC). (2018). *Travel and tourism economic impact 2018 Botswana*. London: WTTC.

6 Navigating community conservancies and institutional complexities in Namibia

Paul Hebinck, Richard Dimba Kiaka and Rodgers Lubilo

Introduction

Since the mid-1990s, a community-based natural resource and development programme has been promoted and implemented by the Namibian state and non-state agencies to simultaneously ensure ecological sustainability and socio-economic empowerment of people living in Namibia's communal rural areas (Nuulimba & Taylor, 2015). The programme is globally known as Community Based Natural Resource Management (CBNRM). It entailed in Namibia the formation of conservancies on communal land, granting members of rural communities restricted usufruct rights over wildlife and other natural resources and to redistribute the benefits derived from such use among its registered members (Boudreaux & Nelson, 2011). Currently, there are 86 registered conservancies, covering 163,017 km^2, which is some 20% of Namibia's territory inhabited by about 227,941 people (www.nacso.org.na). The expansion and operation of communal conservancies are positively associated with the increase in the numbers and diversity of wildlife. Ecotourism and trophy hunting provide the conservancies with monetary resources for cash payouts and other benefits for community development projects (Lapeyre, 2011). The conservancy programme is hence generally lauded in Namibia and beyond as successfully combining conservation (Naidoo et al., 2016a) with the socio-economic development of rural communities (Jones, Diggle, & Thouless, 2015). For this success to materialise, the management of the conservancies is expected to be participatory and democratic in nature (de Vette, Kashululu, & Hebinck, 2012; Lubilo, 2018; Bollig, 2016). The formation of conservancies assumed and implied that new institutional forms and modes of participating in decision-making and sharing of the monetary benefits of nature had to be designed and subsequently introduced in the communities (Ostrom, 1990). This occurred with substantial financial and technical support from international environmental organisations and donors.

We argue here that the formation of conservancies did not unfold as smoothly as planned. We demonstrate here that the conservancy model is contested and at the same time is being remodelled by those social actors who are in a position to exert their power and authority. We base our argument on field research from two conservancies in Namibia: ǂKhoadi ǁHôas and Wuparo in northwest and northeast Namibia respectively (see Figure 6.1). These are amongst the first communal conservancies to be registered in Namibia and are considered in Namibia by the Integrated Rural Development and Nature Conservation (IRDNC) and Ministry of Environment and Tourism (MET) as well as by the World Wildlife Fund (WWF) to be successful and stable (Jones et al., 2015; Naidoo et al., 2016a). The data was collected at different times by the authors between 2016 and 2016; and 2013 and 2015. Data collection methods included participant observation, focus group discussions and key informant interviews with various members of the conservancies, including committee members, chairpersons and ordinary members. We also collected survey data and conducted a review of the available secondary data. Additionally we revisited the conservancies in 2018 and 2019 to observe the changes that had occurred after the fieldwork.

The chapter proceeds as follows: we first zoom in on the institutional complexities in which the communal conservancy programme became enmeshed and how the "modern" and "traditional" forms of organisations configure each other in and during the conservancy formation process. Secondly, we show that the conservancy has created spaces from which many will benefit; however, one does so more than the other. Conservancy actors thus manage to navigate the conservancy but each in their own ways, generating a variety of outcomes.

Theoretical framing

The CBNRM programme initiated in Namibia the formation of conservancies and conceived these as territory-based communities with legal status, governed by a constitution codifying the role of managers and that of elected committees in the day-to-day management. This would democratically regulate the distribution of proceeds from nature such as trophy hunting and tourism to conservancy members. The formation process is triggered by a common-pool resource perspective that asserts that individuals sharing a resource can collectively participate in modifying the rules for resource management (Ostrom, 1990). Users are positioned as motivated by economic incentives to collaboratively and communally manage the (natural) resources sustainably. The premise is that one can design, modify and enforce rules that ensure conservation and regulate income distribution fairly and that the rights to the (natural) resources can be allocated to the people who reside on the land (Murphree, 2005; Saunders, 2014). The communal conservancy model often involves collaborations between private entrepreneurs, NGO's and the state (Jones, 2010), which in the situations pertaining to nature conservation are incentivised by networks of global conservation-minded elites, global conservation NGOs, philanthropists and royalty and tourism entrepreneurs marketing nature for pleasure and profit (Van der Duim, Meyer, Saarinen, & Zellmer, 2012). Patches of communal land in Southern Africa have, as a result, been turned into commoditised conservation areas.

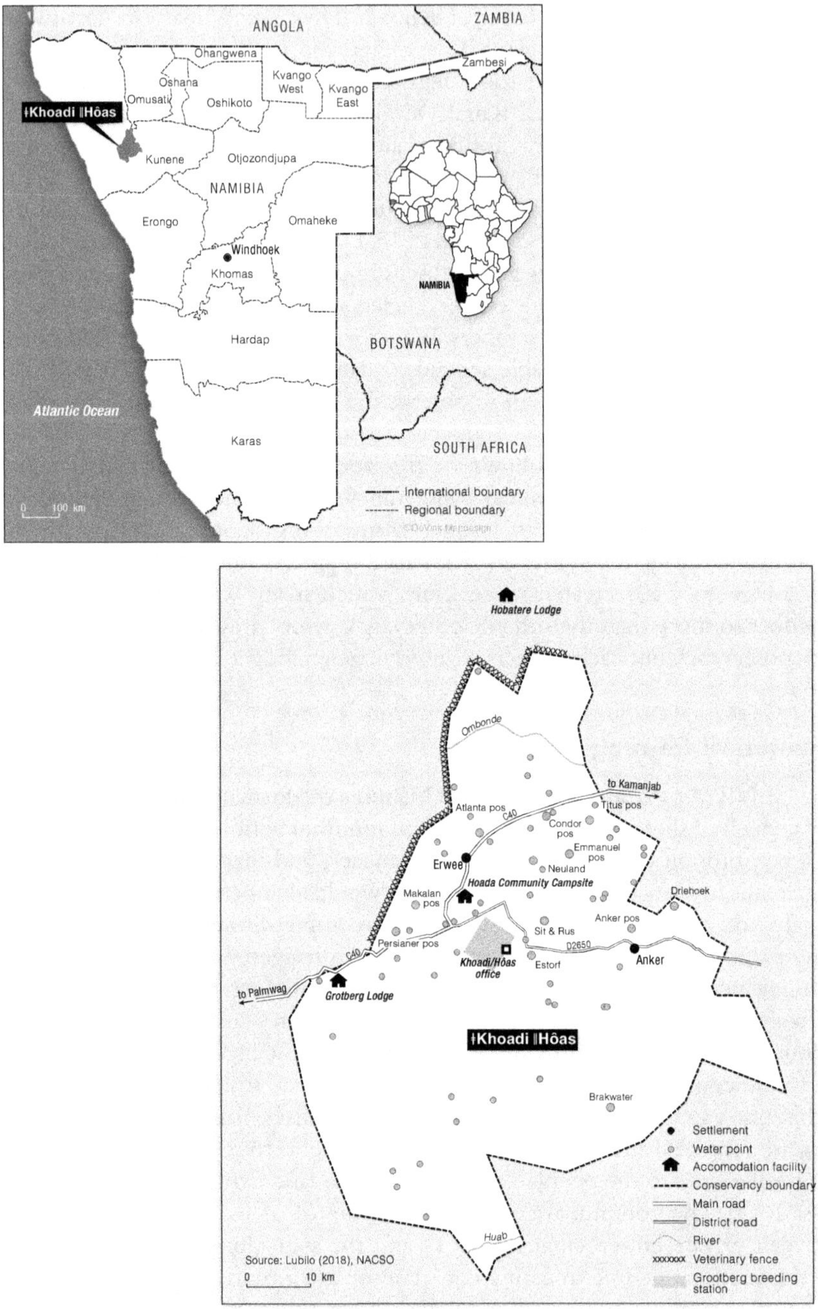

Figure 6.1 Situational map ǂKhoadi ǁHôas Conservancy

Considering that the boundaries of social institutions constituting a conservancy are pervious as a result of multiple ties and functions (Schnegg, 2018), the various social actors potentially (re)model the formation and daily management of the conservancies (Bollig, 2016; Kiaka, 2018; Lubilo, 2018). Some of these operate from "outside" the conservancy (e.g. policy makers, NGOs, labour migrants residing in Windhoek or elsewhere, tourists and hunters, lodge owners and trophy-hunting organisations) and those who work and live in the conservancy (e.g. ordinary members, chiefs and village headmen, managers, game guards). They wield asymmetrical powers and their authority shifts in and during the conservancy formation process. In considering their roles and relative influence, we need to take into account that conservancy formation processes carry elements of one or another form of controlling or disciplining (de Vette et al., 2012; Lubilo & Hebinck, 2019). Becoming registered as a conservancy entails following certain procedures, for example, the formulation of a constitution which is facilitated through training offered by NGOs and consultants, drafting of management and land-use plans and holding democratic elections to appoint committees and managers. These are rituals which have become part of the efforts to transform people into beneficiaries of development interventions (Mosse, 2005; de Vette et al., 2012) and to make them stakeholders in global conservation efforts with a local reach. The emphasis on designing and participation has unintentionally built upon pre-existing power structures, reinforcing these to the advantage of the "new" and "old" elites who hold various but differentiated positions of economic and political power. The old elites consist of chiefs and village headmen who customarily play key roles in distributing land. They nowadays form part of the Traditional Authority (TA) that are represented in the Regional and Communal Land Boards. Namibia's post-independence reforms in land, water and political administration reasserted the authority of chiefs and headmen in natural resource management.

The new elites are those in salaried positions and private businessmen. The new elites have gained in importance, particularly after Independence in 1990. Both types of elites managed well in internalising the conservancy discourse and speaking the conservancy management language. They are thus able to gain (or to maintain) privileged access to development resources. They also attach themselves to the various nodes of power and control over rights and access to land, water and other natural resources as well as benefits of tourism and trophy hunting. Platteau (2004) understands this as elite capture. Critics of CPR (Acheson, 2011; Saunders, 2014) similarly argue that designing common property management is underpinned by assumptions that ignore the complexities that are involved in what we frame here as "navigating conservancy politics and everyday life." Cleaver (2012) coined the term institutional bricolage to describe a process where actors (bricoleurs) navigate and patch institutional arrangements together "from cultural resources available to them in response to changing conditions."

Although the communal conservancy programme has been evaluated as successfully streamlining natural resource management in Namibia (Nelson &

Agrawal, 2008; Naidoo et al., 2016a; Nuulimba & Taylor, 2015), the reality is that conservation efforts have been parachuted into an existing socio-political field of institutional complexities. The institutionalisation of conservancies has empowered a range of new actors, generating, in turn, a series of interfaces and contestations with the pre-existing so-called "traditional" modes and forms of organising the use and access to natural resources.

It is in this emerging and complex socio-political field that conservation efforts were inserted. Therefore, we suggest here to conceptualise conservancies as enmeshed with other forms and modes of power and authority, configuring and reconfiguring each other. We hypothesise, however, that those in positions of power and authority are not the sole beneficiaries of conservancies and that the conservancy provides a space or a room for manoeuvre for the actors to navigate. Building on the work of Long (2001), we conceptualise a conservancy as an emergent "arena." Long (2001, p. 59) defines arena as "social locations or situations where issues, resources, values and representations contest with each other." These arenas unfold either as spaces in which contestations and struggles associated with different practices and values of different domains take place or as spaces within a single domain where attempts are made to resolve discrepancies in value interpretation and incompatibility between the various actors' interests. Moreover, where there exist different interests and values entangled with power imbalances, institutional complexities lead to micro-politics that may manifest in protracted contestations and conflicts over a resource (Saunders, 2014). This happens as the actors involved make use of the ambiguities and contradictions to navigate the complexities and carve out spaces to pursue their own ecological, social, economic and political interests.

Conservancy formation history: the case of ǂKhoadi ǁHôas and Wuparo

ǂKhoadi ǁHôas

The formation of ǂKhoadi ǁHôas Conservancy is well described in Kiaka (2018). Established in 1998 as a conservancy, in what was known as the Grootberg ward, ǂKhoadi ǁHôas has a population of about 4,300 inhabitants and occupies 3,366 km^2 of land (see Figure 6.1). Conservation NGOs consider ǂKhoadi ǁHôas as a success story and it serves as a model for CBNRM in Namibia because it currently rarely relies on donor funding. The Conservancy is well known for being the first to construct and renovate two fully community-owned tourist lodges for which it received a Community Benefit Award at the prestigious World Travel and Tourism Council's "Tourism for Tomorrow." Moreover, since 1999, ǂKhoadi ǁHôas has been a stable player in Namibia's trophy hunting industry and creates employment and cash income for the local community.

The conservancy did not unfold in an institutional vacuum. In the late 1990s, the Grootberg Farmers Union focused on rangeland management and livestock breeding improvement and marketing. The union wielded power from local farmers and became an instrument for mobilising local support for the conservancy programme. Its officials formed the founding committee of the conservancy. Furthermore, as part of the black empowerment policy of post-independence Namibia, more local people were employed and posted as extension workers, teachers and administrators in the Grootberg area. Some of these government workers were asked by their seniors in Windhoek to work with the conservancy programme team and ensure the conservancy was established. Some of them became part of the conservancy management committee or became employees.

The |Gaiodaman TA is in charge of local land administration and allocation of grazing rights. Both land and grazing rights give people residency and allow them the use of communal water points. However, the manner in which water is pumped into a communal reservoir and made available for communities is controlled by community water point committees. These water institutional arrangements have existed since early 2000 as part of a donor- and state-driven rural water decentralisation policy (Kiaka, 2018). The common usage of these resources is also shaped by informal social institutions that mediate everyday life. Although hunting is criminalised and monitored by the conservancy, kinship relations prevent people from reporting each other to the authorities (Kiaka, 2018). Trophy hunting and tourism increased, opening the area to capital investment mainly by international tour operators and hunting agencies. In 2014/2015, the conservancy earned about 54,000 US$ (824,040 Namibian dollars) from tourism and trophy hunting. The desire to control and benefit from these financial proceeds increases both local and external actors' stake in wildlife conservancy affairs. Damages from wild animals in the area, especially elephants, have increased the need for compensation (Schnegg & Kiaka, 2018).

Wuparo

Established in 1999, Wuparo conservancy covers an area of 148 km^2. The conservancy is sparsely populated; residents live in several villages clustered throughout the area (see Figure 6.2). The formation of the conservancy was largely driven by NGOs such as IRDNC and the MET. They worked closely with the local chief and held a series of meetings to convince the local people that they would benefit from the conservancy programme. There are also reports that the formation was resisted by many people, especially the elders. They viewed the conservancy with suspicion, thinking that the government was planning to extend the Nkasa Ruparo National Park and that people would be displaced again, leading to the loss of their livelihoods.

Wuparo is situated in the floodplains, but over time a mosaic of woodlands and grasslands has emerged. Its abundant biodiversity and concentration of large and small mammals create an enabling environment for prime hunting and tourism; both activities contribute significantly to conservancy income. Its tourism potential has been increased as a result of the development of Nsaka

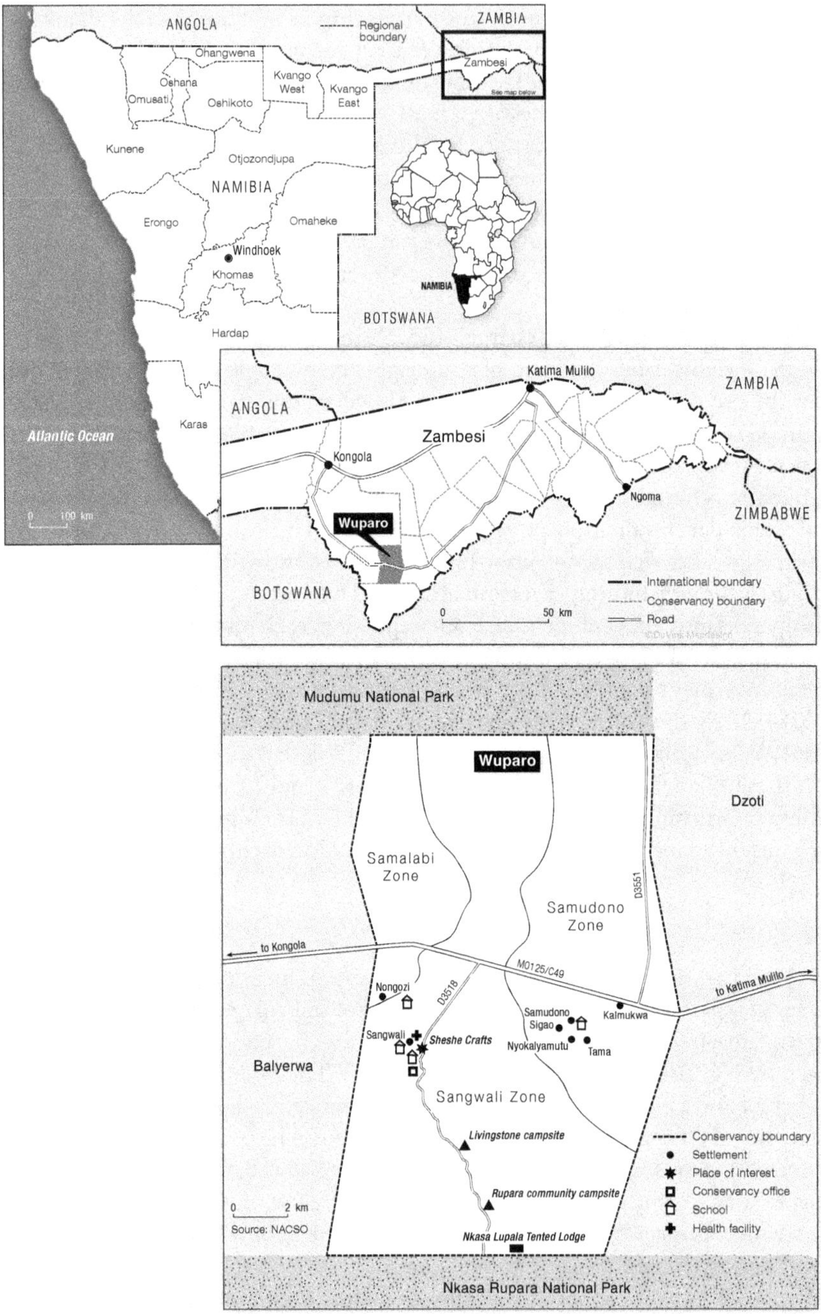

Figure 6.2 Situational map Wuparo Conservancy

Lupara tented lodge, thanks to a grant from the US-funded Millenium Challenge Account. To distribute cash incomes to its members, the conservancy embraced a decentralised governance system which created village zones as recipients of the financial benefits.

Navigating practices and processes within institutional complexities

This section builds on the idea that conservancies are rather differentiated in their capacity and ability to navigate the opportunities that a conservancy model represents. We make a distinction here between new elites, the representatives of the traditional authorities and the ordinary community members, the so-called commoners. We also single out the conservancy bureaucracy from the "new" elite category as they occupy positions that enable them to directly forge decisions.

Elite capture and elite control

In the formation and development of ǂKhoadi ǁHôas conservancy, MET and NGOs were significantly assisted by new local elites who worked for the government as extension workers and teachers in the area. Because of their ability to communicate complex policy information to the communities, they were seen as change agents. In some cases, senior MET officials in the capital directly instructed the extension workers to teach the communities about CBNRM policies and principles and to mobilise them to support the conservancy idea. Translating CBNRM policy into practice was added to their job responsibilities. Their social status of being educated, holding government jobs and regular income accorded them community respect and qualified them as local people who are capable of providing leadership. Consequently, they were elected to key positions in the ǂKhoadi ǁHôas conservancy management committee. Some left their government positions to work for the conservancy in donor-funded positions.

In addition to their elitist advantage, the new elites have a strong political base from two populous urban settlements (Erwee and Anker) which host government offices and is where they live. Many people with an interest in the conservancy leadership, but from the more rural and less-populated settlements, complain that they are unable to unseat elites who reside in the two urban settlements characterised by a larger population. Therefore, the dominant presence of local elites in the ǂKhoadi ǁHôas conservancy has persisted to date.

Being a member of the conservancy management committee means that one gets seating allowances and portions of game meat whenever they attend conservancy meetings. In addition, committee members are also responsible for the recruitment of conservancy staff and employees of its two lodges. These jobs are highly sought after in an area where people live precarious lives characterised by meagre incomes from menial jobs, widespread hunger and inadequate supply of other basic household needs. Employment opportunities, in particular, raise stakes and breed conflicts between conservancy members and officials. Many residents of the conservancy fault the recruitment process of the conservancy and mention situations where the intervention of key conservancy officials may have influenced the employment of their relatives in the conservancy or its tourism ventures. Indeed, the local elites of ǂKhoadi ǁHôas have played an important role in controlling the conservancy formation and development process to its current glory of organisational stability and income generation. But they have also captured its benefits because of their privileged positions.

The MET and IRDNC collaborated in Wuparo closely together with Chief Sifu of the Mayeni people and those close to the royal clan to establish the conservancy. The new elites who slowly emerged through the ranks of the conservancy quickly realised that no decision could be made without endorsement from either the local headman or chief. Decisions on who gets employed or elected became a preserve of these elites (Lubilo, 2018). Those in leadership positions or employment manoeuvred their close relatives and friends into favourable positions. When it came to employing staff at the Nkasa Lupara lodge, those in leadership changed the selection procedure from being appointed at a general meeting to an interview. This was commonly viewed as allowing a secret selection of close relatives and friends. The elites are eager to occupy leadership positions as this gives them access to allowances, training and other kinds of exposure, such as trips abroad, with long-term benefits. For example, a former conservancy manager has been employed by MET after being trained by the conservancy. However, after many years of trial and error, Wuparo embarked on a transformation process to improve participation, consultation and increase benefits. While this improved the communication between those in leadership and members, those occupying key positions used the process to maintain their leadership positions. For example, the chairman and manager kept alternating their positions to make sure they stayed long at the helm.

Traditional authority (TA)

The TA, through the local Mayeyi chief and his village headmen, plays an active role in the management and decision-making processes of the conservancy at Wuparo. They are customarily involved in land allocation and settling disputes. This is still quite common in the Zambesi region (Harring & Odendaal, 2012), whereas the role of chiefs in the ǂKhoadi ǁHôas conservancy is less prominent and dominant. The conservancy opened new opportunities for the traditional authorities to extract levies or rent payments. The chief and village headmen at Wuparo receive an annual grant from the conservancy. They also receive other royalties, including meat. The conservancy committee worked from the start through the traditional authority structure to secure land and to enter into conservation-related arrangements. To strengthen this relationship, the conservancy committees have included the TAs in their decision-making process; they are also involved in selecting those who should be employed by the conservancy. In addition, the conservancy annually allocates a budget for chiefs for their traditional ceremonies. The conservancy has divided itself into three major zones, and in each of these there is an Induna who participates in conservancy management. The chief has a representative in the committee, and sometimes the local indunas take part in meetings. The decisions that are made without the TA are deemed illegitimate. All meetings are held at the *Khuta* (the traditional court) and the Induna is allowed to give some comments at the beginning and end of the meeting. Sometimes this dual governance system collides with the democratic nature of how conservancy decisions are made. This situation is ambiguous for the management committee as members tend to have more trust in the TA than in the conservancy committee. The TA has, after all, a longer history than conservancies' structures, which are still relatively new.

The commoners

The 'commoners' is in itself a problematic category as it is heterogeneous in composition. They have in common though – hence commoners – that they do not or only marginally participate in conservancy politics. They are a necessary ingredient of a conservancy but do not fully share in the benefits. Most withdraw from the conservancy affairs (Schnegg & Kiaka, 2018), while others deviate from its rules by continuing to hunt (Lubilo & Hebinck, 2019).

For ǂKhoadi ǁHôas, the outsider's view considers it a sustainable conservancy and beneficial to its common members. Unfortunately, the glory of this label does not resonate with the commoners of the conservancy. Only 67 people out of a possible adult population of 2,100 are employed by the conservancy and its two lodges. Only those who are employed receive further training benefits. The majority hardly benefits from the conservancy, yet they carry the burden inflicted by elephant destroying communal water points. Many consider this as an unjust outcome of the conservancy process and withdraw from participating in conservancy affairs. They do not attend conservancy meetings, especially if those meetings do not lead to a redistribution of costs of conservation through just compensation. Generally, those who attend the meetings mostly do so because they appreciate the free meals provided by the conservancy. Moreover, in ǂKhoadi ǁHôas, people rarely report incidences of hunting by members of their kin to the conservancy partly because they do not want to betray trust embedded in helpful kinship ties, but also as a way to sabotage the governing process of the conservancy, whose outcome they consider as unjust.

Wuparo Conservancy has a registered membership of about 2,600 members, of which about 50–60 are employed and receive payments from the conservancy. The conversancy has over the years seen a drop in people attending the bi-annual meetings because few benefits trickle down to the ordinary members. The inadequate flow of benefits to ordinary members has also manifested itself into members' apathy towards the management of the conservancy. The hunting of game is quite substantial in the conservancy (Lubilo & Hebinck, 2019). Members voice their frustration and pressurise the management committee to improve the (re)distribution of benefits. This does not mean that Wuparo is a failure. Some people have accessed jobs, human–wildlife conflicts have been reduced and wildlife management and utilisation have improved.

The conservancy bureaucracy

Although the glory accorded to ǂKhoadi ǁHôas does not resonate with the living conditions of its residents and members, the conservancy bureaucracy succeeds to deflect this reality through the creation of success stories. In a first example, after the launch of "State of the Communal Conservancy Report" in 2016, an article with pictures of an elderly woman appeared in a Namibian newspaper describing how communities, especially women, benefit from communal conservancies. The fact that the story was purported to be told by an elderly woman amplified the success of CBNRM in ǂKhoadi ǁHôas. In subsequent interviews, the woman starring in the success story vehemently contested its contents. She insisted that the photos were actually taken in 2014 and used in an earlier newspaper article after a team of conservationists, journalists and senior conservancy staff interviewed her about the challenges of human–wildlife conflicts.

In a second example, a newspaper article generously wrote about a senior employee of ǂKhoadi ǁHôas. "A leading woman in community conservation in Namibia" read the title of the article in the Namibian newspaper in 2015. Her rise as a manager of the conservancy was captured by retracing the roots of a young "farm girl" out of school and facing the gendered vulnerabilities in poor rural communities to the rise to leadership epitomising a classic case of women's empowerment. Nevertheless, the article was silent on the many frustrations the "leading woman in conservation" has over low remuneration and failure to compensate farmers of the loss they incur from human–wildlife conflicts.

Wuparo conservancy, despite its many challenges in meeting the needs of the membership, is well managed. The conservancy has continued to distribute some cash and project benefits to its members. Wuparo annually generates about US$100,000 (some N$1.5 million) from trophy hunting. Part of this income is distributed to individual members as cash benefits. The figure sounds substantial, but when disaggregated amongst members and all other costs accounted for, it only amounts to an average of US$14 (i.e. N$192) per annum per adult. This is far below the US$300 that local elites such as teachers or conservancy staff earn as a monthly salary. Despite the meagre per capita amount, its aggregated value is captured in the conservancy reports as an outstanding success. For example, based on this "success" the conservancy was granted a Millennium Challenge Account award which enabled it to enter into a joint-venture partnership to construct the Nkasa Lupala Lodge, which further opens the area to private tourism entrepreneurs. Success is further created around financing community projects, yet the projects are not so sustainable. Only one out of a total of nine community projects is still operational (Lubilo, 2018). In addition, the conservancy is reported to have supported the training of its staff and local communities, yet only a few new elites are able to use their acquired skills to realise some financial benefits such as through employment.

Navigating conservancy politics and everyday life

The central argument that runs through the chapter is that a conservancy should not be treated as a homogenous entity or as a neutral socio-political space. This aspect and the complexities that arise from the conceptualisation of conservancy as a "contested space" is rather ignored in the CBNRM literature that merely emphasises the success of community-based conservation (Naidoo et al., 2016b; Naidoo, Weaver, De Longcamp, & Du Plessis, 2011a; Naidoo, Weaver, Stuart-Hill, & Tagg, 2011b; Angula et al., 2018). We concur with Blaikie (2006) in his ground-breaking critique of CBNRM, "[t]here are success stories too, although they are stories told by the initiating agencies themselves." We add to this debate by stipulating the importance of issues and relations of power and how that is shaped by the design principles of the conservancy. We have demonstrated here that a conservancy has a designed institutional arrangement with its inbuilt nodes of power, affects and simultaneously is affected by institutional arrangements that draw from different, pre-existing positions of power and authority. The conservancy unfolds as an arena or a socio-political space that consists of an interconnected web of institutions, norms and values that mediate actors' relations with each other concerning the resources that are key to their livelihoods. The so-called extra local actors who drove the designing and enforced its institutionalisation are looped into the same socio-political force field. They continue to expect as per the design that conservancies perform as a community in order to be labelled as success stories of conservation (i.e. reducing poverty and conserving natural resources through sustainable management).

We unpacked a conservancy as an arena where different actor groups navigate and manoeuvre to reap the benefits of community-based conservation. In this way, we do not ignore and simplify but problematise the existing socio-political inequalities of the society that now constitute a conservancy. We distinguished four different categories of conservancy actors: elites, the traditional authorities, the commoners and the conservancy bureaucracy. They each exercise their agency in their own way to navigate the conservancy. While doing so, they employ a variety of discursive means (i.e. adhering to as well as manipulating and contesting the conservation model) and create different versions of the success of the conservancy for their respective constituencies. The elites exhibit their mastery of the conservation discourse, but simultaneously their social status manoeuvres them into positions so that they can capture the benefits of the conservancy. Backed by post-independence political reforms that strengthened their "traditional" power and authority positions, chiefs and headmen managed to extend that into the conservancies, allowing them, in turn, to extract annual land rents and levies and influencing appointments of staff. The commoners only marginally benefit; their non-participation and withdrawal should be read as a political weapon to contest the unequal distribution of benefits. They found alternative avenues to navigate everyday life in the conservancy. The conservancy bureaucracy, on the other hand, deflects local realities by communicating success stories which are palatable to national and international support for community-based conservation.

The conservancy remains an arena in which the conservancy actors navigate in various ways with various discursive means. Under such conditions, serious initiatives need to be taken to redesign the conservancy model, for instance, by taking everyday life as a starting point.

Acknowledgements

Richard Kiaka's research was part of the research project "Local Institutions in Globalized Societies – LINGS" (www.lings-net.de) which is funded by the German Research Foundation. Rodgers Lubilo's fieldwork was part of the collaborative project, "Mobility, Networks and Institutions in the Management of Natural Resources in Contemporary Africa" (2009–2014) which was funded by the Volkswagen Foundation. We thank Stasja Koot for comments on an early draft. Nel de Vink drew the maps based on information obtained from authors' site visits to the website from the Namibian Association of CBNRM support organisations (NACSO; www.nacso.org.na).

Literature references

Acheson, J. (2011). Ostrom for anthropologists. *International Journal of the Commons, 5*, 319–339.

Angula, H., Stuart-Hill, G., Ward, D., Matongo, G., Diggle, R., & Naidoo, R. (2018). Local perceptions of trophy hunting on communal lands in Namibia. *Biological Conservation, 218*, 26–31.

Blaikie, P. (2006). Is small really beautiful? Community-based natural resource management in Malawi and Botswana. *World Development, 34*, 1942–1957.

Bollig, M. (2016). Towards an Arid Eden? Boundary-making, governance and benefit sharing and the political ecology of the new commons of Kunene Region, Northern Namibia. *The International Journal of the Commons, 10*, 771–799.

Boudreaux, K., & Nelson, F. (2011). Community conservation in Namibia: Empowering the poor with property rights. *Economic Affairs, 31*, 17–24.

Cleaver, F. (2012). *Development through bricolage: rethinking institutions for natural resource management*. London: Routledge.

De Vette, M., Kashululu, R., & Hebinck, P. (2012). Conservancies in Namibia: A discourse in action. In B. Arts, S. Van Bommel, M. Ros-Tonen, & G. Verschoor (Eds.), *Forest-people interfaces*. Wageningen: Wageningen Academic Publishers.

Harring, S., & Odendaal, W. (2012). God stopped making land. *Land rights, conflict and law in Namibia's Caprivi region*. Legal Assistance Centre, Windhoek.

Jones, B. (2010). The evolution of Namibia's communal conservancies. In F. Nelson (Ed.), *Community rights, conservation and contested land: The politics of natural resource governance in Africa*. London: Earthscan.

Jones, B., Diggle, R., & Thouless, C. (2015). From exploitation to ownership: wildlife-based tourism and communal area conservancies in Namibia. In R. Van Der Duim, M. Lamers, &

J. VanWijk (Eds.), *Institutional arrangements for conservation, development and tourism in eastern and Southern Africa a dynamic perspective*. Delft: Elbron.

Kiaka, R. (2018). *Environmental (in)justice in Namibia. Costs and benefits of community-based water and wildlife management*. Hamburg: Unpublished PhD Thesis. The University of Hamburg.

Lapeyre, R. (2011). Governance structures and the distribution of tourism income in Namibian communal lands: A new institutional framework. *Tijdschrift voor Economische en Sociale Geografie, 102*, 302–315.

Long, N. (2001). *Development Sociology: Actor Perspectives*, London: Routledge.

Lubilo, R. (2018). *Enactment of 'community' in community-based natural resource management in the Zambezi region, Namibia*. Wageningen: Wageningen University. Unpublished PhD thesis. Sociology of Development and Change.

Lubilo, R., & Hebinck, P. (2019). 'Local hunting' and community-based natural resource management: Resistance and livelihoods in Namibia. *Geoforum, 101*, 62–75.

Mosse, D. (2005). *Cultivating development: An ethnography of aid policy and practice*. London: Pluto Press.

Murphree, M. W. (2005). Congruent objectives, competing interests, and strategic compromise: Concept and process in the evolution of Zimbabwe's campfire, 1984–1996. In J. P. Brosius, A. Lowenhaupt, & C. Zernerr (Eds.), *Communities and conservation: Histories and politics of community-based natural resource management*. Oxford: Altamira Press.

Naidoo, R., Weaver, C., De Longcamp, M., & Du Plessis, P. (2011a). Namibia's community-based natural resource management programme: An unrecognized payments for ecosystem services scheme. *Environmental Conservation, 38*, 445–453.

Naidoo, R., Weaver, C., Stuart-Hill, G., & Tagg, J. (2011b). Effect of biodiversity on economic benefits from communal lands in Namibia. *Journal of Applied Ecology, 48*, 310–316.

Naidoo, R., Weaver, C., Diggle, R., Matongo, G., Stuart-Hill, G., & Thouless, C. (2016a). Complementary benefits of tourism and hunting to communal conservancies in Namibia. *Conservation Biology, 30*, 628–638.

Naidoo, R., Weaver, C., Diggle, R., Matongo, G., Stuart-Hill, G., & Thouless, C. (2016b). Complimentary benefits of tourism and hunting to communal conservancies in Namibia. *Conservation Biology, 30*, 628–638.

Nelson, F., & Agrawal, A. (2008). Patronage or participation? Community-based Natural Resource Management Reform in Sub-Saharan Africa. *Development and Change, 39*, 557–585.

Nuulimba, K., & Taylor, J. (2015). 25 years of CBNRM in Namibia: A retrospective on accomplishments, contestation and contemporary challenges. *Journal of Namibian Studies: History Culture Politics, 18*, 89–110.

Ostrom, E. (1990). *Governing the commons: The evolution of institutions for collective actions*. Cambridge: Cambridge: Cambridge University Press.

Platteau, J. P. (2004). Monitoring elite capturing in community-driven development. *Development and Change, 35*, 223–246.

Saunders, F. (2014). The promise of common pool resource theory and the reality of commons projects. *International Journal of the Commons, 8*, 636–656.

Schnegg, M. (2018). Institutional multiplexity: Social networks and community-based natural resource management. *Sustainability Science, 13*, 1017–1030.

Schnegg, M., & Kiaka, R. (2018). Subsidized elephants: Community-based resource governance and environmental (in)justice in Namibia. *Geoforum, 93*, 105–115.

Van Der Duim, R., Meyer, D., Saarinen, J., & Zellmer, K. (Eds.). (2012). *New alliances for tourism, conservation and development in eastern and Southern Africa*. Delft: Elbron.

7 Fourteen years of tourism and climate change research in Southern Africa

Lessons on sustainability under conditions of global change

Gijsbert Hoogendoorn and Jennifer M. Fitchett

Introduction

The first academic investigation into climate change and sustainable tourism development in Southern Africa was conducted by Preston-Whyte and Watson (2005), published as a book chapter entitled, *Nature Tourism and Climatic Change in Southern Africa* in Hall and Higham's (2005), *Tourism, Recreation and Climate Change* (Channelview). Since the publication of this chapter, research on climate change and tourism has gained significant traction in a number of localities throughout Southern Africa (see Saarinen, Hambira, Atlhopheng, & Manwa, 2012; Tervo-Kankare, Saarinen, Kimaro, & Moswete, 2017; Hoogendoorn & Fitchett, 2018b for local examples and Hoogendoorn & Fitchett, 2018a for a continental overview), but for the most part issues around tourism and climate change remain desperately under researched. This chapter explores the concerns raised by Preston-Whyte and Watson (2005) regarding the threats and opportunities that climate change poses to the sustainability of tourism in the context of more recent research on regional climate change and the contemporary impacts of climatic variability on tourism sectors. We have extracted each of the key climate change threats cited by Preston-Whyte and Watson (2005) and reviewed the contemporary literature on climate change and tourism, as well as the literature on specific biomes and activities, to provide a comparison for each claim. These are subsequently regrouped under larger themes reflecting points of contention and agreement.

Re-evaluating Preston-Whyte and Watson's (2005) concerns regarding climate change and tourism sustainability

Preston-Whyte and Watson (2005) address two primary themes in their chapter: the first synthesises on published research at the time in terms of regional climatic trends; the second outlines the likely impacts of regional warming and aridification on nature-based tourism, categorized by ecozones. Rather than providing empirical evidence quantifying specific climate change threats to tourism, exploring operators' perceptions or experiences of climate change, or interrogating tourists' responses to climate change, this work integrated the outputs of climate change and tourism literature. This provided an important starting point for research on climate change and tourism in the region, highlighting key points of concern for investigation (Hoogendoorn & Fitchett, 2018a).

Preston-Whyte and Watson (2005) explore the impacts of a coupled increase in temperature and aridification on nature-based tourism across six key ecozones in Southern Africa with implications on alien species, malaria and poverty. The climate change basis for their assumptions is largely undisputed; temperature increases are within range, and for most regions a shift to aridity remains projected (SAWS, 2017). Given the absence of any prior research exploring climate change impacts on tourism in the region, this provided a well-structured approach to interrogating the varied impacts of a warming climate and the progressive decline in both precipitation and net moisture availability in Southern Africa. Preston-Whyte and Watson (2005, p. 140) argue that "Southern Africa stands to lose specific nature tourism attractions if, by 2050, the predictions are correct regarding increased global warming trends and associated increased aridification." These nature tourism attractions that they identify as being under threat are numerous, and many of these have in the inter-leading time been proven to be deteriorating as a direct result of climate change.

Recent studies provide empirical evidence confirming the threats of a reduction of the Zambezi River flow on tourism in the region (Dube & Nhamo, 2018a, 2018b); of sea level rise-induced flooding to tourism destinations along the eastern and southern coast of South Africa (Fitchett, Grant, & Hoogendoorn, 2016a; Hoogendoorn, Grant, & Fitchett, 2016); and the impacts of ocean warming on a change in the timing, magnitude and spatial extent of the sardine run (Fitchett, Grab, & Portwig, 2019). Research has also explored the impacts of warming on snow- and ice-related tourism on the African continent (Stockigt, Hoogendoorn, Fitchett, & Saarinen, 2018) and on the continued melting of the Kilimanjaro glaciers, although these are now found to be controlled more directly by changes in the moisture balance than temperature (Mölg, Cullen, Hardy, Kaser, & Klok, 2008). The threats of drought to viticulture in the Western Cape of South Africa (Araujo, Abiodun, & Crespo, 2016), to the Namaqualand daisies (Cowling, Esler, & Rundel, 1999), and to the Fynbos biome, have provided further situational evidence of the contemporary impacts of climate change, although in the instance of Fynbos, mitigated largely by the concurrent warming (Esler, Von Staden, & Midgley, 2015). The damage to coral reefs raised by Preston-Whyte and Watson (2005) has indeed also become a key concern for a range of destinations for which snorkeling and scuba diving provide a key attraction, although more recent science indicates that the damage to coral reefs is induced by climate change-related ocean acidification rather than sea level rise (Hooidonk, Maynard, Manzello, & Planes, 2014). Similarly, Preston-Whyte and Watson's (2005) projections of an expansion in the malaria-prone region of Southern Africa have been supported by recent empirical studies, although this increase in a malaria-suitable climate is related to an increase in temperature and precipitation, rather than of aridity (Caminade et al., 2014; Ryan et al., 2015).

In light of these numerous significant threats of climate change to tourism that are detailed within their chapter, it is surprising that Preston-Whyte and Watson (2005, p. 140) conclude with such a positive outlook for nature-based tourism in the region:

> in general, a warmer and drier climatic scenario should enhance the promotion of popular images of the 'African bush' experience given that the expansion of the eutrophic savannas favours flat-topped thorn trees, large ungulate herds and the "Big Five." Even a red hue to the African sunset could be enhanced by increased atmospheric aerosol loading from fires needed to regulate bush encroachment and dust from eroded soils.

These benefits of climate change to tourism in the region are not clearly communicated in the ecozone classified analysis of the regional impacts and rely heavily on the capacity for the existing fenced national parks to facilitate such changes in large mammal composition. Indeed, in the analysis of the savanna ecoregion, Preston-Whyte and Watson (2005) argue the contrary, that an extension of the eutrophic expansion of Kruger National Park under climate change would reduce the diversity of large mammals to the detriment of Big Five viewing, while the bush encroachment in the Hluhluwe-Umfolozi would hamper game sightings. While for the regions surrounding the Zambezi River, Preston-Whyte and Watson (2005) argue for a prolonged presence and concentration of large mammals around water sources, they caution this with concerns about the sustainability for vegetation and water resources.

For many of the ecoregions, Preston-Whyte and Watson (2005) suggest a change in faunal composition, with neither assemblage particularly stronger or weaker in their tourist attraction (e.g. shift from zebra to buffalo in the lowland tropical coastal forest) and thus is difficult to argue as a net benefit to tourism. Preston-Whyte and Watson (2005) accurately highlight that critical to the adaptation of ecosystems to climate change is the potential for range shifts (Root et al., 2003) to higher altitudes and latitudes, which the development of transfrontier parks certainly facilitates. However, the political tensions around these parks, the severe impacts of individual storm and drought events and numerous instances of poaching are likely to prevent these natural migrations from taking place at the magnitude or scale required to avoid climate-related stressors and eventual death (Fitchett, Robinson, & Hoogendoorn, 2016; Morais, Bunn, & Hoogendoorn, 2018). Moreover, even if temperature increases and changes in precipitation patterns were to favourably affect tourism in the short term, these gains would be limited within the thermal thresholds of each of the large mammals in question, which would under continued warming migrate polewards or upslope until this was impossible, leading to eventual extirpation.

The two outright benefits of climate change to tourism, as outlined by Preston-Whyte and Watson (2005), relate to the Nama Karoo becoming more desolate and thus playing into its current attraction, and an improvement in the sunsets. Yet, Kaján and Saarinen (2013) argue that while desolation because of drought may possibly improve tourism attractions (i.e. desert tourism), communities in these areas may face serious socio-economic troubles as a result. Without empirical evidence to quantify the importance of these factors to the tourist market, it is difficult to argue outright that these are less important than the losses incurred for other forms of nature-based tourism.

In the case of both the Drakensberg and Table Mountain, Preston-Whyte and Watson (2005) argue that the impacts of climate change on the natural environment are unlikely to reduce the destination image, as this is related more strongly to the geomorphology. This raises an important point regarding the importance of nature in a tourist attraction and the degree to which this is contingent on the weather at the time, a factor of the mean climatic conditions of the destination (Giddy, Fitchett, & Hoogendoorn, 2017a). A key determinant is the amount of time spent outdoors for a particular activity, but associated to this are the climatic expectations of the tourist and their baseline climatic sensitivity (Du Bois, Ceron, Gössling, & Hall, 2016; Steiger, Abegg, & Jänicke, 2016; Fitchett & Hoogendoorn, 2018b). Indeed a significant portion of climate change and tourism literature both locally and internationally has explored the primary impacts of climate change on a reduced climatic suitability of a destination for tourism (cf. Mieczkowski, 1985; Perch-Nielsen, Amelung, & Knutti, 2010; Fitchett et al., 2016a; Fitchett, Hoogendoorn, & Swemmer, 2016; Fitchett, Robinson, & Hoogendoorn, 2017b). On this basis we therefore expand the definition of nature-based tourism beyond the predominantly large-mammal wildlife considered by Preston-Whyte and Watson (2005) to include nature-dependent tourism in Southern Africa for which climate change and tourism research has been conducted, namely, adventure tourism, beach tourism, cultural and heritage tourism.

What we know now: regional climate threats to tourism sub-sectors

The main focus of Preston-Whyte and Watson's (2005) chapter was on national parks considered through the lens of specific biogeographic ecozones. While this still remains important to sustainable tourism development in Southern Africa, emerging research on climate change and tourism in Southern Africa spans a greater array of tourism attractions. Notably, these too rely heavily on the natural environment and the outdoor climate and consequently face many of the same challenges that plague the more narrowly defined nature-based tourism. This broad-based impact will be discussed below.

Nature-based tourism and national parks

Despite the clear research agenda put forward by Preston-Whyte and Watson (2005), relatively little empirical research has been conducted regarding the climate change impacts on nature-based tourism and national parks in Southern Africa. Hambira (2017) conducted the most comprehensive research in this domain, exploring tour operators' and governmental policy makers' awareness and responsiveness to the impact of climate change on tourism in Botswana. Centered in the Okavango Delta and the Kgalagadi Transfrontier Park regions, this study focusses specifically on the vulnerability and adaption mechanisms of the previously mentioned stakeholders. The study confirms the concerns raised by Preston-Whyte and Watson (2005), as increasing temperatures are already found to be affecting tourist activities in Botswana (Hambira, Saarinen, Manwa, & Atlhopheng, 2013; Hambira, 2017). Tervo-Kankare, Saarinen, Kimaro, and Moswete (2017) explored tourism business perceptions at the Tsiseb Conservancy in Namibia, finding that tour operators are aware of how recent climate variability has affected their business but are skeptical about the impact of future climate change on the tourism industry. This may be due to the potential benefits that drought poses through desert tourism (Kaján & Saarinen, 2013). In South Africa, the only analyses of climate threats directly to the tourists engaged in nature-based tourism comprises the computing of Tourism Climatic Index scores for destinations that primarily comprise nature-based tourism attractions, such as the Pilanesberg (Fitchett et al., 2017b) and an investigation into the TripAdvisor reviews of tourists visiting these destinations (Fitchett & Hoogendoorn, 2018a). These studies indicate that at present, the climate is "ideally suited" to tourism and that tourists are largely satisfied with their experiences of the weather (Fitchett et al., 2017b; Fitchett & Hoogendoorn, 2018a).

A second key threat to nature-based tourism and national parks in Southern Africa, particularly in the northeastern regions, is storm damage inflicted by tropical cyclones (Southon & Van Der Merwe, 2018). It is worth mentioning that Preston-Whyte and Watson (2005) do not reflect on the impacts of tropical cyclones on nature-based tourism, nor the threats of the changing storm dynamics under climate change. Increases in the intensity and changes in landfall position of these storms has been documented in the inter-leading years (Malherbe, Engelbrecht, & Landman, 2013; Fitchett & Grab, 2014; Fitchett, 2018). The impacts on nature-based tourism arise from the temporary flooding, damage to tourism and transportation infrastructure, and longer-term destruction of habitat for the large mammal fauna (Fitchett et al., 2016). An increased incidence of southward tracking storms and a heightened intensity pose serious threats to nature-based tourism associated with Kruger National Park and the broader Great Limpopo Transfrontier Park, and with a greater footprint of these high-intensity storms, effects will likely be felt as far afield as Botswana, Mozambique and Malawi. Evidence of this is being observed with the landfall of Tropical Cyclone Idai in March 2019.

Preston-Whyte and Watson (2005) were instead concerned with drought and aridity. Drought events do pose significant threats to nature-based tourism, stressing ecosystem functions and reducing the aesthetic quality of the park. TripAdvisor reviews reveal the sensitivity of tourists to drought conditions, particularly within the eastern regions of South Africa (Fitchett & Hoogendoorn, 2018a). The particularly dry conditions in 2016 were responsible for widespread water shortage in the Kruger National Park and in turn the death of large mammals, including hippopotamus and buffalo (Van Vuuren, 2016). A primary adaptation response to the drought was the opening and refilling of a number of artificial waterholes across the park (Grové-Shrader, 2016). However, such artificial waterholes have been found to change the herbivore distribution and behaviour, posing threats to both predator-prey interactions and game viewing opportunities (Smit, Grant, & Devereux, 2007; Purdon & Van Aarde, 2017).

For both flood and drought events, a relatively short-lived climatic perturbation can have long-lasting impacts on the ecosystem viability of the national park and in turn its ability to continue to attract tourists. As both drought and flood events are projected to intensify in the region under climate change (SAWS, 2017), proactive adaptation strategies with a long-term view to sustainable ecosystem structure, game viewing and user experience are required to prevent significant losses. As the Kruger National Park is responsible for more than 40% of the South African National Parks (SANParks), revenue earnings and 26.9% of the total visitor numbers for the 2016/2017 reporting period (SANParks, 2017), the SANParks' business model is highly reliant on the Kruger National Park for sustained revenue and the cross-subsidisation of other parks (Ferreira & Harmse, 2014). The wide array of climate change threats to this park in particular threaten the sustainability of nature-based tourism in South Africa more broadly.

Adventure tourism

Adventure tourism in Southern Africa often spatially overlaps with nature-based tourism in the same or adjacent destinations. Adventure tourism activities rely heavily on the natural environment as a drawcard (Giddy & Webb, 2016). The weather is therefore paramount to daily operation, and thus in turn any major changes in the climatic suitability of the destination under climate change may yield the destination and activities financially unsustainable (Giddy, Fitchett, & Hoogendoorn, 2017b). Giddy et al. (2017a) explored the effects of extreme weather events on whitewater tourism in southern Africa, finding that both drought and flooding events severely affected tour operators, compromising the economic stability of these operations, which were all small-sized enterprises. The tour operators are responding to the threats of climate change by moving away from a reliance on a single activity (Giddy et al., 2017a). A similar adaptation strategy is being adopted in the ski tourism industry at Afriski in Lesotho through a diversification of offerings and marketing of off-season events (Stockigt et al., 2018). Critical to the successful adaptation to climate change across the adventure tourism spectrum is the capacity for high-resolution climate modelling and forecasting, given the diversity of climatic conditions required for each activity. Given the reliance on the natural environment as a drawcard for adventure tourism (Giddy & Webb, 2018), the secondary impacts of climate change on the regional fauna and flora, many of which have been outlined by Preston-Whyte and Watson (2005), should also be considered in adaptation plans.

Beach tourism

Beach tourism is a focal point of tourism and climate change research in South Africa and relies heavily on the natural environment and the climatic conditions of a destination because of the outdoor nature of beach-related activities (Hoogendoorn et al., 2016). In many instances, the beach provides the primary tourist attraction, while nature-based activities such as birdwatching, game viewing and hiking contribute as a secondary attraction to destinations (Priskin, 2001). Empirical research that has been conducted on the climate change threats to beach tourism explore the impacts of sea level rise on beach recession and flooding, storm surges and changes in the climatic suitability of coastal destinations for prolonged periods outdoors (Fitchett et al., 2016a, 2017b; Hoogendoorn et al., 2016). The impacts of sea level rise are particularly problematic in the stretches of the South African coastline that are characterised by shallow, sloping, sandy beaches, such as St Francis Bay and Cape St Francis, where properties along the coastlines face severe threats of flooding by 2050 (Fitchett et al., 2016). However, the greatest concern for the sustainability of the beach tourism sector relates to the disjuncture in perceptions of climate change threats, both amongst local and international tourists, and tourists and accommodation establishment proprietors in terms of the nature and extent of climate change threats (Fitchett et al., 2016; Hoogendoorn et al., 2016). Accommodation establishment proprietors are particularly concerned about small-magnitude fluctuations in day-to-day weather, reducing these through the provision of air conditioning, heaters and blankets, and indoor activities (Hoogendoorn et al., 2016). The guests, by contrast, are concerned about sea level rise, storm surges and beach recession, and indicate concern that accommodation establishment proprietors are not responding with even low-cost solutions such as localised retaining walls (Fitchett et al., 2016; Hoogendoorn et al., 2016). Although Tourism Climatic Index scores indicate concerns of high temperatures, high humidity and year-round rainfall as factors which currently reduce the climatic suitability of these destinations, and are likely to worsen in the future, the scores remain high by international standards (Fitchett et al., 2016).

Cultural and heritage tourism

Cultural and heritage tourism has recently come under the spotlight in the context of climate change threats to tourism, yet remains under studied across the globe (Hall & Ram, 2016). While Hall, Baird, James, and Ram (2016) mention a variety of environments that could be affected by climate change, including cultural landscapes, built environments, buried archaeology, parks and gardens, these cases have not been considered in the African context. Southern Africa has a variety of heritage and cultural landscapes that can be accessed in urban areas and often indoors (van der Merwe, 2013; Manwa, Moswete, & Saarinen, 2016), and which thus are unlikely to be affected directly by climate change. However, a large number of Southern Africa's heritage and cultural tourism products are outdoors, such as the Cradle of Humankind and Vredefort Crater (South Africa), Victoria Falls (Zambia and Zimbabwe) and the Petrified Forest (Namibia), and thus expose tourists to the outdoor climate and in turn rely on favourable weather to operate. Moreover, climatic events may directly impact on the heritage or cultural asset itself: flooding, erosion and damage to culture and heritage artefacts under increased temperatures, changes in rainfall and humidity, changes in wind speed and direction can have a severe effect on established cultural and heritage sites as well as newly discovered heritage such as rock art (see Duval & Smith, 2012). That is not to say that these changing conditions do not affect cultural and heritage landscapes in the Global North (see Fernandes, 2015), but we would argue that once again the issue of low adaptive capacity comes to the fore in the Southern African context where socio-economic needs will in all likelihood overshadow the protection of heritage and cultural landscapes. Although a growing body of literature considers cultural and heritage tourism in Southern Africa (see van der Merwe, 2016), issues around the climatic preferences of tourists visiting heritage and cultural sites, threats of climate hazards to artefacts, and adaptation plans within this sector to longer-term climate change threats have not been investigated. Empirical research is therefore urgently required in this domain to set the agenda for successful adaptation, both to protect the source of tourism and more importantly to protect these important artefacts.

Conclusion

Preston-Whyte and Watson (2005) provided a crucial starting point for climate change and research on sustainable tourism in Southern Africa. In the absence of empirical research that explored the direct contemporary impacts and future threats of climate change to tourism in the region, the chapter synthesised the knowledge at the time regarding climate change projections for the region and the key nature-based tourism products within each ecozone (Preston-Whyte & Watson, 2005, summarised in Table 7.1). In the subsequent 14 years, this has driven a key research agenda for tourism in Southern Africa, and a large number of empirical studies have been conducted, spanning South Africa, Zimbabwe, Botswana, Namibia and Lesotho. Notably, much of the work conducted in this period confirms the hypothesised impacts that Preston-Whyte and Watson (2005) had listed.

Following a considerable list of threats that climate change poses to nature-based tourism in Southern Africa by 2050, Preston-Whyte and Watson (2005) surprisingly concluded on a very positive note, arguing that climate change may offer some opportunities for tourism in the region. In the years following this publication, a growing awareness of climate change threats to tourism among both tourists and tourism operators has been documented. A greater issue now is in the prioritisation of resources to facilitate successful adaptation strategies to prevent some of the harms of climate change to the tourism sub-sectors that rely heavily on the natural environment and, by virtue of their outdoor nature, the daily weather and long-term climate. Issues of prioritisation at a national level centre on challenges of meeting the immediate needs of service and infrastructure delivery in a chronically unequal region. These immediate needs often preclude the development of infrastructure to address longer-term challenges. Prioritisation issues are, however, also driven by uncertainties, denialism and misinformation within the tourism sector. Often, although tourists are well aware of the climate change threats to the sector, operators are unwilling to implement the necessary adaptation mechanisms or respond with adaptations that are inappropriate to the primary challenges and interrogate the sustainability of the sector (Hoogendoorn et al., 2016). There is also a critical lack of engagement with the local communities to ensure that tourism adaptations are in line with local strategies and that these will encourage rather than hinder community involvement in this emerging sector. The question is no longer whether climate change will have a positive or negative impact on the sustainability of tourism, but how best to adapt to threats with minimal capital investment.

References

Araujo, J. A., Abiodun, B. J., & Crespo, O. (2016). Impacts of drought on grape yields in Western Cape, South Africa. *Theoretical and Applied Climatology, 123*(1–2), 117–130. doi.10.1007/s00704-014-1336-3.

Caminade, C., Kovats, S., Rocklov, J., Tompkins, A. M., Morse, A. P., Colón-González, F. J., . . . Lloyd, S. J. (2014). Impact of climate change on global malaria distribution. *Proceedings of the National Academy of Sciences, 111*(9), 3286–3291. doi.10.1073/pnas.1302089111.

Cowling, R. M., Esler, K. J., & Rundel, P. W. (1999). Namaqualand, South Africa – an overview of a unique winter-rainfall desert ecosystem. *Plant Ecology, 142*(1–2), 3–21. doi.10.1023/A:1009831308074

Dube, K., & Nhamo, G. (2018a). Climate variability, change and potential impacts on tourism: Evidence from the Zambian side of the Victoria Falls. *Environmental Science and Policy, 84*, 113–123. doi.10.1016/j.envsci.2018.03.009.

Dube, K., & Nhamo, G. (2018b). Climate change and potential impacts on tourism: Evidence from the Zimbabwean side of the Victoria Falls. *Environment, Development and Sustainability*. doi. 10.1007%2Fs10668-018-0118-y.

Du Bois, G., Ceron, J. P., Gössling, S., & Hall, C. M. (2016). Weather preferences of French tourists: Lessons for climate change impact assessment. *Climatic Change, 136*(2), 339–351. doi.10.1007/s10584-016-1620-6.

Duval, M., & Smith, B. (2012). Rock art tourism in the Ukhahlamba/Drakensberg world heritage site: Obstacles to the development of sustainable tourism. *Journal of Sustainable Tourism, 21*(1), 134–153. doi.10.1080/09669582.2012.699060.

Esler, K. J., Von Staden, L., & Midgley, G. F. (2015). Determinants of the Fynbos/succulent karoo biome boundary: Insights from a reciprocal transplant experiment. *South African Journal of Botany*, *101*, 120–128. doi.10.1016/j.sajb.2015.02.006.

Fernandes, F. (2015). Built heritage and flash floods: Hiking trails and tourism on Madeira Island. *Journal of Heritage Tourism*, *11*(1), 88–95. doi.10.1080/1743873X.2015.1082574.

Ferreira, S. L. M., & Harmse, A. (2014). Kruger national park: Tourism development and issues around the management of large numbers of tourists. *Journal of Ecotourism*, *13*(1), 16–34. doi.10.1080/14724049.2014.925907.

Fitchett, J. M. (2018). Recent emergence of CAT5 tropical cyclones in the South Indian Ocean. *South African Journal of Science*, *114*(11/12), 83–88. doi.10.17159/sajs.2018/4426.

Fitchett, J. M., & Grab, S.W. (2014). A 66-year tropical cyclone record for South-East Africa: Temporal trends in a global context. *International Journal of Climatology*, *34*(13), 3604–3615. doi.10.1002/joc.3932.

Fitchett, J. M., Grab, S.W., & Portwig, H. (2019). Advances in the timing of Sardine migration in the South-West Indian Ocean: Ocean-atmospheric considerations under global change. *South African Journal of Science*, In Press.

Fitchett, J. M., Grant, B., & Hoogendoorn, G. (2016). Climate change threats to two low-lying South African coastal towns: Risks versus perceptions. *South African Journal of Science*, *112*(5–6), 86–94. doi.10.17159/sajs.2016/20150262.

Fitchett, J. M., & Hoogendoorn, G. (2018a). Exploring the climate sensitivity of tourists to South Africa through tripadvisor reviews. *South African Geographical Journal*. doi.10.1080/03736245.2018.1541022.

Fitchett, J. M., & Hoogendoorn, G. (2018b). An analysis of factors affecting tourists' accounts of weather in South Africa. *International Journal of Biometeorology*, *62*(12), 2161–2172. doi.10.1007/s00484-018-1617-0.

Fitchett, J. M., Hoogendoorn, G., & Swemmer, A. (2016). Economic costs of the 2012 floods on tourism in the Mopani district municipality, South Africa. *Transactions of the Royal Society of South Africa*, *71*, 187–194. doi.10.1080/0035919X.2016.1167788.

Fitchett, J. M., Robinson, D., & Hoogendoorn, G. (2017b). Climate suitability for tourism in South Africa. *Journal of Sustainable Tourism*, *25*(6), 851–867. doi.10.1080/09669582.2016.1251933.

Giddy, J. K., Fitchett, J. M., & Hoogendoorn, G. (2017a). A Case Study on the Preparedness of White-Water Tourism to Severe Climatic Events in Southern Africa. *Tourism Review International*, *21*(2), 213–220. doi.10.3727/154427217X14984977561754.

Giddy, J. K., Fitchett, J. M., & Hoogendoorn, G. (2017b). Insight into American tourists experiences with weather in South Africa. *Bulletin of Geography: Socio-Economic Series*, *38*, 57–71. doi.10.1515/bog-2017–0034.

Giddy, J. K., & Webb, N. L. (2016). The influence of the environment on adventure tourism: From motivations to experiences. *Current Issues in Tourism*, *21*(18), 2124–2138. doi.10.1080/13683500.2016.1245715.

Giddy, J. K., & Webb, N. L. (2018). The influence of the environment on adventure tourism: from motivations to experiences. *Current Issues in Tourism*, *21*(18): 2124–2138.

Grové-Schrader, H. (2016). Kruger Park opening waterholes to save animals from drought. Retrieved from www.News24.com/green/news/Kruger-Park-Opening-Waterholes-To-Save-Animals-From-Drought-20160307 [accessed 31 May 2018].

Hall, C. M., Baird, T., James, M., & Ram, Y. (2016). Climate change and cultural heritage: Conservation and heritage tourism in the anthropocene. *Journal of Heritage Tourism*, *11*(1), 10–24. doi.10.1080/1743873X.2015.1082573.

Hall, C. M., & Higham, J. (2005). *Tourism, recreation and climate change*. Clevedon: Channelview.

Hall, C. M., & Ram, Y. (2016). Heritage in the intergovernmental panel on climate change assessment reports: A Lexical assessment. *Journal of Heritage Tourism, 11*(1), 96–104. doi.10.1080/1743873X.2015.1082572.

Hambira, W. L. (2017). Botswana tourism operators' and policy makers' perceptions and responses to the tourism-climate change nexus: Vulnerabilities and adaptations to climate change in Maun and Tshabong areas. *Nordia Geographical Publications, 46*(2), 59.

Hambira, W. L., Saarinen, J., Manwa, H., & Atlhopheng, J. R. (2013). Climate change adaptation practices in nature-based tourism in Maun in the Okavango Delta Area, Botswana: How prepared are tourism businesses? *Tourism Review International, 17*, 19–29. doi.10.3727/154427213X13649094288025.

Hoogendoorn, G., & Fitchett, J. M. (2018a). Tourism and climate change: A review of threats and adaptation strategies for Africa. *Current Issues in Tourism, 21*(7), 742–759. doi.10.1080/13683500.2016.1188893.

Hoogendoorn, G., & Fitchett, J. M. (2018b). Perspectives on second homes, climate change and tourism South Africa. *African Journal of Hospitality, Tourism and Leisure*, 7(2), 1–18.

Hoogendoorn, G., Grant, B., & Fitchett, J. M. (2016). Disjunct perceptions? Climate change threats in two low-lying South African coastal towns. *Bulletin of Geography: Socio-Economic Series, 31*, 59–71. doi.10.1515/bog-2016–0005.

Hooidonk, R., Maynard, J. A., Manzello, D., & Planes, S. (2014). Opposite latitudinal gradients in projected Ocean acidification and bleaching impacts on coral reefs. *Global Change Biology, 20*(1), 103–112. doi.10.1111/gcb.12394.

Kaján, E., & Saarinen, J. (2013). Tourism, climate change and adaptation: A review. *Current Issues in Tourism, 16*(2), 167–195. doi.10.1080/13683500.2013.774323.

Malherbe, J., Engelbrecht, F. A., & Landman, W. A. (2013). Projected changes in Tropical Cyclone Climatology and Landfall in the Southwest Indian Ocean Region under enhanced anthropogenic forcing. *Climate Dynamics, 40*(11–12), 2867–2886. doi.10.1007/s00382-012-1635-2.

Manwa, H., Moswete, N., & Saarinen, J. (Eds.). (2016). *Cultural tourism in Southern Africa*. Bristol: Channel View.

Mieczkowski, Z. (1985). The tourism climatic index: A method of evaluating world climates for tourism. *Canadian Geographer, 29*(3), 220–233. doi.10.1111/j.1541–0064.1985.tb00365.x.

Mölg, T., Cullen, N. J., Hardy, D. R., Kaser, G., & Klok, L. (2008). Mass balance of a slope glacier on Kilimanjaro and its sensitivity to climate. *International Journal of Climatology, 28*(7), 881–892. doi.10.1002/joc.1589.

Morais, D. B., Bunn, D., Hoogendoorn, G., & KC, B. (2018). The potential role of tourism microentrepreneurship in the prevention of rhino poaching. *International Development Planning Review, 40*(4), 443–461.

Perch-Nielsen, S. L., Amelung, B., & Knutti, R. (2010). Future climate resources for tourism in Europe based on the daily tourism climatic index. *Climatic Change, 103*(3–4), 363–381. doi.10.1007/s10584-009-9772-2.

Preston-Whyte, R. A., & Watson, H. K. (2005). Nature tourism and climatic change in Southern Africa. In C. M. Hall & J. Higham (Eds.), *Tourism, recreation and climate change* (pp. 130–142), Bristol: Channelview Publications.

Priskin, J. (2001). Assessment of natural resources for nature-based tourism: The case of the central coast region of Australia. *Tourism Management, 22*, 637–648.

Purdon, A., & Van Aarde, R. J. (2017). Water provisioning in Kruger national park alters elephant spatial utilisation patterns. *Journal of Arid Environments*, *141*, 45–51. doi.10.1016/j.jaridenv.2017.01.014.

Root, T. L., Price, J. T., Hall, K. R., Schneider, S. H., Rosenzweig, C., & Pounds, J. A. (2003). Fingerprints of global warming on wild animals and plants. *Nature*, *421*(6918), 57. doi.10.1038/nature01333.

Ryan, S. J., Mcnally, A., Johnson, L. R., Mordecai, E. A., Ben-Horin, T., Paaijmans, K., & Lafferty, K. D. (2015). Mapping physiological suitability limits for malaria in Africa under climate change. *Vector-Borne and Zoonotic Diseases*, *15*(12), 718–725. doi.10.1089/vbz.2015.1822.

Saarinen, J., Hambira, W. L., Atlhopheng, J., & Manwa, H. (2012). Tourism industry reaction to climate change in Kgalagadi south district, Botswana. *Development Southern Africa*, *29*(2), 273–285. doi.10.1080/0376835X.2012.675697.

SAWS. (2017). *A climate change reference atlas*. Pretoria: Water Research Commission.

Smit, I. P., Grant, C. C., & Devereux, B. J. (2007). Do artificial waterholes influence the way herbivores use the Landscape? Herbivore distribution patterns around rivers and artificial surface water sources in a large African Savanna park. *Biological Conservation*, *136*(1), 85–99. doi.10.1016/j.biocon.2006.11.009.

South African National Parks. (2017). Annual report. Retrieved from www.sanparks.org/assets/docs/general/annual-report-2017.pdf [accessed 26 April 2018].

Southon, M. P., & Van Der Merwe, C. M. (2018). Flooded with risks or opportunities: Exploring flooding impacts on tourist accommodation. *African Journal of Hospitality, Tourism and Leisure*, 7(1), 1–27.

Steiger, R., Abegg, B., & Jänicke, L. (2016). Rain, rain, go away, come again another day. Weather preferences of summer tourists in mountain environments. *Atmosphere*, 7(5), 63. doi.10.3390/atmos7050063.

Stockigt, L., Hoogendoorn, G., Fitchett, J. M., & Saarinen, J. (2018). Climate sensitivity and snow-based tourism in Africa: An investigation of Tripadvisor reviews on Afriski, Lesotho. *Proceedings of the Biennial Conference of the Society of South African Geographers*, Bloemfontein 1–5 October, pp. 207–224.

Tervo-Kankare, K., Saarinen, J., Kimaro, M. E., & Moswete, N. N. (2017). Nature-based tourism operators' responses to changing environment and climate in Uis, Namibia. *African Geographical Review*, *37*(3), 273–282. doi.10.1080/19376812.2017.1286246.

Van Der Merwe, C. D. (2013). The limits of urban heritage tourism in South Africa: The case of constitutional hill, Johannesburg. *Urban Forum*, *24*(4), 573–588. doi.10.1007/s12132–013–9197-x.

Van Der Merwe, C. D. (2016). Tourists guides' perceptions of cultural heritage tourism in South Africa. *Bulletin of Geography: Socio-Economic Series*, *34*, 117–130. doi.10.1515/bog-2016–0039.

Van Vuuren, L. (2016). Drought-nature's lessons in overdrive in Kruger national park: Drought management-feature. *Water Wheel*, *15*(5), 34–37.

Part II

Natural resource-based tourism development and growth in Southern Africa

Policy issues, challenges and practices

8 The analysis of conflict and coexistence of traditional and contemporary land uses

Patricia K. Mogomotsi, Melville Saayman and Andrea Saayman

Introduction

Although tourism as an economic sector has played a considerable role in growing various economies around the world, it has negatively affected social and economic development as well as the environment (Kim, Jun, Walker, & Drane, 2015). With specific reference to agriculture, it has been observed that since most tourist resorts and activities are located in rural areas, tourism development may affect local agricultural production (Liu, Liu, Hu, Wu, & Dai, 2008). This leads to conflicts over land use between tourism and agricultural activities. The relationship between tourism and agriculture is often referred to as ranging from the opportunity to conflict (Telfer & Wall, 1996). The Okavango Delta in Botswana is not an exception.

Over the years, the Okavango Delta has experienced negative natural resource dynamics, including water pollution, biodiversity loss and some cases of natural resource depletion (Darkoh & Mbaiwa, 2005). There has also been increased competition over natural resources in the Delta, resulting in land use conflicts (Darkoh & Mbaiwa, 2014). Notably, there are conflicts between livestock and wildlife. For instance, a veterinary fence has been erected around the Delta to contain and protect the growing livestock by blocking the migratory routes of wildlife. Consequently, the fence has contributed to the death and decline of wildlife species around the Delta (Darkoh & Mbaiwa, 2005). This decline has potential adverse effects on tourism, especially wildlife-based tourism, in the area.

Wildlife resources play significant ecological and economic roles in the landscapes of the Okavango Delta. One of the most common income-generating activities in the Delta is wildlife and safari tourism, which are also called nature-based tourism (Mogende & Moswete, 2018). Tourism and wildlife management are some of the contemporary land uses that are taking place in what was historically considered as the tribal land. However, there are a few studies that define traditional and contemporary land use patterns (see Kgathi, 2002; Mbaiwa, Ngwenya, & Kgathi, 2008). Further, there are even fewer studies that discuss the conflicts between contemporary and traditional land uses on the livelihoods of farmers (see Darkoh & Mbaiwa, 2005; Okavango Delta

Management Plan (ODMP), 2007). Therefore, this chapter aims to address the following objectives: (i) to analyse the farmer's perceptions on land uses in the Okavango Delta, (ii) to analyse the conflicts between contemporary land use and arable farming, and (iii) to discuss the conflicts between contemporary land use and farming.

Description of study site and methods

The study site is the Okavango Delta, which is located in North West District, Botswana (Figure 8.1). The Okavango Delta is an inland delta covering 16,000 square kilometres and is fed by the Okavango River (Kgathi et al., 2006). It sustains numerous livelihood activities such as agriculture and fishing as well as tourism development. The Okavango Delta has been declared a Wetland of International Importance and Ramsar Site. It has also been declared United

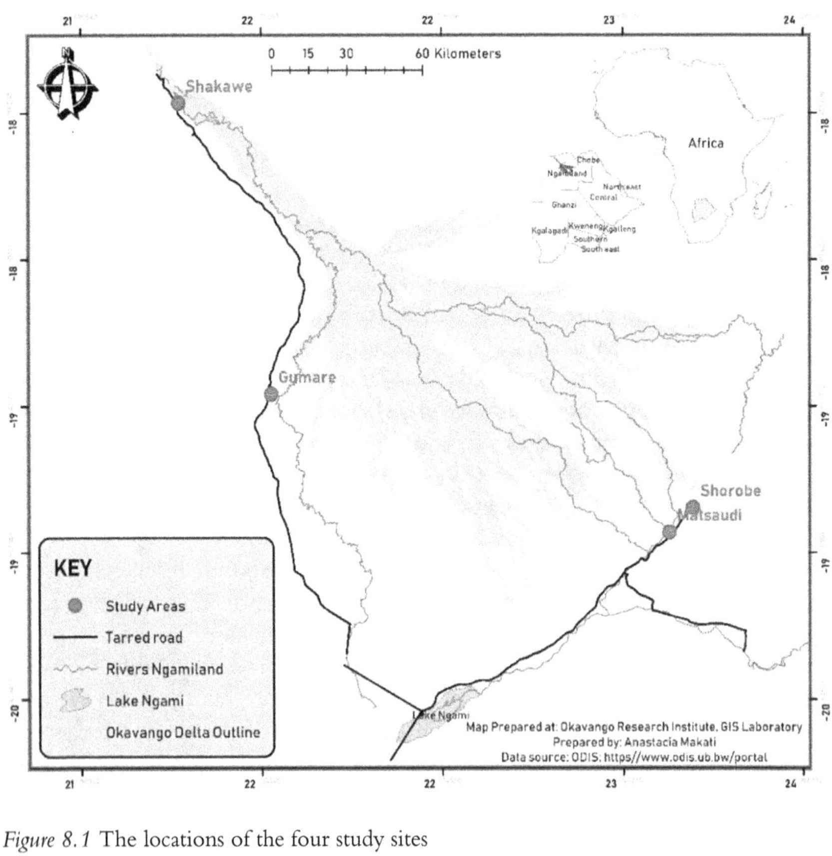

Figure 8.1 The locations of the four study sites

Source: (Okavango Research Institute GIS Lab, 2019)

Nations Educational, Scientific and Cultural Organization (UNESCO)'s 1,000th World Heritage Site.

This study generated empirical data through three primary data methods, namely, households surveys, key informant interviews and focus group discussions. Villages reflective of land use conflicts and coexistence within the Okavango Delta were identified from literature sources and through site visits. After identifying the villages with land use conflicts, four villages were conveniently sampled. These four villages are Shorobe, Matsaudi, Gumare and Shakawe (Figure 8.1).

Using the 2011 Botswana Population and Housing Census results, the total sample size for the household survey obtained for this study was 228 out of 2,976 estimated farming households at a 95% confidence level and 80% estimated response rate. This was calculated using a Raosoft sample size calculator, which is survey software. Simple random sampling was used to select households in each village. In total, 230 households were sampled. However, 221 responded, while nine of the sampled households did not participate. Data from the households' survey were collected using household survey questionnaires. The questionnaires contained a variety of open- and closed-ended questions.

The key informant interviewees were purposively selected using the expertise-oriented approach. The selection of respondents was based on their knowledge of socio-economic issues reflecting tensions between agriculture and tourism, such as farmers' compensations for crop raiding. It was also based on the respondents' expertise in documents and policies that inform agriculture and tourism. Eighteen (18) key informants with knowledge on land use conflicts and socio-economic issues reflecting tensions between agriculture and tourism in the Okavango Delta were selected. The respondents had at least five years of experience in the study area. The details of the key informants are summarised in Table 8.1.

Table 8.1 Details of the key informants

Characteristic	*Description*	*Number of representatives*
Gender	Male	13
	Female	5
	Total	**18**
Organisation or designation	*Dikgosi*	2
	Department of Wildlife National Parks	5
	Village Development Committee (VDC)	5
	Department of Animal Health and Production	2
	Botswana Tourism Organisation	2
	Department of Crop Production	2
	Total	**18**
Years of experience	5–10	8
	Above 10	10
	Total	**18**

Four (4) focus groups were held, one in each village. The groups were comprised of ten participants conveniently selected with the help of the Village Development Committee (VDC) and other community members. In this study, the focus group participants were members of the communities aged 21 years and above. The exclusion criteria were comprised of members of communities with no farming experience and those younger than 21 years. The focus group discussion guide/protocol was comprised of questions that necessitated farming experience of three years at the minimum, or at least knowledge of the conflicts between agriculture and tourism in the study area. In order to eliminate dominant member syndrome and patriarchal society biases, each group was comprised of an equal number of males and females. Therefore, 20 females and 20 males participated in the focus groups.

For data analysis, the concept-driven approach and thematic content analysis were used side by side in this study because, as Rubin and Rubin (1995, p. 26) argues, "you discover themes and concepts embedded throughout your interviews." On the one hand, a concept-driven approach is defined as the identification of data using predetermined concepts or ideas (Gibbs, 2007). On the other hand, thematic content analysis "provides a purely qualitative, detailed, and nuanced account of data" (Vaismoradi, Turunen, & Bondas, 2013, p. 400). The themes were aligned to the objectives of the study, and they represented patterned responses in the dataset.

Both qualitative and quantitative data were analysed using descriptive and inferential analyses. Pearson's Chi-square ($\chi 2$) test of independence was used to determine if there were significant relationships between various independent variables and dependent variables. The $\chi 2$ tests were preferred in this study because they provide details on the significance of any observed differences as well as on the categories that account for any differences found.

Results and discussion

Land use in the Okavango Delta

In this study, the uses of land in the Okavango Delta were categorised into two broad groups, namely, traditional land and contemporary land uses. The significant traditional land uses identified by focus groups in all of the four villages were residential, arable farming, livestock farming, fishing harvesting of natural resources and livestock grazing. The key informants substantiated the data obtained from the focus group discussions. A key informant from the Department of Wildlife National Parks highlighted, however, that the traditional land was used and still is, mainly for subsistence purposes.

Various studies in the Okavango Delta have noted that the traditional land uses and livelihood activities in the delta are similar to the ones identified in this study (see Kgathi et al., 2007; Darkoh & Mbaiwa, 2009; Ngwenya, Mosepele, & Magole, 2012). However, Kgathi et al. (2007, p. 294) note that over the years, the importance of traditional land uses and their corresponding contribution to household livelihoods in the delta have been reduced in part "because the land they depended on (and its resources) is now used for other activities such as tourism and conservation."

The contemporary land uses identified through focus group discussions were ranches and commercial establishments associated with tourism such as camps, lodges and hotels; other commercial establishments such as shops, national parks and wildlife management areas (WMAs); controlled hunting areas (CHAs); and the erection of veterinary fences. They highlighted that the Ministry of Agriculture erected the veterinary fences in response to a contagious bovine pleuropneumonia (CBPP) pandemic and foot and mouth disease. The fences serve as barriers for isolating livestock in case of a disease outbreak and for preventing the transmission of diseases. The CBPP fences erected in the delta are illustrated in Figure 8.2.

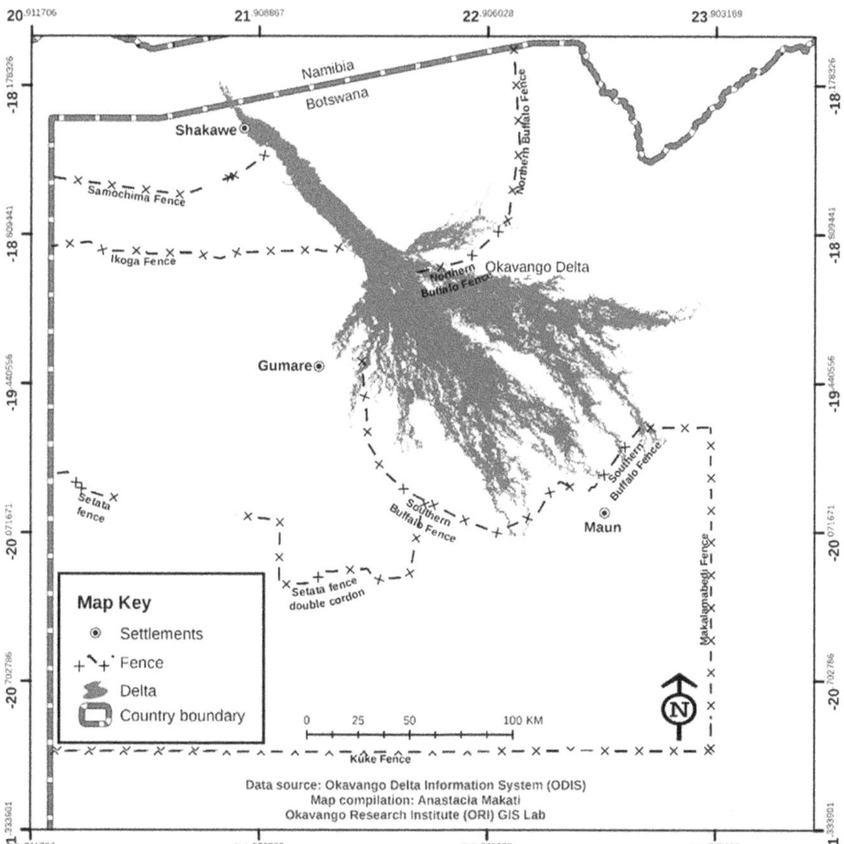

Figure 8.2 CBPP fences in the Okavango Delta

Source: (Okavango Research Institute, 2019)

The key informants supported the results obtained from focus group discussions on contemporary land uses. In addition to the reasons for the erection of fences highlighted by the focus group participants and key informants in this study, a study by Darkoh and Mbaiwa (2009) argues that veterinary fences were also constructed to protect the country's beef production and export, predominantly to the European markets. According to the results from both the focus group discussions and key informant interviews, the emergence of contemporary land uses in the Okavango Delta have led to conflicts between land uses and livelihood activities. The following section discusses the results on conflicting land uses in the Okavango Delta.

Land-use conflicts in the Okavango Delta

Conflicts between contemporary land-use with arable farming

Data from household surveys reveal that 72.4% of the respondents reported that they own *masimo* in various parts of the Okavango Delta (Figure 8.3). While dryland farming is widely practised in the region, some farmers in this study

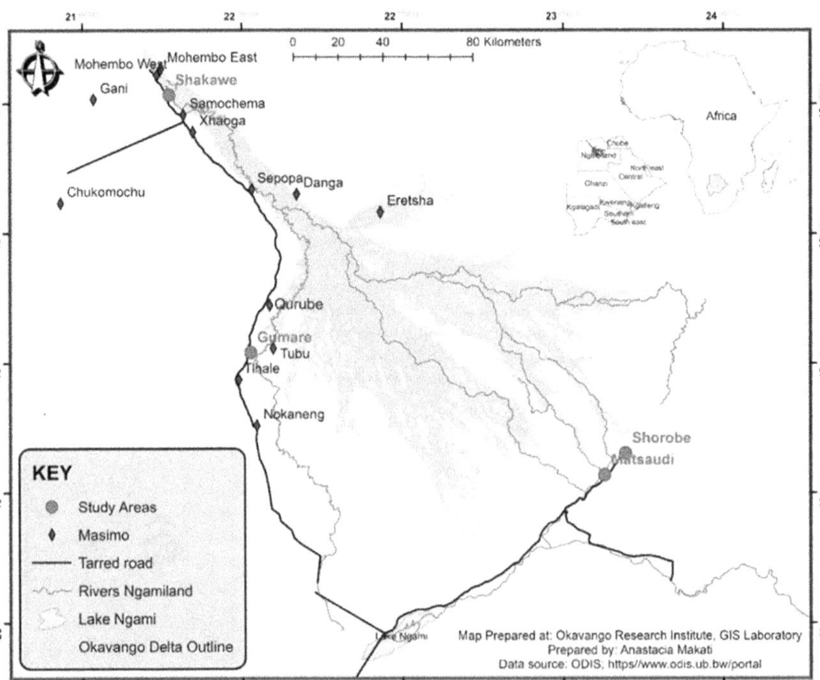

Figure 8.3 Study areas and respondents' ploughing fields (*Masimo*)

Source: (Okavango Research Institute, 2019)

use flood recession farming, locally known as *molapo* arable farming in the Okavango Delta. Correspondingly, it has been established that 75% of arable land is used as dryland fields, while 25% is used for flood recession farming in the delta (Mfundisi & Petros, 2015). It is argued that "there is higher organic matter accumulation in flood recession farms which [makes] the *molapo* farms more fertile than dryland farms" (Mfundisi & Petros, 2015, p. 148).

In this study, 76.3% of the respondents indicated that they have experienced raiding in the last three years as illustrated in Table 8.2. Only 10% of the respondents noted that such animals as warthogs, monkeys or baboons and hippopotamuses raided their farms. The majority (90%) of the farmers' whose farms have been raided reported that elephants raided the farms.

Some of the respondents lamented that when wildlife, especially elephants, raid their farms, they do not only damage crops, but also damage their dwellings and other properties on the farm. For example, a farmer in Matsaudi reported that his water tanks were broken by the elephants during a raiding experience in 2017. Similarly, a farmer in Shorobe reported that elephants destroyed her water pumps in April 2018, while a farmer in Shakawe indicated that one of her huts collapsed due to raiding by elephants in 2016. All of the respondents who have experienced raiding reported that their fences were destroyed in the processes.

The focus group discussions substantiated the results from household surveys and key informant interviews, wherein elephants were cited as animals that cause the most damage to crops and other property. According to the Okavango Delta Management Plan (ODMP), in addition to damaging crops, the other destructions caused by elephants to farms include destroying boreholes and water storage, threatening human life as well as damaging houses and roads (ODMP, 2007).

Although the damages caused by elephants and other wild animals to farms and crops are significant, some (23.3%) of the farmers whose farms were raided by wildlife in the past three years indicated that they did not report the raiding to relevant authorities (Table 8.2). A key informant from the Department of Wildlife and National Parks, based in Shakawe, also noted the incidences of non-reporting of wildlife raiding. The key informant noted that,

> We have registered numerous cases of crop raiding. Wildlife destroys the fences of *masimo* and the crops. We have recorded between 800 and 1000

Table 8.2 Crop-raiding experiences and reporting

		Reported raiding to authorities		*Total*
		Yes	*No*	
Experienced raiding	Yes	100	31	131
Total				131

> cases in Shakawe in the last two years. Elephants and hippos are frequent culprits. We are also aware that some cases go unreported. We always encourage farmers to report, but some choose not to.

Through the household survey questionnaires, farmers were asked to state the reason(s) for not reporting damages caused by wildlife to *masimo*. The reasons are summarised in Table 8.3. The commonly cited reasons for non-reporting were "authorities do not attend" (29%) and "no need because the compensation is low" (29%). In support of the former, a respondent from Gumare decried that,

> I reported to relevant authorities two years ago. The feedback after I reported was that "we will visit the scene tomorrow." It has been two full years. Surely their "tomorrow" has not arrived yet. Maybe it will arrive when I'm dead. I am not the only victim of "tomorrow," that takes months and years.

Equally the respondents noted that the reason for not reporting was that the compensation offered by the government is often "too low." By implication, the respondents perceive that the compensation does not match the damage caused by wildlife. Therefore, for such respondents, there is no incentive to report to the Department of Wildlife and National Parks. Some studies have also noted the discrepancies between the amount paid as compensation for damage caused by wildlife and the amount of damage caused by the wildlife (see ODMP, 2006, 2007; Mmopelwa & Mpolokeng, 2008).

About 23% of the respondents indicated that they did not report crop raiding because animals not listed for compensation caused the damages. Subsequent to recognising and appreciating human–wildlife conflicts in the country, the government of Botswana, through the Department of Wildlife and National Parks, has established a Problem Animal Control (PAC) unit (Republic of Botswana, 2005). The unit aims to ensure that farmers who suffered damage, inconvenience and loss due to wildlife are compensated (Republic of Botswana, 2005). However, compensation is not paid for damage caused by some wildlife species.

Table 8.3 Reasons for not reporting crop raiding

	Frequency	*Percent*
Authorities do not attend	9	29.0
Did not cause much damage	4	12.9
Damage caused by a species not listed	7	22.6
No compensation for unfenced fields	2	6.5
No need because the compensation is too low	9	29.0
Total	31	100.0

It is limited to damage caused by seven species, namely, cheetah, crocodile, elephant, leopard, lion, wild dog and hippopotamus (Republic of Botswana, 2005; Mmopelwa & Mpolokeng, 2008).

According to 6.5% of the respondents, the other reason for not reporting incidents of crop raiding is that the government does not offer compensation if the raiding occurred in unfenced fields. In a study by Sitati and Walpole (2006), it is argued that in rural Kenya, non-electrified fences serve as one of the traditional mitigation measures to deter wildlife from raiding the fields. However, fencing has financial and non-financial costs that prevent some low-income households from erecting it timeously (Sitati & Walpole, 2006).

Focus group discussions echoed similar sentiments to household surveys. In addition to the reasons summarised in Table 8.3 above, the discussions with focus groups yielded two more reasons. The first one is that the payments of compensation by the appropriate offices take too long. One of the participants in Matsaudi indicated that they had waited to receive compensation since 2015. The other reason is that in some areas, there are no Agricultural Extension Officers or field demonstrators, and the officers are untrained in other areas.

The ODMP (2007) made similar observations regarding the incidence of non-reporting of crop raiding experiences. According to the ODMP (2007), 24% of the households in the sample did not report crop-raiding experiences to the Department of Wildlife and National Parks. The reasons for not reporting given by the respondents in the ODMP are similar to the ones noted by the respondents in this study.

The majority (58%) of the respondents who reported crop-raiding incidents by wildlife to relevant authorities alleged that they had not been compensated yet. The respondents cited various reasons for not receiving compensation. One of the commonly cited reasons for delayed compensations is that "it takes too long to receive compensation" (32.7%) as reflected in Figure 8.4.

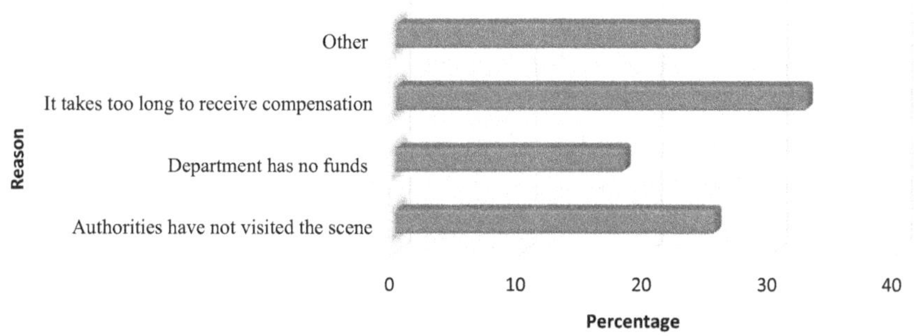

Figure 8.4 Reasons for not getting compensation

Source: Authors field work

About 25% of the respondents reported that they had not received compensation yet because "authorities have not visited the scene" as illustrated in Figure 8.4. As one of the respondents explained,

> We acknowledge that the offices that we are reporting to have other duties and that they have limited staff at times. However, they should start treating our cases as urgent. I reported last year March; they are yet to visit the scene.

The ODMP (2007) report also notes the cases of delayed assessments of fields. According to the report, the Department of Wildlife and National Parks did not assess at least 19% of the fields. Further, "only 59% of the households who reported damage had received a compensation payment at the time of the household survey (or 73% of those whose fields were assessed)" (ODMP, 2007, p. 37).

In this study, only 33.3% of the respondents remarked that they were satisfied with the amount of compensation they had received. The majority (66.7%) of the respondents noted various reasons for their dissatisfaction. A significant proportion (75%) of such respondents commented that the compensation payments were too low.

The occurrences of crop raiding have an adverse effect on the household livelihoods of arable farmers. They affect the farmers' financial capital by disallowing them to earn cash from the sale of surplus harvests, and they have a negative impact on the farmers' physical capital through the destruction of property such as fences.

More than 35% of the land in the Okavango Delta is designated as national parks, game reserves and WMAs. These areas are unfenced (Mogomotsi, 2019). Therefore, wildlife moves freely into grazing lands, leading to conflicts between wildlife conservation as contemporary land use and livestock farming as traditional land use. The following section discusses the conflicts between contemporary land uses with livestock farming in the Okavango Delta.

Conflicts between contemporary land-use with livestock farming

According to focus group discussions, the erection of veterinary fences is one of the primary sources of conflict between contemporary land uses and livestock farming. The participants pronounced that the fences have significantly reduced the area of grazing lands, thereby affecting the productivity of the livestock negatively. Generally, the participants linked the veterinary fences to wildlife conservation and tourism. They noted that their livestock in the Okavango Delta has a lower value compared to livestock in other areas of the country due to foot and mouth disease spread by buffaloes. In a study by Stone and Nyaupane (2018) at Chobe, it is argued that buffaloes carry the virus responsible for the foot and mouth disease within the region.

Consequently, cattle from the Chobe cannot be sold to the European Union (EU) beef market. "This problem has brought some dilemma to both the community and government as to how to secure farmers' livelihoods through the beef industry while at the same time safeguarding the tourism sector" (Stone & Nyaupane, 2018, p. 317). The Okavango Delta is faced with a similar dilemma.

A key informant from the Department of Animal Health and Production (DAHP) in this study noted that the fences had promoted negative attitudes of farmers towards both wildlife conservation and tourism. The informant highlighted that the farmers' negative attitudes have a basis and "their hatred towards wildlife conservation and tourism is not misplaced." In addition to the reduction of grazing land by fences, the other primary source of conflict between contemporary land uses and livestock farming identified in this study is livestock predation. Table 8.4 summarises the results of the responses regarding livestock predation between the years 2015 and 2018. These were obtained through household surveys. The majority (70.5%) of the respondents indicated that they had experienced livestock predation. The types of livestock killed included cattle, goats, sheep, donkeys and horses.

About 51% of the respondents whose livestock have been killed indicated that they had not reported the predation incidences to the relevant offices. Their decisions for not reporting are supported by the reasons summarised in Figure 8.5. The majority (52.1%) of the respondents narrated that they have not reported because the authorities rarely ever visit the scenes timeously. Their concerns were similar to those of arable farmers discussed in the previous section. This arguably reflects a vacuum in the organisation in place that serves as an institutional apparatus. The wildlife management institutions in place make proscriptions and prescriptions for curbing, mitigating, controlling and managing human–wildlife conflicts that exist in Botswana and the Okavango Delta. However, the disincentive to report on account of non-attendance to the scene by those in positions reflect the vacuum in the enforcement of the institutions' proscriptions and prescriptions.

The second highest commonly cited reason for not reporting occurrences of livestock predation is that the compensation payments for donkeys are too low, and at the time, no compensations are paid at all (20.8%). There seems to be a mismatch between the economic value attached to donkeys by the government

Table 8.4 Responses about livestock predation

	Response	*Frequency*	*Percent*
Livestock killed between 2015 to 2018	Yes	91	70.5
	No	38	29.5
	Total	129	100

Source: Authors fieldwork

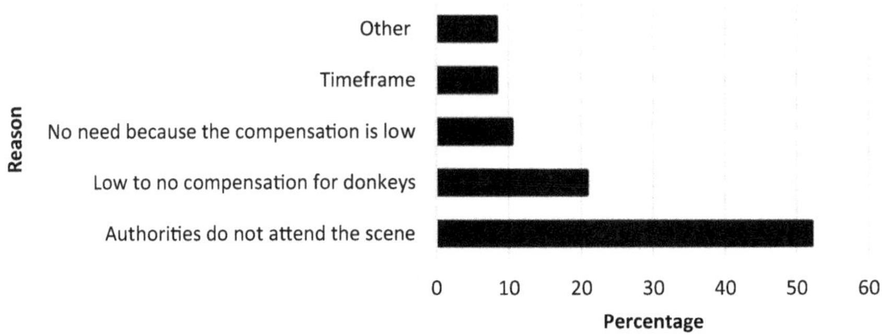

Figure 8.5 Reasons for not reporting livestock predation

Source: Authors fieldwork

and the communities. In 2013, the government noted some changes in the rates of compensation payments for livestock as follows:

> in a case where a farmer's cow was killed by a lion, it would be compensated with P3 000, which will be an increase from the current P1 050. [The] compensation for a donkey killed by a lion had increased from P70 to P200 . . . in a case where a farmer's bull had been killed by a lion, compensation was raised from P1 900 to P5 500.
>
> (Galeragwe, 2013).

The compensation payments for donkeys are significantly lower than those for cattle. However, in the Okavango Delta, the loss of donkeys due to predation represents an economic loss to farmers, not only regarding direct losses but also regarding the increased costs and work effort. According to the respondents in this study, the government has undervalued the importance of donkeys to their livelihoods. In the Okavango Delta, donkeys play an essential role in the livelihoods of subsistence farmers (Geiger & Hovorka, 2015). They serve as an affordable means of transport and draught power for subsistence farmers (Geiger & Hovorka, 2015). They also contribute towards household incomes and food security for small-scale farmers (Geiger & Hovorka, 2015).

The mismatch in the values attached to donkeys by the government and the farmers in the Okavango Delta has not only created a disincentive in reporting incidences of donkey predation by lions and other wildlife but has also led to the retaliatory killing of the predators by the farmers. It has been noted that the failure of government institutions and their corresponding institutional arrangements often leave farmers with no alternative but retaliation (Minnie, 2009).

The other noteworthy reason for not reporting livestock predation incidences to relevant offices is the "timeframe" (Figure 8.5). According to 8.3%

of the farmers, they often discover the carcasses of their livestock days after the incidences. In the Okavango Delta, as in other parts of the country, households widely practice the traditional three-tier land-use system (Darkoh & Mbaiwa, 2005). The three-tier system encompasses the village, the lands (*masimo*) and the cattle post (*moraka*). As evidenced by the above quote, the household has to divide time and other resources among the three-tier land use. However, the evidence of livestock predation is lost as time elapses. The farmers noted that the livestock could be killed by any of the large carnivores in the Okavango Delta, such as the lion (*Panthera leo*), leopard (*Panthera pardus*), cheetah (*Acinonyx jubatus*), African wild dog (*Lycaon pictus*) and spotted hyena (*Crocuta crocuta*). However, some of these carnivores, such as spotted hyena, are not on the list of predator species identified for compensation payments. It, therefore, becomes difficult to identify the species that killed the livestock after several days have passed.

Of the 49% of farmers who reported livestock predation to the relevant authorities, only 39% were not compensated. In this study, a χ^2 test of independence was used to determine if there is a significant relationship between socio-demographic features of the farmers, namely, age, gender and highest qualification, and compensation payments. The null hypothesis of the test assumed that there is no significant association between the payment of compensation and the farmers' socio-demographic features. At $p > 0.05$ for all the variables, no association was found between the payment of compensation and the farmers' socio-demographic profiles. The null hypothesis is therefore accepted.

Although the majority (about 61%) of the farmers received compensation payments, only one respondent (3.6%) reported that he was satisfied with the payment. One of the highly quoted reasons for the farmers' degree of dissatisfaction is "compensation was too low" (74.1%). The key informant from the PAC unit acknowledged the concerns about the low rates of compensation payments. In his words, "the money is not meant to be more than or equal to the value of the livestock. It aims to help the farmer in the events of conflict with wildlife." He further noted that his office had not made follow-ups to enquire whether the complainants were satisfied with the compensation payments they had received between 2015 and 2018.

Conclusion

This chapter analysed the land-use conflicts in the Okavango Delta and highlighted the inherent conflicts between tourism and agriculture in the region. The emergence and development of contemporary land use in the region, such as WMAs and tourism, have created conflicts between traditional land uses and modern ones. The majority of the arable farmers (76.3%) indicated that they had experienced raiding by wildlife in the last three years. However, 58% of the respondents who reported crop-raiding incidents by wildlife to relevant authorities noted that they had not been compensated. Similarly, the majority

(70.5%) of the livestock farmers indicated that they had experienced livestock predation in the last three years, the majority (51%) indicated that they had not reported the predation incidences to the relevant offices.

Although the government offers compensation to farmers who have suffered a loss due to wildlife, some farmers choose not to report the incidents of crop raiding and livestock predation. In cases where farmers report, some receive no compensation while those who receive compensation are generally not satisfied with the payments for various reasons. One of the most common reasons for the dissatisfaction is that the compensation payment is far exceeded by the loss suffered. Therefore, compensation payments are ineffective in the study areas as the rates paid by the government to farmers often do not match the extent of the damages incurred by wildlife to livestock and crops. In the process, the livelihoods of the farmers are negatively affected. Therefore, the institutional instruments should be revised and mechanisms of calculating the value of the damage caused and value lost should be developed. The mechanisms should use the market rates and other objective measures to calculate the costs related to wildlife damages. Furthermore, there is a need for the adoption of adaptive management strategies in the Okavango Delta specifically, and in Botswana broadly. The strategies will not only help reconcile different land uses but will also promote a balance between contemporary and traditional land uses in order to simultaneously improve the livelihoods for farmers and develop the business environment for tourism-related enterprises.

References

Albrechts, L. (2004). Strategic (spatial) planning reexamined. *Environment and Planning B: Planning and Design, 31*, 743–758.

Darkoh, M., & Mbaiwa, J. E. (2005). *Natural resource utilisation and land use conflicts in the Okavango Delta, Botswana*. START.

Darkoh, M. B., & Mbaiwa, J. E. (2009). Land-use and resource conflicts in the Okavango Delta, Botswana. *African Journal of Ecology*, *47*, 161–165.

Darkoh, M. B. K., & Mbaiwa, J, E. (2014) Okavango Delta – A Kalahari Oasis Under Environmental Threats. *Journal of Biodiversity & Endanger Species*, *2*:138. doi:10.4172/2332-2543.1000138.

Galeragwe, M. (2013). *Government increases compensation*. Retrieved from Botswana Daily News: www.dailynews.gov.bw/news-details.php?nid=6652.

Geiger, M., & Hovorka, A. J. (2015). Animal performativity: Exploring the lives of donkeys in Botswana. *Environment and Planning D: Society and Space*, *33*(6), 1098–1117.

Gibbs, G. R. (2007). *Qualitative research kit: Analyzing qualitative data*. London: Sage Publications.

Kgathi, D. L. (2002). *Natural resources, tenure and access in Botswana's Okavango Basin*. Maun: HOORC.

Kgathi, D. L., Kniveton, D., Ringrose, S., Turton, A. R., Vanderpost, C., & Lundqvist, J. (2006). The Okavango: A river supporting its people, environment and economic development. *Journal of Hydrology*, *331*(1–2), 3–17.

Kgathi, D. L., Ngwenya, B. N., & Wilk, J. (2007). Shocks and rural livelihoods in the Okavango Delta, Botswana. *Development Southern Africa*, *24*(2), 289–308.

Kim, W., Jun, H. M., Walker, M., & Drane, D. (2015). Evaluating the perceived social impacts of hosting large-scale sport tourism events: Scale development and validation. *Tourism Management*, *48*, 21–32.

Liu, G., Liu, Z., Hu, H., Wu, G., & Dai, L. (2008). The impact of tourism on agriculture in Lugu Lake region. *International Journal of Sustainable Development and World Ecology, 15*, 3–9.

Mbaiwa, J. E. (2004). The socio-cultural effects of tourism development in the Okavango Delta. *Journal of Tourism and Cultural Change, 2*(3), 163–184.

Mbaiwa, J. E. (2011). Changes on traditional livelihood activities and lifestyles caused by tourism development in the Okavango Delta, Botswana. *Tourism Management, 32*(5), 1050–1060.

Mbaiwa, J. E., Ngwenya, B. N., & Kgathi, D. (2008). Contending with unequal and Privileged access to natural resources and land in the Okavango Delta, Botswana. *Singapore Tropical Geographical Journal, 29*(2), 155–172.

Mfundisi, K. B., & Petros, O. (2015). A comparative assessment of soil fertility on flood recession (Molapo) and dryland farms: The case of Xhobe settlement in the Okavango Delta world heritage site. *Botswana Notes and Records, 47*, 148–157.

Minnie, L. (2009). *Socio-economic and ecological correlates of leopard-stock farmer conflict in the Baviaanskloof mega-reserve, Eastern Cape*. Port Elizabeth: Unpublished masters thesis, Nelson Mandela Metropolitan.

Mmopelwa, G., & Mpolokeng, T. (2008). Attitudes and perceptions of livestock farmers on the adequacy of government compensation scheme: Human-carnivore conflict in Ngamiland. *Botswana Notes and Records, 40*, 147–158.

Mogende, E., & Moswete, N. (2018). Perspectives on the environmental impacts of wildlife-based tourism at the Chobe national park river front, Botswana. *PULA: Botswana Journal of African Studies, 32*(1), 96–125.

Mogomotsi, P. K. (2019). *An institutional framework for the sustainable co-existence of tourism and agriculture in Botswana*. Potchefstroom, South Africa: Unpublished PhD thesis, North West University.

Mowforth, M., & Munt, I. (2015). *Tourism and sustainability: Development, globalisation and new tourism in the third world*. London: Routledge.

Ngwenya, B. N., Mosepele, K., & Magole, L. (2012). A case for gender equity in governance of the Okavango Delta fisheries in Botswana. *Natural Resources Forum, 36*, 109–122.

Noga, S. R., Kolawole, O. D., Thakadu, O. T., & Masunga, G. S. (2018). 'Wildlife officials only care about animals': Farmers' perceptions of a Ministry-based extension delivery system in mitigating human-wildlife conflicts in the Okavango Delta, Botswana. *Journal of Rural Studies, 61*, 216–226.

ODMP. (2006). OKAVANGO DELTA MANAGEMENT PLAN PROJECT SECRETARIAT. Draft final framework management plan. Maun, Botswana.

ODMP. (2007). *Okavango delta management plan: Component 5: Wildlife management – human elephant conflict*. Gaborone, Botswana: Natural Resources & People (NRP).

Rubin, H. J., & Rubin, I. S. (1995). *Qualitative interviewing: The art of hearing data* (2nd ed.). London: Sage Publications.

Sitati, N. W., & Walpole, M. J. (2006). Assessing farm-based measures for mitigating human-elephant conflict in Transmara District, Kenya. *Oryx, 40*(3), 279–286.

Stone, M. T., & Nyaupane, G. P. (2018). Protected areas, wildlife-based community tourism and community livelihoods dynamics: Spiraling up and down of community capitals. *Journal of Sustainable Tourism, 26*(2), 307–324.

Telfer, D. J., & Wall, G. (1996). Linkages between tourism and food production. *Annals of Tourism Research, 23*(3), 635–653.

Vaismoradi, M., Turunen, H., & Bondas, T. (2013). Content analysis and thematic analysis: Implications for conducting a qualitative descriptive study. *Nursing Health Sciences, 15*(3), 398–405.

9 Local people's perspectives on wildlife conservation, ecotourism and community livelihoods

A case study of Lusaka National Park, Zambia

Vincent R. Nyirenda, Castro Milimo and Ngawo Namukonde

Introduction

Establishment of protected areas on land originally under customary regime may result in negative people-park relations (Bragagnolo, Malhado, Jepson, & Ladle, 2016). The people's negative positions and attitudes towards protected areas breed intolerance to wildlife (Snyman, 2014). Attitudes can be expressed as characteristic responses over public environmental concerns (Huddart-Kennedy, Beckley, McFarlane, & Nadeau, 2009). Local people's perceptions of protected areas may depend on a number of factors, inter alia their perceived cost and benefits, and levels of trust about governance institutions (Mutanga, Muboko, & Gandiwa, 2017).

Local perceptions influence attitudes and practices for public participation and undermine public participation in agenda setting, decision-making, policy formulation and implementation (Ward, Holmes, & Stringer, 2017). Historically, establishment of many African protected areas coerced inhabitants to relocate and alienated them from accessing natural resources from such protected areas (Mombeshora & Le Bel, 2009; Fischer, Muchapondwa, & Sterner, 2011). Consequently, local people resented protected areas where "fortress conservation" was prevalent as a conservation protectionist approach (Child, 2009; Strickland-Munro, 2010). When local people are excluded from the resources they depend on, they also change their resource use behaviour (Ebua, Agwafo, & Fonkwo, 2011). However, compounding factors for local perceptions, such as wildlife crop damages, livestock predation and insufficient compensation benefits complicate the local people's behaviour towards wildlife conservation (Mutanga, Vengesayi, Muboko, & Gandiwa, 2015). Negative perceptions, attitudes and practices tend to variedly undermine long-term biodiversity conservation and tourism development goals and efforts (Imran, Alam, & Beaumont, 2014; Allendorf, Aung, Swe, & Songer, 2017). Forcibly internalized costs by local people compromise their cooperative attitudes (Martin, Myers, & Dawson, 2018). As the interactions and actions by local people are influenced by perceptions and attitudes, sustenance of management strategies and conservation effectiveness depend on them (Mamo, 2015).

Sustainable tourism development plays a critical role in internalising and increasing the net benefits and reducing the costs among the beneficiaries (i.e. local people) by securing their livelihoods and promoting wildlife conservation (Andereck et al., 2005; Snyman, 2014). The sustainable tourism development assumes promotion of a socio-ecological system through the protection of natural resources on which tourism depends while harnessing benefits from the resources (Berno & Bricker, 2001). Sustainable tourism development has the potential to generate a flow of benefits to local people and improve their socio-economic lives, livelihoods and positively contribute to wildlife conservation through their sustained support in the long run (Emptaz-Collomb, 2009; Mbaiwa, 2018; Snyman, 2014), largely influenced by positive social exchanges among the residents (Nyirenda, Kaoma, Nyirongo, Lwali, & Chomba, 2017; Mbaiwa, 2018). Thus, tourism strategies may result in unexpected outcomes because of nature and complexities associated with tourism development. Therefore, there is a need to approach the analysis of sustainable tourism development from multiple perspectives (Fodness, 2017).

Local people acquaint and interact with protected areas through resource use, traditional ecological knowledge and environmental awareness initiatives (Gandiwa, 2012). In order to evaluate the conservation projects' successes or failures, there is a need to understand local people's perceptions (Sundaresan et al., 2012). Local people abutting protected areas are important in identifying and implementing community livelihoods and biodiversity conservation interventions (Whitham, Kun, & Riordan, 2015).

There is now increasing consideration by the wildlife researchers and conservation practitioners for enhanced capacity in local livelihoods strategies (Chechina, Neveux, Parkins, & Hamann, 2018); Ward, Stringer, & Holmes, 2018). Considering that perceptions, attitudes and practices among local people are contextual and site-specific, we focus on the wildlife conservation, ecotourism and community livelihoods associated with the peri-urban Lusaka National Park. We explore two research questions: (i) What are the factors underlying the local people's perceptions, attitudes and practices towards the peri-urban Lusaka National Park? and, (ii) How do Shantumbu people interact with their neighbouring peri-urban Lusaka National Park? (iii) And whether the interaction behaviours with the park are predicted by livelihoods effects?

Methods and materials

Study area

The Shantumbu is a customary land abutting a fenced Lusaka National Park (LNP; 6,715 hectares) see Figure 9.1. The park was established in 2011 and declared a National Park in 2015, principally for wildlife conservation and

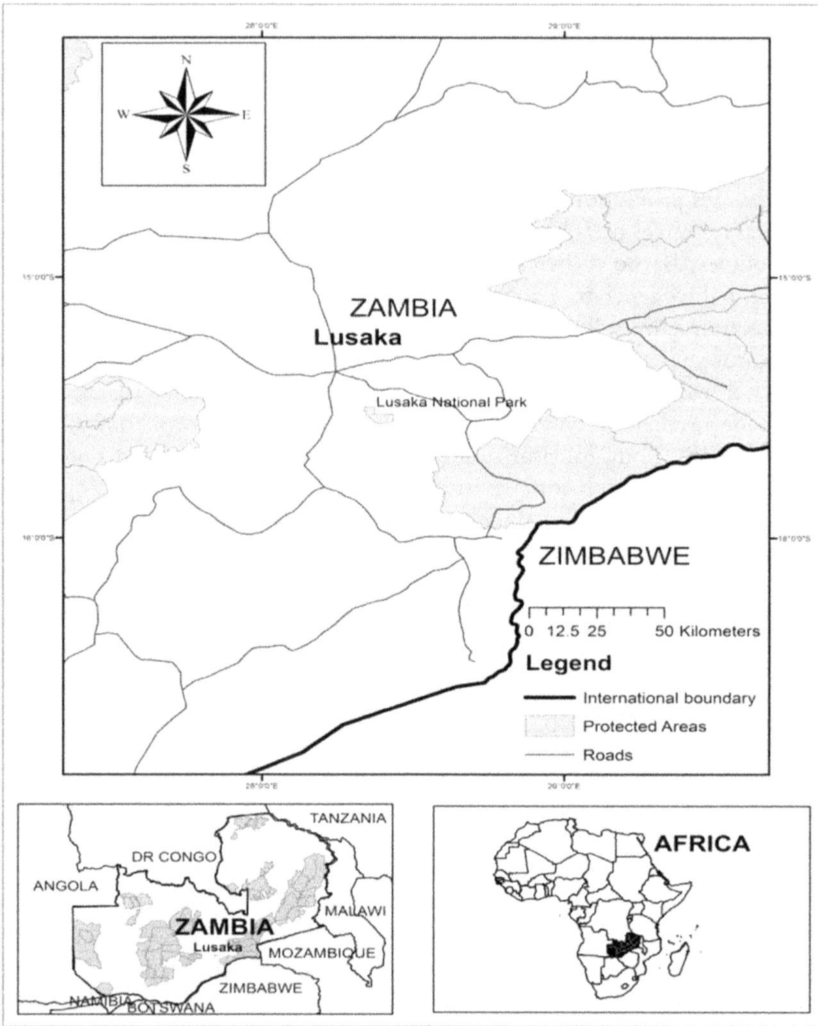

Figure 9.1 Lusaka National Park and its surrounding areas, Zambia

Source: created by V. R. Nyirenda, 2019

tourism development. It is located in the southeast peripherals of Lusaka city – Zambia's capital. Antecedent to creation of the park, the Shantumbu's predominately Soli-speaking people utilised the area, which constituted designated Forests No. 26 and No. 55 and previously was converted from a customary land for firewood collection, charcoal burning, timber harvesting, crop cultivation and sand quarrying (Nyirenda, Siamudaala, & Kaula, 2014). The park has been inhabiting several wildlife species that include white rhinoceros (*Ceratotherium simum* Burchell, 1817), giraffe (*Giraffa camelopardalis* Linnaeus, 1758), common eland (Taurotragus oryx Pallas, 1766), hartebeest (*Alcelaphus buselaphus* Pallas, 1766), zebra (*Equus quagga* Boddaert, 1785), sable antelope (*Hippotragus niger* Harris, 1838), great kudu (*Tragelaphus strepsiceros* Pallas, 1766), blue wildebeest (*Connochaetes taurinus* Burchell, 1823), common waterbuck (*Kobus ellipsiprymnus ellipsiprymnus* Ogilbyi, 1833), black lechwe (*Kobus leche smithemani* Lydekker, 1900), impala (*Aepyceros melampus* Lichtenstein, 1812), puku (*Kobus vardonii* Livingstone, 1857), bushbuck (*Tragelaphus scriptus* Pallas, 1766), southern reedbuck (*Redunca arundinum* Boddaert, 1785), warthog (*Phacochoerus africanus* Gmelin, 1788), pangolin spp., and axis deers (*Cervus axis* Erxleben, 1777) (Nyirenda et al., 2014). The numbers accompanying the species scientific names and authorities are representing the year of species designation. These animal species were translocated to the park from elsewhere, within and outside Zambia.

Data collection

In February, 2018, a total of 82 randomly sampled elderly respondents (≥18 years old) were drawn from 150 households of Shantumbu village based on the village register and subsequently interviewed. Data was collected using a semi-structured questionnaire which was administered through face-to-face interviews, where the interviewers read out the questions from the questionnaires to the respondents. Where necessary, the questions were translated into a local language to ensure that the respondents fully comprehended them. Such interview protocols provided participants with freedom to effectively participate and researchers with freedom to promptly evaluate and probe responses (Dimitrakopoulos et al., 2010).

Data analysis

Data was analysed using R statistical software version 3.5.1 and ggplot2 package for statistical analyses(R-Core Team, 2018). The statistics were calculated in accordance with Zuur, Ieno, Walker, Saveliev, and Smith (2009). Knowledge, attitudes and practices (KAP) framework of analysis was adopted because of its thematic rigor (Kaliyaperumal, 2004; Gemeda, Minstro, Feyssa, Sima, & Gutema, 2016). Using qualitative content analysis (Guest, 2012), in-depth insights of responses by the interviewees were synthesised by thematic classification of issues raised. The Likert scale was adopted to provide a degree of agreement or disagreement in the context of measure (cf. Dunlap, 2008; Hawcroft & Milfont, 2010) and coded *a priori* analyses. Further, we adopted mixed methods to analyse sustainable tourism development in the study area due to its complex nature and multiple perspectives (Fodness, 2017).

Results

Social factors influencing perceptions, attitudes and practices

Socio-demographics

Out of a total of 82 respondents, 59.8% (n=49) were male and 40.2% (n=33) were female. The age of the respondents ranged from 18–72 years. A large number of interviewees (43.9%) were aged between 26–35 years. Most of the respondents were married (65.9%, n=54) while the rest were single (15.9 %, n=13), widows (7.3%, n=6), widowers (2.4%, n=2) and "other" that included participants divorced or engaged (8.5%, n=7). The respondents' household sizes ranged from 2 to 10 members per household (median=6). A higher number of the respondents (53.7%, n=44) had household sizes ranging from 6–10 (large families) while the rest (46.3%, n=38) had household sizes of 5 and below (small families). The majority of the respondents were small- to medium-scale farmers (96.3%, n=79), while the rest were students at secondary schools (3.7%, n=3). Maize was the most commonly cultivated crop, and occasionally groundnuts and beans were cultivated. All of the respondents were Christians. The highest level of education attained among respondents was secondary school. A higher number of the respondents (63.4%, n=52) merely reached primary school than the rest (36.6%, n=30), who reached secondary school. Out of female respondents, 72.7% (n=24) reached primary school, while 27.3% (n=9) reached secondary school. From male respondents, 57.1% (n=28) reached primary school, while 42.9% (n=21) reached secondary school. The minimum length of residence of the respondents was 1 year and the maximum was 50 years (median=15 years). Most of the respondents (61%, n=50) lived in Shantumbu for more than 10 years and only 39.0% (n=32) lived in Shantumbu for 10 years and below. The number of livestock owned by the respondents ranged from 1 to 50 (median=5). A higher number of the respondents (58.5%, n=48) had numbers of livestock at 10 and below. Goats and chickens were the most commonly kept livestock and, in rare cases, pigs, ducks and guinea fowls were reared.

Knowledge: physical factors influencing perceptions, attitudes and practices

Size and boundary

All of the 82 respondents did not know the size of the LNP, but the majority of them (59.8%, n=49) knew its physical boundary. Only a few of the female respondents (42.4%, n=14) knew the boundary of the LNP as compared to male counterparts (71.4%, n=35). There was a significant difference between gender and knowledge on the park boundary (χ^2=6.898, df=1, p=0.009). More elderly respondents were aware of the park boundary compared to younger ones. However, there was an insignificant difference between age and knowledge on the park boundary (χ^2=8.910, df=4, p=0.063). There was a significant association between household sizes and knowledge on the park boundary (χ^2=9.175, df=1, p=0.002). Respondents with larger (>5) household sizes were more aware of the park boundary compared to those with smaller (≤5) household sizes. Most of the respondents who reached secondary school (76.7%, n=23) knew the park boundary as compared to those who reached primary school (50%, n=26). There was a significant difference between attained level of education and knowledge on the park boundary (χ^2=5.626, df=1, p=0.018). Further, there was also a significant association between length of residence and knowledge on the park boundary (χ^2=21.853, df=4, p<0.001). Respondents who have resided in Shantumbu village for a longer time were aware of the park boundary than those who have resided there for a relatively shorter time. There was also significant association between the number of livestock and knowledge on the park boundary (χ^2=8.710, df=3, p=0.033). Respondents with a larger number of livestock were more aware of the park boundary than respondents with fewer livestock.

Animal species in LNP

Most respondents (95.1%, n=78) knew the type of wild animals in LNP, while only a small fraction of respondents (4.9%, n=4) were unaware. Those that were aware of the animal types did not only name them but gave additional information about them. For example, one male respondent indicated: "I know the wild animals that are fenced and kept in the park, such as giraffes, hartebeests, warthogs, sables, zebras, wildebeests, rhinos, waterbuck, elands, impalas, puku, lechwes, kudu, bushbuck and reedbuck." There was an insignificant difference between gender and knowledge on the wild animals (χ^2=2.112, df=1, p=0.146). However, there was a significant association between length of residence and knowledge on the type of wild animals (χ^2=10.708, df=1, p=0.030). Respondents who resided in Shantumbu village for a longer time were more aware of the types of wild animals in the park than those who had resided there for a relatively shorter period of time.

Perceptions, attitudes and practices among the Shantumbu people

Residents' perceptions

The major predictors of perceptions among the Shantumbu people were level of education attained, gender, marital status, occupation and number of livestock owned (Table 9.1). The majority of the respondents (69.5%, n=57) did not perceive any losses or costs from the park, while 30.5% (n=25) of the respondents perceived them. The most frequently perceived losses were crop damage by warthogs. The warthogs were reported to evade the park electric wire fence and wildlife staff at night by use of trenches to predate on crops such

Table 9.1 Major predictors of perceptions among Shantumbu people

Predictors	*Df.*	*SS*	*RSS*	*AIC*	*F-Value*	*P-value*	*Level of significance*
Education	1	1.980	5.978	-194.73	35.155	<0.001	***
Gender	1	0.521	4.519	-217.67	9.247	0.003	**
Length of residency	2	0.397	4.637	-218.02	4.567	0.002	*
Marital status	4	0.803	4.801	-218.70	3.565	0.010	*
Occupation	1	0.237	4.236	-222.98	4.216	0.044	*
No. of livestock	1	0.224	4.223	-223.23	3.987	0.050	*
Age	1	0.011	4.010	-227.48	0.198	0.658	NS
Size of household	1	0.001	3.998	-227.71	0.001	0.987	NS

NB: Df.-Degree of freedom; SS-Sum of squares; RSS-Residual sum of squares; AIC-Akaike information criterion; ***-Highly statistically significant; **-Very statistically significant; *- Statistically significant.

as maize (*Zea mays*) in Shantumbu village. Other forms of losses included loss of pasturelands for livestock as a result of park fencing and the exclusion of local people from accessing resources in the park, such as firewood and mushrooms. For instance, one female respondent whose crop field was located adjacent to the park lamented like many others: "The 'wild pigs come out and destroy some of our crops and return to the park in the night. On [the] other hand, we no longer take our animals to the park for grazing [and] can neither cultivate nor collect firewood, fruits and mushrooms from there as we used to do prior to the park fencing."The 51.2% (n=42) of respondents perceived conservation of wild animals in LNP as important, while 48.8% (n=40) perceived it as not important. The respondents outlined tourism as the main reason why conservation of the park's wild animals is important. A significant proportion of men (63.3%, n=31) compared to women (33.3%, n=11) agreed that conserving the park's wild animals is vital (χ^2=7.071, df=1, p=0.008). One male respondent remarked: "Wildlife-based tourism can help us improve our lives through sales of curios to tourists and visitors to the park like it is the case in other places." Further, most of the respondents who reached secondary school (93.3%, n=28) perceived conservation of wild animals to be important compared to those who reached primary school (26.9%, n=14). There was a significant association between level of education and perception of conservation of the park's wild animals (33.533, df=1, p<0.001). However, the level of education attained, gender, marital status and occupation had varied relationships with the number of livestock (Figure 9.2), which was an apparent source of conflict over limited pastureland. Table 9.1: shows the major predictors of perceptions among Shantumbu people adjacent to Lusaka National Park, Zambia, February 2018, based on multiple logistic regression.

Half of the respondents (n=41) indicated that the park was created to benefit the local people while the other half disagreed. Those who submitted that the park was created to benefit the local people were optimistic that the government would generate income from the park through tourism which will in turn benefit the local people through provision of social amenities such as

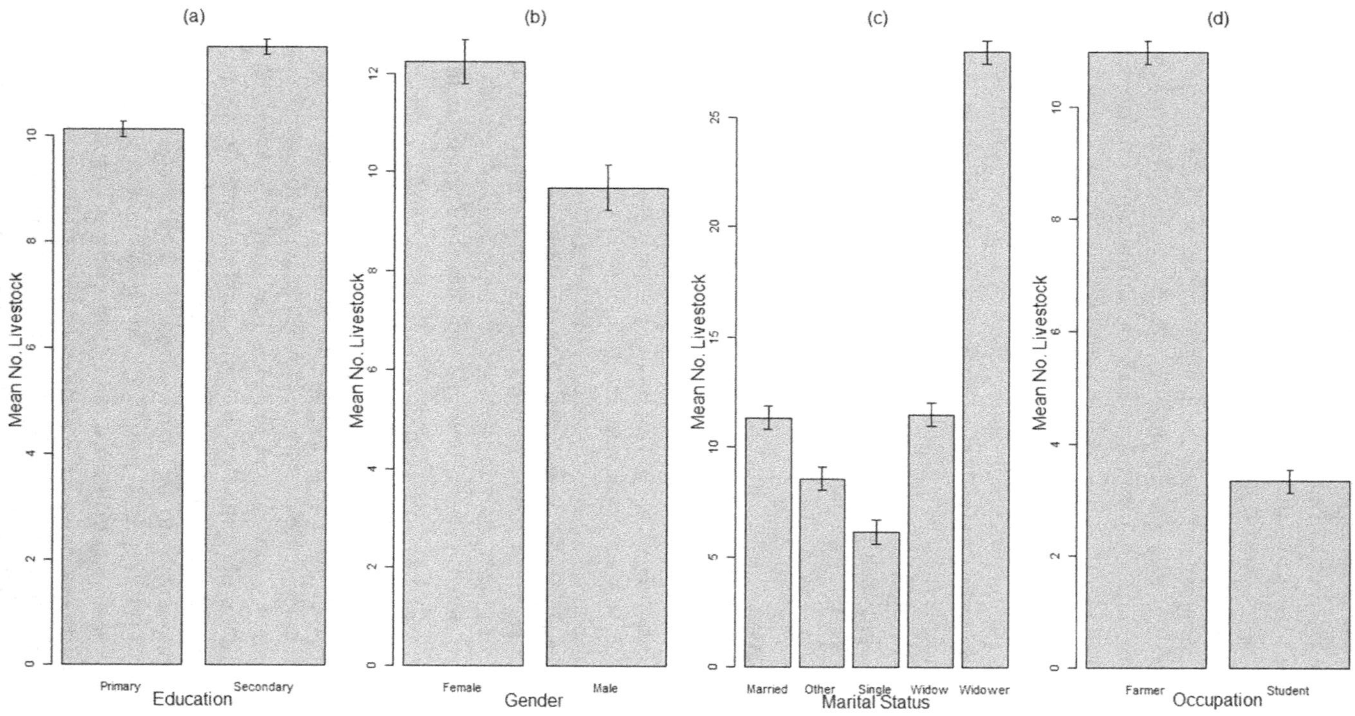

Figure 9.2 Relationship between level of education attained, gender, marital status and occupation

schools and health centres. Figure 9.2: The relationship between four factors (level of education attained, gender, marital status and occupation) influencing perception towards Lusaka National Park, and the number of livestock.

There was a significant difference between the number of men and women who perceived local benefits from the park (χ^2=6.136, df=1, p=0.013). However, some respondents perceived more tourism-related benefits accruing to individual entrepreneurs than the community as a whole in the form of social amenities. For example, a female respondent added: "The park was created [so] that we make money from small tourism business." Further, most of the respondents (64.6%, n=53) declined greater freedom to "control" (legally cull) straying animals from the park on the basis of lack of local capacity and also feared there would be on-slaughtering of the animals. Many submissions recorded from respondents can be summarised by one from a female respondent: "Wildlife from the park will finish if anyone is free to kill them." The numbers of respondents (51.9%, n=27) who reached primary school who agreed that local people should be given greater freedom to control stray animals from the park was significantly different from those holding the same perception who reached secondary school (6.7%, n=2)(χ^2=17.046, df=1, p<0.001).

RESIDENTS' ATTITUDES

The majority of the respondents (92.7%, n=76) denied LNP having improved their lives, because they had not received direct benefits from the park ever since it was declared a national park in 2015. At least half of the respondents (58.5%, n=48) were dissatisfied that their households lived close to the LNP. One of the male respondents urged that: "We suffer more costs than we gain such that our lives have not improved by living close to the park." A significantly greater number of male respondents (53.1%, n=26) were dissatisfied that their households lived close to the park than females (24.2%, n=8) (χ2=6.748, df=1, p=0.009). Their dissatisfaction was attributed to crop damages by warthogs and a dearth of benefits from the park. Further, significantly more respondents who reached secondary school (90.0%, n=27) were satisfied that their households lived close to the LNP than those who reached primary school (13.5%, n=7) (χ2=45.917, df=1, p<0.001).

Only 48.8% (n=40) of the respondents favoured the park while 51.2% (n=42) disliked its existence near them, principally due to the perceived absence of financial benefits accrued from the park. The number of male respondents (59.2%, n=29) who liked the park is significantly different from that of their female counterparts (33.3%, n=11) (χ^2=5.274, df=1, p=0.022). A higher number of the respondents who are married liked the LNP than those who were not married (χ^2=11.024, df=4, p=0.024). One male indicated that: "My family and I enjoy seeing animals from this nearby park, as I cannot afford to go elsewhere for game viewing." A significantly higher number of respondents who reached secondary school (93.3%, n=28) liked the park than those who reached primary school (23.1%, n=12) (χ^2=37.584, df=1, p<0.001). Viewing the animals from the park was the reason advanced by park enthusiasts. However, respondents who reached primary school (100%, n=52) agreed that the size of the LNP should be reduced against a few secondary school attendees (10.0%, n=3). One of the female respondents' similar submissions reported: "We can have more land for agriculture and other developmental activities if the park size was further reduced. Currently, we are not allowed pasture for our livestock in the park." There was a significant association between length of residence and respondents' attitudes in favour of reducing the size of the park (χ^2=8.349, df=4, p=0.050). People with a longer residency in Shantumbu village favoured reduction of the park for agriculture and settlements.

RESIDENTS' PRACTICES AND ACTIONS

Most respondents (93.9%, n=77) took no "control" methods but reported to the village headman while the rest (6.1%, n=5) reported to the wildlife agency for action against straying animal incidences out of the park. Like other respondents, a man justified actions by the community members: "We report to the village headman because we respect him and [he] has to know everything that goes on in his village. Upon getting our complaints, he can then report to relevant authorities for corrective actions." There was no compensation policy for crop damages, loss of property and other incursions. In the event that they came across a wild animal carcass, some respondents (14.6%, n=12) buried the carcass themselves while others (69.5%, n=57) reported it to the village headman and the rest (15.9%, n=13) reported it to the wildlife agency which then disposed it off. On encountering a member of their community in possession of illegal wildlife products, most respondents (66.5%, n=54) reported it to the village headman while others (3.4%, n=3) to the wildlife agency and the rest (30.1%, n=25) preferred to do nothing about it. Although there had never been any physical attack on a local person by a wild animal from the park, most of the respondents (97.6%, n=80) indicated that, if such an incident happened, reports would be made to the village headman.

Discussion

This chapter posits that level of education attained, gender, marital status, occupation and number of livestock owned influence local perceptions towards an ecotourism-oriented urban national park (Table 9.1). Shantumbu residents held mixed perceptions towards the LNP based on their individual situations, such as level of education attained, gender, marital status, occupation and number of livestock owned, which impacted upon their interaction with the peri-urban park. Higher education, public awareness and occupation reinforce positive perceptions about biodiversity conservation, which may lead to support for local decisions and management efforts (Gandiwa et al., 2014). However, limited awareness and communication with authorities demoralise local people and hinder their participation in conservation and tourism activities (Ciocănea, MirelaIanoşi, & Bagrinovschi, 2016).

Better educated residents tend to access better employment, providing alternative livelihood strategies that contribute to reducing dependency on natural resources from protected areas and increase their tolerance towards the park. Similar studies elsewhere have shown that positive perceptions and attitudes can develop among the local people living adjacent to peri-urban national parks if they have a strong "sense of place" in respect to it (Tanner-McAllister, Rhodes, & Hockings, 2014; Rossi, Pickering, & Byrne, 2016). The more educated and affluent respondents tend to hold ecocentric values in favour of the park and its tourism activities, while the less educated ones would be more anthropocentric and supportive of utilitarian activities (Pickering & Rossi, 2016).

Further, gender and marital status may influence local people's perceptions on conservation based on the level and nature of interactions with the natural environment (Bragagnolo et al., 2016). Where men and women find roles to participate in tourism, the resource base is less likely to be undermined and individuals work to resolve any conservation conflicts that may arise, thereby contributing to growth of tourism development. Understanding of the gender and other social elements in local people in societies that have strong traditions can be an important ingredient of tourism sustainability (Chiu, Chan, & Marafa, 2016). In addition, households with large livestock herds in the study area are likely to be more negative towards protected areas and often less supportive of conservation efforts than those with fewer livestock due to the scarcity of pasture caused by access and use restrictions of a protected area (Osunsina & Fagbeyiro, 2015).

Ecocentrism minimises conservation conflicts and supports sustainable tourism development and biodiversity conservation (Nyirenda et al., 2017). Sustainable tourism development requires the support and participation of local people. Further, effective conservation planning strategies need to integrate stakeholder participation, including local people, taking into account various local socio-economic elements (Andrade & Rhodes, 2012; Mutanga et al. 2015). The biodiversity conservation, sustainable tourism and community livelihoods related developmental concepts should be generated with the involvement of local stakeholders to ensure their sustained support (Carter et al., 2014). There is need to also establish effective collaborative networks to sustain tourism in peri-urban parks (Cruz, Albrecht, & Briones, 2016).

Although the park was entirely game fenced, warthogs in particular used trenches to invade Shantumbu crop fields. Previous studies have depicted the negative impacts of fences on both wildlife and neighbouring communities (Lindsey, Masterson, Beck, & Romanăch, 2012). The strong sense of exclusion of local people from the park contributes to negative local perceptions. Tension and conflicts emanating from exclusionary tendencies can protract and potentially derail conservation and tourism interventions (Strickland-Munro, 2010; Holmes, 2013). Therefore, there is a need to effectively manage the tourism developments among the stakeholders, particularly with the involvement of local people.

Length of residency around the LNP increased the knowledge base of the environment among the Shantumbu people. Elsewhere, it has also influenced conservation perceptions of the local people, as the longer the people stayed in a village, the more attached they became to their environment (Arjunan, Holmes, Puyravaud, & Davidar, 2006). However, new residents are likely to be less dependent on the natural resources and therefore are likely to be more positive about newly established protected areas than older residents (Garekae, Thakadu, & Lepetu, 2016). Where benefits such as tourism products exceed costs such as crop damage, local people are likely to be more cooperative and supportive of biological conservation (Mfunda, Holmern, & Røskaft, 2012). Proximate to Lusaka City, the park has the potential for elevated benefits to the local people. Although, ever-increasing urban sprawl due to the growth of Lusaka city puts pressure on Shantumbu people, effective conservation planning can guarantee conservation and ecotourism benefits to them.

Conservation agencies should direct extra effort to managing negative perceptions contributing to negative attitudes, given that perceptions are attitude-forming processes (Allendorf et al., 2017). There is a need to nurture positive perceptions and address determinants of negative perceptions in order to improve community appreciation of biodiversity conservation and ecotourism interventions. If harnessed, strong relational social capital can be essential to natural resource management (Nath, Inoue, & Pretty, 2010; Alexander & Armitage, 2014). Further, there is also a need to consider community heterogeneity in the conservation planning.

Conclusion

A combination of socio-demographic and park attributes helped to shape local perceptions, attitudes and practices towards LNP and provided the basis for the improvement of protected area management and sustainable tourism development. The level of education attained, gender, length of residency, marital status, occupation and number of livestock owned influenced the levels of perceptions, attitudes and practices towards tourism and biodiversity conservation in the park. The perceived low benefits to local people were the underlying reason for reduced participation in the park's tourism development by the local people. Understanding such elements and implementing appropriate interventions can ensure sustainable tourism development in the area. There is a need for more integration of local people into decision-making and the park's tourism activities. The collaborative action will potentially create a sense of place and stimulate local participation and sustainable tourism development through social networks. Local people have a critical role to play in sustainable tourism development. To foster positive partnerships with adjacent local people and successfully implement conservation and tourism strategies, it is important to consider various factors that impact the park management. The conservation strategies should also support local people's livelihoods in order to shift their negative perceptions, attitudes and practices towards positive ones in the long run.

References

Alexander, S. M., & Armitage, D. (2014). A social relational network perspective for MPA science. *Conservation Letters, 8*(1), 1–13.

Allendorf, T., Aung, M., Swe, K. K., & Songer, M. (2017). Pathways to improve park-people relationships: Gendered attitude changes in Chatthin Wildlife Sanctuary, Myanmar. *Biological Conservation, 216*, 78–85.

Andereck, K. L., Valentine, K. M., Knopf, R. C., & Vogt, C. A. (2005). Residents' perceptions of community tourism impacts. *Annals of tourism research, 32*(4), 1056–1076.

Andrade, G. S. M., & Rhodes, R. (2012). Protected areas and local communities: An inevitable partnership toward successful conservation strategies? *Ecology and Society, 17*(4), 14. doi:10.5751/ES-05216-170414

Arjunan, M., Holmes, C., Puyravaud, J. P., & Davidar P. (2006). Do developmental initiatives influence local attitudes toward conservation? A case study from the Kalakad – Mundanthurai Tiger Reserve, India. *Journal of Environmental Management, 79*(2), 188–197.

Berno, T., & Bricker, K. (2001). Sustainable tourism development: The long road from theory to practice. *International Journal of Economic Development, 3*(3), 1–18.

Bragagnolo, C., Malhado, A. C., Jepson, P., & Ladle, R. J. (2016). Modelling local attitudes to protected areas in developing countries. *Conservation and Society, 14*(3), 163–182.

Carter, N. H., Viña, A., Hull, V., McConnell, W. J., Axinn, W., Ghimire, D., & Liu, J. (2014). Coupled human and natural systems approach to wildlife research and conservation. *Ecology and Society, 19*(3), 43. doi:10.5751/ES-06881-190343

Chechina, M., Neveux, Y., Parkins, J. R., & Hamann, A. (2018). Balancing conservation and livelihoods: A study of forest-dependent communities in the Philippines. *Conservation and Society, 16*, 420–430.

Child, B. (2009). The emergence of parks and conservation narratives in Southern Africa. In H. Suich, B. Child, & A. Spenceley (Eds.), *Evolution and innovation in wildlife conservation: Parks and game ranches to transfrontier conservation areas* (pp. 19–33). London, Earthscan.

Chiu, H.Y., Chan, C.-S., & Marafa, L. M. (2016). Local perception and preferences in nature tourism in Hong Kong. *Tourism Management Perspectives, 20*, 87–97.

Ciocănea, C. M., MirelaIanoşi, C. S., & Bagrinovschi, V. (2016). Assessing public perception on protected areas in Iron Gates Natural Park. *Procedia Environmental Science, 32*, 70–79.

Cruz, E., Albrecht, H., & Briones, A. (2016). Collaborative networks and tourism management of peri-urban forests. *International Journal of Sustainable Development Planning, 11*(2), 172–181.

Dimitrakopoulos, P. G., Jones, N., Losifides, T., Florokapi, I., Lasda, O., Paliouras, F., & Evangelinos, K. I. (2010). Local attitudes on protected areas: Evidence from three Natura 2000 wetland sites in Greece. *Journal of Environmental Management, 91*(9), 1847–1854.

Dunlap, R. E. (2008). The new environmental paradigm scale: From marginality to worldwide use. *Journal of Environmental Education, 40*(1), 3–18.

Ebua, V. B., Agwafo, T. E., & Fonkwo, S. N. (2011). Attitudes and perceptions as threats to wildlife conservation in the Bakossi area, South West Cameroon. *International Journal of Biodiversity and Conservation, 3*(12), 631–636.

Emptaz-Collomb, J.G.J. (2009). Linking tourism, human wellbeing and conservation in the Caprivi Strip (Namibia). PhD Dissertation, University of Florida, USA.

Ficko, A., & Bončina, A. (2019). Public attitudes toward environmental protection in the most developed countries: The environmental concern Kuznets curve theory. *Journal of Environmental Management, 231*, 968–981.

Fischer, C., Muchapondwa, E., & Sterner, T. (2011). A bio-economic model of community incentives for wildlife management under CAMPFIRE. *Environmental and Resource Economics, 48*(2), 303–319.

Fodness, D. (2017). The problematic nature of sustainable tourism: Some implications for planners and managers. *Current Issues in Tourism, 20*(16), 1671–1683.

Gandiwa, E. (2012). Local knowledge and perceptions of animal population abundances by communities adjacent to the northern Gonarezhou National Park, Zimbabwe. *Tropical Conservation Science, 5*(3), 255–269.

Gandiwa, E., Zisadza-Gandiwa, P., Muboko, N., Libombo, E., Mashapa, C., & Gwazani, R. (2014). Local people's knowledge and perceptions of wildlife conservation in southeastern Zimbabwe. *Journal of Environmental Protection, 5*, 474–485.

Garekae, H., Thakadu, O. T., & Lepetu, J. (2016). Attitudes of local communities towards forest conservation in Botswana: A case of Chobe Forest Reserve. *International Forest Review, 18*(2), 180–191.

Gemeda, D. O., Minstro, A. A., Feyssa, D. H., Sima, A. D., & Gutema, T. M. (2016). Community knowledge, attitude and practice towards black crowned crane (*Balearica pavonina*, L.) conservation in Chora Boter district of Jimma Zone, Ethiopia. *Journal of Ecology and the Natural Environment, 8*(4), 40–48.

Guest, G. (2012). *Applied thematic analysis*. Thousand Oaks, CA: Sage Publications.

Hawcroft, L. J., & Milfont, T. L. (2010). The use (and abuse) of the new environmental paradigm scale over the last 30 years: A meta-analysis. *Journal of Environmental Psychology, 30*, 143–158.

Holmes, G. (2013). Exploring the relationship between local support and the success of protected areas. *Conservation and Society, 11*(1), 72–82.

Huddart-Kennedy, E., Beckley, T. M., McFarlane, B. L., & Nadeau, S. (2009). Rural-urban differences in environmental concern in Canada. *Rural Sociology, 74*(3), 309–329.

Imran, S., Alam, K., & Beaumont, N. (2014). Environmental orientations and environmental behaviour: Perceptions of protected area tourism stakeholders. *Tourism Management, 40*, 290–299.

Kaliyaperumal, K. (2004). Guideline for conducting a knowledge, attitude and practice (KAP) study. *Community Ophthalmology, 4*(1), 7–9.

Lindsey, P. A., Masterson, C. L., Beck, A. L., & Romanãch, S. (2012). Ecological, social and financial issues related to fencing as a conservation tool in Africa. In M. J. Somers & M. Hayward (Eds.), *Fencing for conservation* (pp. 215–234). New York: Springer.

Mamo, Y. (2015). Attitudes and perceptions of the local people towards benefits and conflicts they get from conservation of the Bale Mountains National Park and Mountain Nyala (*Tragelaphus buxtoni*), Ethiopia. *International Journal of Biodiversity and Conservation, 7*(1), 28–40.

Martin, A., Myers, R., & Dawson, N. M. (2018). The park is ruining our livelihoods. We support the park! Unravelling the paradox of attitudes to protected areas. *Human Ecology, 46*(1), 93–105.

Mbaiwa, J. E. (2018). Effects of the safari hunting tourism ban on rural livelihoods and wildlife conservation in Northern Botswana. *South African Geographical Journal, 100*(1), 41–61.

Mfunda, M., Holmern, T., & Røskaft, E. (2012). Benefits and access to natural resources influence conservation perceptions and relationship between local people and other stakeholders: The case of Serengeti ecosystem, Tanzania. *International Journal of Biodiversity and Conservation, 4*(14), 535–554.

Mombeshora, S., & Le Bel, S. (2009). Parks-people conflicts: The case of Gonarezhou National Park and the Chitsa community in south-east Zimbabwe. *Biodiversity and Conservation, 18*(10), 2601–2623.

Mutanga, C. N., Muboko, N., & Gandiwa, E. (2017). Protected area staff and local community viewpoints: A qualitative assessment of conservation relationships in Zimbabwe. *PLOS ONE, 12*(9), e0184779.

Mutanga, C. N., Vengesayi, S., Muboko, N., & Gandiwa, E. (2015). Towards harmonious conservation relationships: A framework for understanding protected area staff-local community relationships in developing countries. *Journal for Nature Conservation, 25*, 8–16.

Nath, T. K., Inoue, M., & Pretty, J. (2010). Formation and function of social capital for forest resource management and the improved livelihoods of indigenous people in Bangladesh. *Journal of Rural and Community Development, 5*(3), 104–122.

Nyirenda, V. R., Kaoma, C., Nyirongo, S., Lwali, C. A., & Chomba, C. (2017). Social exchange and structuralist-constructivism approaches for enhanced ecotourism and food security in wildlife-agrarian mosaic landscapes: Insights from eastern Zambia. *International Journal Tourism Policy*, 7(2), 93–109.

Nyirenda, V. R., Siamudaala, V. M., & Kaula, M. (2014). Management effectiveness and potential for tourism of peri-urban Lusaka National Park, Zambia: A preliminary assessment. *Environment and Natural Resources Research, 4*(1), 117–129.

Osunsina, I. O. O., & Fagbeyiro, A. O. (2015). Local community perception and attitude towards the non-utilisation of natural resources in old Oyo Natural Park, Oyo State, Nigeria. *Journal of Agriculture and Environment for International Development, 109*(2), 291–306.

Pickering, C. M., & Rossi, S. (2016). Mountain biking in peri-urban park: Social factors influencing perceptions of conflicts in three popular national parks in Australia. *Journal of Outdoor Recreation and Tourism, 15*, 71–81.

R Core Team. (2018). *R: A language and environment for statistical computing*. Vienna: R Foundation for Statistical Computing.

Rossi, S. D., Pickering, C. M., & Byrne, J. A. (2016). Not in our park! Local community perceptions of recreational activities in peri-urban national parks. *Australasian Journal of Environmental Management, 23*(3), 245–264.

Snyman, S. (2014). Assessment of the main factors impacting community members' attitudes towards tourism and protected areas in six Southern African countries. *Koedoe, 56*(2), 1–12.

Strickland-Munro, J. (2010). *Understanding the interactions among local communities, protected areas and tourism: Case studies of Kruger National Park and Purnululu National Park*. Doctoral dissertation, Murdoch University.

Sundaresan, S., Bruyere, B., Parker, G., Low, B., Stafford, N., & Davis, S. (2012). Pastoralists' perceptions of the endangered Grevy's Zebra in Kenya. *Human Dimensions of Wildlife, 17*(4), 270–281.

Tanner-McAllister, S., Rhodes, J., & Hockings, M. (2014). Community and park manager's perceptions of protected area management: A southeast Queensland study. *Australasian Journal of Environmental Management, 21*, 320–336.

Ward, C., Holmes, G., & Stringer, L. (2017). Perceived barriers to and drivers of community participation in protected area governance. *Conservation Biology, 32*(2), 437–446.

Ward, C., Stringer, L. C., & Holmes, G. (2018). Protected area co-management and perceived livelihood impacts. *Journal of Environmental Management, 228*, 1–12.

Whitham, C. E. L., Kun, S., & Riordan, P. (2015). People and protected areas: Understanding attitude alignment for more effective conservation. *Journal of Resources and Ecology, 6*(5), 281–292.

Zuur, A. F., Ieno, E. N., Walker, N. J., Saveliev, A. A., & Smith, G. M. (2009). *Mixed effects models and extensions in ecology with R*. New York: Springer.

10 Tourism and poverty alleviation in the Global South

Emerging corporate social responsibility in the Namibian nature-based tourism industry

Mary-Ellen Kimaro and Jarkko Saarinen

Introduction

The growing tourism industry represents one of the main income sources for local people in many countries, especially in the Global South. Recently, the tourism industry has been framed as a tool for poverty alleviation. In this respect, the declaration by the United Nations of 2017 as the International Year of Sustainable Tourism for Development provided a unique opportunity to rethink and advance the contribution of the tourism sector to wider development (United Nations, 2017; World Bank Group, 2017). Tourism is also seen as a complementary activity to the United Nations Sustainable Development Goals (SDGs) of the 2030 Agenda for Sustainable Development (United Nations, 2015). This connection between the SDGs and tourism underlines the responsibility of tourism as one of the world's biggest industries to contribute and make a difference to sustainable development (see Saarinen, 2019).

Sustainability in tourism development is often considered as the target goal for tourism planning and management (Frey & George, 2010; Saarinen, Rogerson, & Hall, 2017). However, implementing sustainable tourism has encountered a fair amount of challenges due to an ambiguity of the concept and challenges in its implementation to practice (Sharpley, 2000, 2009). Recognising the gaps of sustainable tourism and the need to address these, the Cape Town Declaration on Responsible Tourism in Destinations was signed in 2002. Since then responsible tourism has become an important element of tourism operations with a focus on the responsibility of individuals and businesses (Constanti et al., 2014; Saarinen, 2014). According to Bramwell and Lane (2008), the essence of responsible tourism is to alter industry's behaviour towards more sustainable development through the assumption that ethical responsibility will supersede self-interest. Thus, responsible tourism aims to ensure that all stakeholders involved in tourism are accountable for their actions and behavior (see Frederick, 1986).

Currently, one of the key tools to create and practice accountability in tourism is the idea of corporate social responsibility (CSR) (Coles, Fenclova, & Dinan, 2013). At a general level, the CSR principles and practices refer to the private sector's interests to respond to externally and/or internally created pressures and motivations concerning ethical obligations towards their operational environment (Holcomb, Upchurch, & Okumus, 2007). Hitherto, there is an emerging evidence of CSR practices in tourism in the Global South (Lapeyre, 2011a; Novelli & Hellwig, 2011). However, relatively little research has been undertaken on the local views on the CSR practices, especially in Namibia. Against this backdrop, this chapter attempts to discuss and analyse the views of a specific local stakeholder group in the Namibian context: the views of local employees on the CSR processes and practices in the selected tourism businesses they work with and their perceptions of the impact of the CSR initiatives towards poverty alleviation in their home communities.

Role of CSR in the tourism sector

According to Spenceley and Goodwin (2007), the tourism sector's local engagement needs to be carefully considered, particularly in relation to community needs, expectations and constraints. This is crucial in natural resource management and community livelihoods contexts as the tourism sector utilises the landscapes and resources of a destination to make a profit in tourism. Hence, the industry should recognise its responsibility to create positive impacts to destination communities and their livelihoods (see Timothy & Ioannides, 2002). Indeed, according to Bowen's (1953, p. 6) original thinking, the idea of social responsibility refers to "the obligations of entrepreneurs to pursue those policies, to make those decisions, or to follow those lines of action which are desirable in terms of the objectives and values of our society." Davis (1960), however, argued that social responsibility is a nebulous idea that should be seen in a specific managerial context. For him, the business is and must remain fundamentally an economic institution, but it also does have responsibilities to help society achieve its basic goals.

A rationale for businesses to make a commitment to the economic as well as social development of surrounding communities and societies where they operate is partly based on the idea of reciprocity. According to Sahlins (1972) there are different forms of reciprocity: generalised, balanced and negative (see Svensson, 2015). Shortly, the generalised reciprocity refers to a solidarity extreme, while the balanced involves direct mutual exchange and the negative reciprocity is often one-dimensional utilisation of relations and resources. In tourism contexts, the negative reciprocity ultimately creates sanctions and other hostile attitudes towards the businesses which makes its operations and utilisation of local natural resources difficult in practice. Thus, while we may not be able to expect an unconditional solidarity extreme between the industry and communities, the mutual exchange and negative reciprocity can explain why the industry would and should be interested in benefitting the localities in which they operate. In addition, the World Business Council for Sustainable Development (2001) has argued that the business case for poverty alleviation is ultimately a straightforward matter, as any business cannot expect to succeed in societies that are dysfunctional or eventually fail. Consequently, a failing society in the tourism context would refer to the economic leakages through badly governed and unsustainable development and management of the tourism industry (Saarinen, 2017), for example, resulting in social and economic suffering of communities where (transnational) tourism businesses thrive. Hence, there is a need to adopt a framework which guides the tourism private sector on voluntary or mandatory compliance to CSR practices.

The involvement of employees in internal and external CSR initiatives

CSR practices may be implemented for employees (internal CSR) and/or by employees (external CSR), who in turn influence the image of the accommodation establishments. In many respects, the key to successful CSR strategies is the involvement of employees, and these kind of strategies have demonstrated to have long-term benefits to accommodation establishments, for example, (Bohdanowicz & Zientara, 2009). Makower (1994) has noted that socially responsible businesses are stimulated by individual employee behavior and that well-motivated employees perform better in terms of productivity and innovation. As a result, a business can create valuable and positive changes in an employee's mindset and behaviour (Ho, Seoki, & Chang, 2010). It is within this context that tourism sector employees' views on the business's internal and external CSR practices become crucial. Internal CSR practices include leadership and human resource processes in terms of strategy, management and working conditions. These cover areas such as local employee recruitment, training and career guidance and development.

External CSR encompasses philanthropy and community contributions and also reflects the way in which the business interacts with the physical environment and its ethical stance towards consumers and other external stakeholders (Carroll, 1979). Since CSR is concerned with those actions that exceed the legal minimum, business contributions in this field are largely discretionary (Carroll, 1979); for example, involvement in some type of a social project, purchase of local products or services, development of partnerships with rural communities and marketing of Fair Traded Products (DEAT, 2002; FENATA, 2004).

CSR has shown evidence of benefits such as competitive advantage through added value (Weeden, 2002) as well as positive changes among employees in retention and low turnover, employee commitment through empowerment (Kazlauskarite, Bucuiniene, & Turauskas, 2006), increased job satisfaction and employee productivity leading to an overall success of the lodge operation (Gurerk, Irlenbusch, & Rockenbach, 2009). In Ditlev-Simonsen and Midttun (2010) motivation of employees to participate in CSR projects internally within the lodge or externally in the community requires a long-term investment. Factors such as incentives and rewards, making sure that all employees share in the success and encouraging employees to voice their complaints and ideas play a vital role in motivating employees to participate.

According to Gartner and Cukier (2012), most tourism lodge owners adhered to labour standards of minimum wage and voluntarily offered fringe benefits such as medical insurance, social security, retirement plans, room and board, funeral and education expenses of lodge employees. These authors noted that fringe benefits had a positive influence on poverty alleviation because it relieves pressure on employees' income and allows them to cope with stress and shocks of daily life. However, despite adherence to labour standards, there is little evidence that lodge employees and their households experience an improvement in poverty conditions. Indeed, several authors (Grant, 2004; Crosby & Bryson, 2005; Cooper, Scott, & Baggio, 2009) have argued that the lack of coordination acts as a hindrance to successful implementation of CSR practices. Hence, there is a need for an integrative system that maximises results and involves employees in decision-making and implementation as stakeholders.

Corporate social responsibility and poverty alleviation: theoretical underpinnings

Carroll (1991) has stated that the concept of corporate social responsibility may from time to time be superseded by various other focuses, such as social responsiveness, social performance, poverty alleviation, climate change and equity or stakeholder management. In respect to this, he has used the idea of "the pyramid of corporate social responsibility" as a framework for understanding the evolving nature of the organization's economic, legal, ethical and philanthropic practices (Figure 10.1). These four practices of the framework provide management with a skeletal outline of the nature and kinds of their CSR (Carroll, 1991).

In other words, the framework helps to expand our awareness and understanding of the concept and scope of CSR practices and what it can entail by calling upon businesses to be profitable (economic), obey the law (legal), be ethical (ethics) and be a good corporate citizen (philanthropic) (See Figure 10.1). Visser (2006) challenged the accuracy and relevance of Carroll's CSR Pyramid in the African context and suggested that the relative priorities of CSR in Africa are likely to be different from the classic, North American ordering. He also argued that Carroll's CSR Pyramid may not be the best model for understanding CSR in general and CSR in Africa in particular and suggested the complexity theory to be a much more useful model than Carroll's CSR Pyramid. De Jongh and Prinsloo (2005) concur by emphasising that the challenges facing CSR in Africa involve messy, "on the edge of chaos" scenarios.

Despite some of its limitations in conceptual clarity and descriptive accuracy, Carroll's CSR Pyramid thinking has proven to be both a durable and useful model for defining and exploring CSR in a specific governance context (see

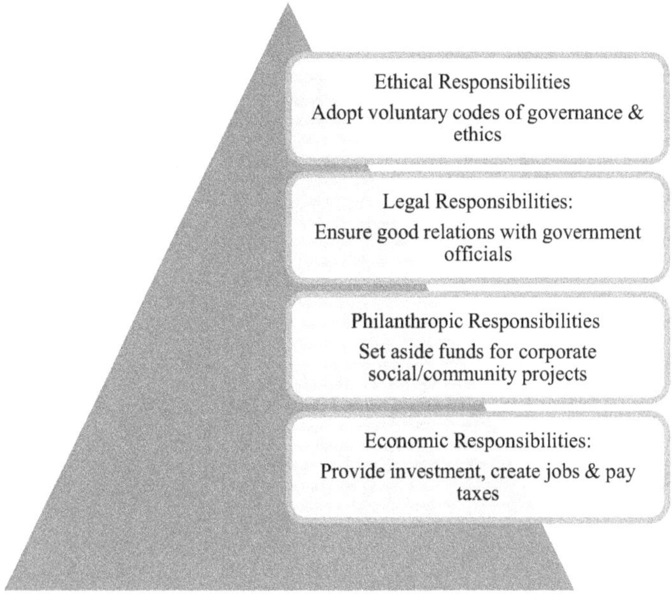

Figure 10.1 Africa's corporate social responsibility pyramid

Source: Adapted from Visser (2006)

Visser, 2006). In this respect, Brejning (2012) has identified five CSR discourses emphasising that corporate social responsibility is an idea which is perceived and conceptualised differently depending on the national context. Table 10.1 presents the identified five CSR discourses and maps out a continuum of the CSR discourses. Here, this chapter links with the liberal and social democratic CSR discourse by coming up with a framework that will provide a business case for CSR (liberal) while at the same time calling for social contracts between a business and its stakeholders. Additionally, this study uses the stakeholder theory, which is the theoretical foundation of liberal CSR discourse. The stakeholder management perspective provides a basis for the coordination of CSR practices internally and externally, which can be used for an integrative system that maximizes results and involves employees in decision-making and implementation as stakeholders (Carroll, 1991; Cooper et al., 2009).

Carroll's CSR Pyramid was built towards the notion of an improved ethical organizational climate manifested by moral management, which can be defined and described through a contrast of immoral/amoral management. In this chapter, the conceptual framework is built toward responsible tourism manifested by poverty alleviation by using Zhao and Ritchie's (2007) integrative *Framework for Anti-Poverty Tourism (APT)*. Zhao and Ritchie's (2007) used

Table 10.1 Mapping the CSR discourses

	Conservative	*Liberal*	*Social democratic*	*Radical*	*Skeptical*
For/against CSR	Against/ Reluctantly for	For	For	For	Against
CSR objectives	None	Business gains. Aligning business and stakeholder interests	Social gains. Involving businesses in social issues including solving social problems.	Bring about global fundamental change. Eliminate global poverty. Justice & sustainability	None. CSR is a scam, a disguised way of continuing business exploitation of society
Contract negotiation	None	Role of business in a mixed economy	Role of business in a mixed economy	Global social contract	Global social contract
Acceptable level of governmental involvement in CSR	None	None. CSR should be a voluntary matter	Some. Main role: to encourage, facilitate rather than legislate	As much as possible. Legislation where practical; otherwise, governmental pressure	None. Instead the power of corporations should be restricted by law
Nongovernmental means of promoting CSR	None	Example setting within the business community through indices and prizes	Example setting within a business community through indices and prizes	NGO activism Academic action Research	None. The media, NGOs and academics must instead expose business misconduct and abuse of CSR for business gains
Means of putting CSR into practice	None	No universal CSR model. Each company must match its CSR efforts with its business profile	Partnerships between business, public sector and/or non-profit sector	Internationally agreed codes of conduct. CSR reporting	None.
Theoretical foundation	Neo-liberal economics	Stakeholder theory	Social contract, Communitarianism, New public management	Critical theory, Neo-Marxism, Action-research	Critical theory, Neo-Marxism
Ideological foundation	Neo-liberal	Liberal	Social democratic	Socialist	Socialist

Source: Adapted from Brejning (2012)

the APT framework (Figure 10.2) as a unifying concept to refer to any tourism development in which poverty alleviation is set as the central or one of the central objectives. They recognized the multidimensional nature of poverty described in the World Bank's (2001) "attacking poverty" framework, which posits that for poverty alleviation to be successful, there is a need to create economic and social opportunities first and then facilitate empowerment and finally enhance security.

In Zhao and Ritchie's (2007) framework the determinants of poverty alleviation include three components, namely, opportunity, empowerment and security, which collectively reflect the guiding principles of contemporary development practices under the leadership of the World Bank (2001) on poverty alleviation. Opportunity means the poor must have access to economic opportunities of which they can take advantage to change their livelihoods, while empowerment aims to enhance the capacity of the poor to influence the state and social institutions and thus strengthen their participation in business processes and local decision-making (World Bank, 2001). Lastly, security is concerned with reducing the vulnerability of the poor to various risks such as ill health, economic shocks and natural disasters. Since poor households have fewer assets and less diversified sources of income to manage any crisis, they could be easily thrown into despair when these adverse impacts happen to them (Dhanani & Islam, 2002).

Zhao and Ritchie (2007) posit that measuring the appropriateness and efficacy of a certain poverty alleviation approach can be as straightforward as to determine how it contributes to the opportunity, empowerment and/or security of the poor. They believe that this approach reveals valuable information that can help to specify the functioning mechanisms of Anti-Poverty Tourism (APT), i.e. poverty alleviation. Additionally, the incorporation of stakeholder orientations in both the APT framework and the CSR Pyramid provides

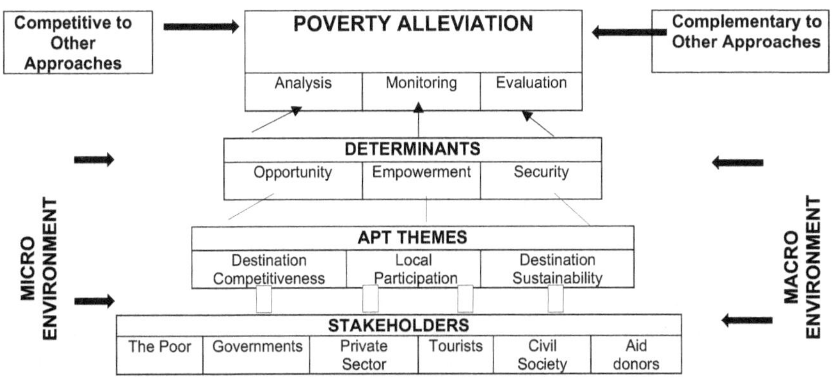

Figure 10.2 An integrative framework for anti-poverty tourism research

Source: Adapted from Zhao and Ritchie (2007)

potentially a practical way of linking moral management to poverty alleviation and consequently unraveling the different categories or types of CSR practices that may be found in tourism businesses and other organizations (Carroll, 1991; World Bank, 2001). Thus, moral management provides a model leadership perspective that enhances security in the poverty alleviation continuum. However, there is limitation on the framework, which Carroll (1991) estimated to result in managers thinking that just by rejecting immoral management, they will be providing ethical leadership. The multidimensional nature of poverty (World Bank, 2001; Zhao & Ritchie, 2007) and the complexity of CSR in Africa (Visser, 2006) can best be addressed through moral management focusing on poverty alleviation, responsible tourism and eventually sustainable development in tourism. These are the basis of the conceptual framework for poverty alleviation through CSR towards responsible tourism (see Table 10.2).

The Namibian case study

CSR initiatives by Namibia's tourism private sector

In overseas markets, Namibia is a relatively new destination in Southern Africa. In addition to regional tourism, the country has received a substantial number of overseas tourists after independence in 1990 (Ministry of Environment and Tourism (MET), 2011). Due to its current relatively favourable macroeconomic environment and political stability, Namibia has been able to maintain a steady number of tourist arrivals. As indicated in the national development plans NDP 1(1995/1996 to 1999/2000) – NDP 5 (2017/2018–2021/2022) (National Planning Commision, 2017) and Vision 2030 (Republic of Namibia, 2004), the tourism industry is regarded as a priority sector in driving sustainable development. The Namibian government considers tourism capable of contributing to the creation of employment opportunities leading to poverty alleviation. In addition, the sector is seen as having the ability to improve the income distribution and living standards of people, and the promotion of small-scale enterprises and informal sector economic activities can reduce income inequalities (Nicanor, 2001).

Recently, the Republic of Namibia has faced a number of social problems such as a high unemployment rate (34%), mostly among the unskilled and youth (NSA, 2019), and unprecedented poverty, especially in the rural areas (Novelli & Hellwig, 2011). In addition, the role of certain ethnic groups, such as the Ovahimbas, in tourism may have been marginalised despite their high visibility in national and regional tourism promotion (Lapeyre, 2011b; Ndlovu, Nyakunu, & Kavita, 2011; Saarinen & Niskala, 2009;). Consequently, at the rural community level, the government and the NGOs have spearheaded the development of community-based tourism in Namibia (Ashley, 2000). Legislation and supporting programmes, such as the National Community Based Natural Resource Management (CBNRM) provide structures that aim for sustainable natural resource management, which can result in social development

Table 10.2 Conceptual framework for poverty alleviation through CSR towards responsible tourism

Poverty alleviation framework (World Bank, 2001; Zhao & Ritchie, 2007)	*Orientation toward employee stakeholders*	*CSR Practices (Carroll, 1991; Visser, 2006, 2011)*			
	Type of management (Carroll, 1991)	*Economic responsibilities*	*Legal responsibilities*	*Philanthropic responsibilities*	*Ethical responsibilities*
Enhancing Security	**Moral Management**	**Competitive CSR (competition-driven, opportunity-based)**	**Mandatory CSR (legally driven, contract-based)**	**Charitable CSR (altruism-driven, philanthropy-based)**	**Discretionary CSR (people driven, socially-based)**
	Employees are a human resource that must be treated with dignity and respect. The goal is to use a leadership style such as consultative/participative that will result in mutual confidence and trust. Commitment is a recurring theme. Employees' rights to due process, privacy, freedom of speech and safety are maximally considered in all decisions. Management seeks out fair dealings with employees	Turning CSR-driven opportunities into a competitive advantage. Its a proactive approach toward taking stakeholder (employees) needs into account, which is more effective than a reactive one because proactivity is more likely to lead to achieving superior value for both employees and the organisation	These are CSR practices which are typically limited and are undertaken only if and when it can be shown that shareholder value will be protected as a result. Examples include employee volunteer programmes that show evidence of improved staff motivation, commitment and productivity	CSR practices that fulfil the organisation's philathropic responsibilities, going beyond preventing possible harms to help alleviate public welfare deficiencies, for example, HIV/AIDS awareness programs	Morally mandatory CSR practices with social gains. Involving businesses in social issues including solving social problems. For instance, building a school for their employees' kids

(*Continued*)

Table 10.2 (Continued)

Poverty alleviation framework (World Bank, 2001; Zhao & Ritchie, 2007)	*Orientation toward employee stakeholders*	*CSR Practices (Carroll, 1991; Visser, 2006, 2011)*			
	Type of management (Carroll, 1991)	*Economic responsibilities*	*Legal responsibilities*	*Philanthropic responsibilities*	*Ethical responsibilities*
Facilitating empowerment	**Amoral management**	**Promotional CSR (image-driven, PR-based)**	**Defensive CSR (compliance-driven, incentive-based)**	**Philathropic CSR (empowerment-driven, sustainability-based)**	**Voluntary CSR (sustainability-driven, code-based)**
	Employees are treated as law requires. Attempts to motivate focus on increasing productivity rather than satisfying employees' growing maturity needs. Employees still seen as factors of production but remunerative approach used. Organization sees self-interest in treating employees with minimal respect. Organization structure, pay incentives, rewards are all geared toward short- and medium-term productivity	CSR is seen mainly as a public relations opportunity to enhance the brand, image and reputation of the company. An example is advertisements of women in positions of authority	These are CSR practices which are typically limited and are undertaken only if and when it can be shown that shareholder value will be protected as a result. Examples include employee volunteer programmes that show evidence of improved staff motivation, commitment and productivity	CSR practices which are voluntary activities that appear to further some social good, beyond the interests of the company, towards sustainability. For instance, the lodge managers are usually a husband and wife team with the spouse empowered in order to keep the family unit whole	CSR practices based on voluntary sustainability standards. For instance, seeking accreditation from certification programs such as Eco Awards or the Sustainable Tourism Certification Alliance Africa

Promoting opportunities	**Immoral management**	**Economic CSR (profit-driven, value-based)**	**Defensive CSR (compliance-driven, risk-based)**	**Corporate philathropy (crisis-driven, survival-based)**	**Strategic CSR (product driven, code-based)**
	Employees are viewed as factors of production to be used, exploited, manipulated for gain of the individual manager or company. No concern is shown for employees' needs/rights/ expectations. Short-term focus. Coercive, controlling, alienating	No CSR policy or objectives. CSR is a scam, a disguised way of continuing business exploitation of society. Employees only benefit through employmentwith minimum to no benefits	Ad hoc CSR investments. As a way to avoid legislation where practical; otherwise, governmental pressure. Characteristic of European companies where CSR is legislated (Brejning (2012). Employees receive minimum to above-minimum wage and legislated benefits	CSR practices which are voluntary activities that appear to further some social good beyond the interests of the company and that which is required by law. Examples include the adoption of advanced human resource management programs	Relates CSR activities to the company's core business and guided by industry code or charter. Examples include discount or free vouchers for employees and their families to enjoy the company's tourist resorts

Source: Adapted from Zhao & Ritchie, 2007; Carroll, 1991; Visser, 2006, 2011.

and economic growth for rural communities. These aims are envisioned with suitable partnership between local communities, government, and increasingly, the private sector (NACSO, 2016).

Indeed, with the initial impetus from the government and NGOs on the development of tourism within the rural communities, the private sector has gradually participated in joint venture partnerships in rural communities, offering their expertise in areas such as lodge management and tour operating. Within these efforts from the private sector, responsible tourism initiatives have evolved. According to Frey and George (2010), responsible tourism emphasizes promoting community involvement with an aim to ensure that economic benefits are gained by host communities. It spreads beyond the management of natural resources alone and, thus, alludes to the social and moral responsibility of tourism businesses towards host societies and the environment. Hence, Dodds and Kuehnel (2008) have argued that core to the concept of responsible tourism is corporate social responsibility (CSR), as it focuses on the accountability of businesses in terms of their impacts on their decisions and actions.

Methods and data

A questionnaire was used for an explorative case study to assess employees' views on the private sectors' practices of CSR based on Carroll's four component practices. Section 1 covered the demographic profile, such as job category, years of employment, level in company and level of education. Section 2 had questions structured on the nature of legal and economic CSR responsibilities (Carroll, 1991), focusing on the level of moral management and poverty alleviation as outlined in the conceptual framework. Section 3 had questions on ethical and philanthropic responsibilities. The close-ended questions were constructed with a positivist approach, with most of the questionnaire responses using the Likert scale and the others number coded to allow the respondents to indicate their attitudes by checking how strongly they agree or disagree with carefully constructed statements. The questionnaire responses were designed to allow the respondents to choose from multiple response alternatives, such as strongly agree, agree, neutral, disagree and strongly disagree (Zikmund & Babin, 2010).

The scale results reveal the respondent's attitude, ranging from very positive to very negative and with a tally total of 32 for each of the four components of CSR responsibilities. These scores were then used to analyse the employees' views or perceptions on the Poverty Alleviation CSR practices dominant in their organisations (Table 10.3). The tool was used to carry out a comparative analysis of the perceptions and views of 126 employees from two of the largest tourism chain lodges in Namibia, with establishments in the most remote and rural areas of the country. The selected companies are geographically spread with branches mainly in the northwest, southern and northern parts of the country. They also operate in proximity to the main tourist attractions sites in Namibia. They volunteered to participate in this research for two reasons. First,

Table 10.3 Poverty alleviation through CSR towards responsible tourism

Tally or score (out of a maximum score of 32)	*Economic responsibilities*	*Legal responsibilities*	*Philanthropic responsibilities*	*Ethical responsibilities*
Scores above 22	Competitive CSR (competition-driven, opportunity-based)	Mandatory CSR (legally driven, contract-based)	Charitable CSR (altruism-driven, philanthropy-based)	Discretionary CSR (people-driven, socially based)
Scores between 11 and 22	Promotional CSR (image-driven, PR-based)	Defensive CSR (compliance-driven, incentive-based)	Philanthropic CSR (empowerment-driven, sustainability-based)	Voluntary CSR (security-driven, code-based)
Scores equal to or less than 10	Economic CSR (profit-driven, value-based)	Defensive CSR (compliance-driven, risk-based)	Corporate Philanthropy (crisis-driven, survival-based)	Strategic CSR (product-driven, code-based)

Source: Authors design

they realised that they have emerged as the leaders in the tourism industry concerning CSR and poverty alleviation initiatives. Thus, there is a certain level of responsibility and need to be transparent in their operations and how they have worked with their employees and adjacent communities. Secondly, both companies are convinced that CSR is a fundamental pillar sustaining any business lifecycle in the social and economic sense.

CSR towards poverty alleviation

The data collection focused on two privately owned tourism businesses, which represent Namibian (company A, n= 75) and South African (company B, n=51) operators. A combination of stratified and purposive sampling was used to select a representative sample of the employees.

The employees represented are from different departments, such as the food and beverage, front office, housekeeping and the back office. The years of employment indicate relatively low employee retention percentages, which varied from 17% with company A for five to ten years, while company B had 10%. Literature indicates that employee retention can be improved through well-implemented CSR initiatives (Kazlauskarite et al., 2006). Most respondents (73%) from Company A and (74 %) from company B had not acquired an educational level higher than Grade 12 (secondary school exit level); this is reflected in the level (within the hierarchy of the company) that the majority of the employees occupied (Table 10.6).

As shown in Table 10.4, the dominant CSR practices by responsibilities in Company A are: Promotional CSR (economic), Defensive CSR (legal), Charitable CSR (philanthropic) and Voluntary CSR (ethical). This implies that Company A has a relatively good capacity and understanding of and embraces the liberal CSR. The company facilitates empowerment of its employees in addressing its economic and legal responsibilities. However, some of the employees still view the company's CSR practices as only creating opportunities.

The overall finding shows that company A is inclined to the liberal CSR practices. It is likely exercising amoral management, where there is no universal CSR model and the company matches its CSR efforts with its business profile. Its ethical neutrality tends to facilitate the empowerment of a few employees in the poverty continuum, who are then paraded for promotional purposes. This might bring success in the short term, but it may not be sustainable in the longer term.

Table 10.5 shows that the dominant CSR practices by responsibilities in Company B are Promotional CSR (economic), Defensive CSR (legal), Charitable CSR (philanthropic) and Voluntary CSR (ethical). Company B, though using the same CSR practices as Company A, its score card seems to suggest it is leaning toward the social contract CSR, where the immoral management is being rejected. The company scores also show a growing focus on sustainability and is capitalizing on Competitive CSR opportunities, which will lead to successful CSR practices, where employees feel involved and initiatives involving employees' personal participation have long-term and sustainable benefits (Bohdanowicz & Zientara, 2009).

It was noted that the employees did not have clarity on management's decision and policy on CSR. However, when questioned about their relationship with the company (legal CSR) as a means of empowerment, 82% from Company A and 76% from Company B indicated that their form of participation

Table 10.4 Company A dominant CSR practices

Economic responsibilities	*Legal responsibilities*	*Philanthropic responsibilities*	*Ethical responsibilities*
Competitive CSR (competition-driven, opportunity-based)	Mandatory CSR (legally driven, contract-based)	Charitable CSR (altruism-driven, philanthropy-based)	Discretionary CSR (people-driven, socially based)
4 (12.1%)	**2 (6.1%)**	**23 (69.7%)**	**2 (6.1%)**
Promotional CSR (image-driven, PR-based)	Defensive CSR (compliance-driven, incentive-based)	Philanthropic CSR (empowerment-driven, sustainability-based)	Voluntary CSR (security-driven, code-based)
28 (84.8%)	**27 (81.8%)**	**10 (30.3%)**	**23 (69.7%)**

Source: Adapted from Visser (2008) and Zhao and Ritchie (2007)

in decision-making was through dialogue with managers conducted mainly through both informal and formal meetings to address employment-related matters and not external CSR initiatives. Open forums with senior managers based in the capital city of Windhoek and the use of trade unions was rare. Although it was felt that the salaries were low, the companies main economic CSR initiative leading to poverty alleviation in society was viewed to be employment creation (Company A – 72% and Company B – 70%) compared to only 4% who viewed that their respective company contributed to overall benefits created in their home communities around them. This was mostly viewed to be in the form of involvement in local community projects, donation of gift aids during natural disasters, e.g. old company blankets, donations to schools, and so on (Company A 21% and Company B 29 % respectively). In addition to the limited awareness and direct involvement in the above-mentioned CSR process and practices, none of the employees were aware of any funds set aside for poverty alleviation through CSR community projects or wider livelihood development initiatives. Only 17% in Company A and 11% in Company B strongly agreed that they felt their employment was secure in the long term. The majority felt vulnerable through lack of security (medical insurance, retirement plan, limited to no coverage of educational expenses) and perceived this as an impediment to poverty alleviation through CSR.

The employees were requested to provide suggestions for their needs towards poverty alleviation through CSR. Their suggestions ranged from sustainable salary increments, more training and awards with nationally recognised certification that would enable them to further their education should they

Table 10.5 Company B dominant CSR practices

Economic responsibilities	*Legal responsibilities*	*Philanthropic responsibilities*	*Ethical responsibilities*
Competitive CSR (competition-driven, opportunity-based)	Mandatory CSR (legally-driven, contract-based)	Charitable CSR (altruism-driven, philanthropy-based)	Discretionary CSR (people-driven, socially based)
19 (25.3%)	**19 (25.3%)**	**53 (70.7%)**	**12 (16%)**
Promotional CSR (image-driven, PR-based)	Defensive CSR (compliance-driven, incentive-based)	Philanthropic CSR (empowerment-driven, sustainability-based)	Voluntary CSR (security-driven, code-based)
55 (73.3%)	**56 (74.7%)**	**22 (29.3%)**	**62 (82.6%)**
Economic CSR (profit-driven, value-based)	Defensive CSR (compliance-driven, risk-based)	Corporate philanthropy (crisis-driven, survival-based)	Strategic CSR (product-driven, code-based)
1 (1.4%)	0	0	**1 (1.4%)**

Source: Adapted from Visser (2008) and Zhao and Ritchie (2007)

Table 10.6 Percentages of those employed at different levels and their educational attainment

Hierarchical positions in the company from highest to lowest	*Company A n=75*	*Company B n=51*	*Educational level attained*	*Company A n=75*	*Company B n=51*
Manager or CEO of branch	11%	6%	Diploma & degree	11%	16%
Level 1 from CEO/ Manager	24%	12%	Vocational	15%	8%
Level 2 from manager	8%	26%	Grade 11 & 12	28%	25%
Level 3 from manager	28%	18%	≤ Grade 10	45%	49%
Level 4 or more from manager	29%	36%	No education	1%	2%

desire to do so. Job rotation and job promotion were mentioned as a means for empowerment, motivation and job satisfaction that could lead to opportunities to explore other potential departments of choice for career development. Employees expressed the need for a retirement plan as a means to attain future security as well as the need for the company to support the community with a clinic and school for their children to avoid being separated from them because of their employment with the company. They were also interested in increased awareness on CSR policy and initiatives.

5. Discussion and conclusions

The chapter reviewed the idea of CSR and how it has been conceptualised and implemented in tourism development policies and discussion. In general, the idea of CSR is understood as a highly contested idea with several existing "competing discourses" which were identified in the chapter. Special focus was given to the social contract discourses surrounding the idea and thoughts of CSR within the context of tourism development, anti-poverty tourism, poverty alleviation, moral management and recently initiated UN SDGs. The chapter sought to increase the level of awareness and understanding of the concept and scope of CSR practices and what they entail by developing a conceptual model for private tourism business in the Global South.

A conceptual framework for poverty alleviation through CSR towards responsible tourism was developed and tested on 126 employees from the two largest tourism chain lodges in Namibia. The framework presented an improvement to the previous CSR pyramid approach, which has a better capacity to deal with the complexity of the CSR system in practice. The framework is expected to adequately address the problem of what should happen when two or more CSR responsibilities are in conflict. CSR in the Global South is highly complex and often chaotic, and it is hoped that poverty alleviation through the CSR framework can provide moral compass for CSR practitioners in their endeavor to help the companies toward responsible tourism.

Based on the results of an explorative case study, philanthropic CSR remains a contested area in CSR in Namibia. Companies tend to use a reactionary approach towards disaster and calamity around their communities rather than safeguarding funds proactively. Thus, philanthropic efforts on poverty alleviation still remain relatively elusive.

In order to achieve poverty alleviation through legal CSR, it is noted that empowerment of employees in the participation of company CSR processes, practices and decision-making is crucial, as it can create a positive mindset and behaviour towards CSR (Ho et al., 2010; see Ditlev-Simonsen & Midttun, 2010). As stated by Visser (2006), the CSR pyramid in Namibia has a major leaning towards the economic CSR, prioritising the provision of employment creation with which the majority of the interviewed employees could also identify. Notwithstanding there is a need for the companies to be more inclusive in the transformational processes towards poverty alleviation. These inclusive measures could include a creation of economic opportunities for employees to access natural and financial resources that enhance and sustain community livelihoods. Although voluntary codes of governance were vague to employees, some elements of ethical CSR were perceived and noted. In this respect, concerted communicative efforts by senior and middle management are required to inform and involve employees. By doing so, poverty alleviation through the CSR framework could offer companies a better direction towards an effective CSR implementation with its employees that would activate local community development and sustainable natural resource utilisation.

Disclosure statement

No potential conflict of interest was reported by the authors.

References

Ashley, C. (2000). *The impact of tourism on rural livelihoods: Namibia's experience*. London: Overseas Development Institute.

Bohdanowicz, P., & Zientara, P. (2009). Hotels companies contribution to improving the quality of life of local communities and the well -being of their employees. *Tourism and Hospitality Research*, *9*(2),147–158.

Bowen, H. (1953). *Social responsibilities of the businessman*. New York: Harper & Row.

Bramwell, B., & Lane, B. (2008). Priorities in sustainable tourism research. *Journal of Sustainable Tourism*, *16*(1), 1–4.

Brejning, J. (2012). *Corporate social responsibility and the welfare state: The historical and contemporary role of CSR in the mixed economy of welfare*. Surrey: Ashgate Publishing.

Carroll, A. B. (1979). A three – dimensional conceptual model of corporate performance. *Academy of Management Review*, *4*(4),497–505.

Carroll, A. B. (1991, July–August). The pyramid of corporate social responsibility: Toward the moral management of organizational. *Business Horizons*. Rohan College of Business Administration Article Reprint. Retrieved from http://www-rohan.sdsu.edu/faculty/dunnweb/rprnts.pyramid.html [accessed 20 January 2015,].

Coles, T., Fenclova, E., & Dinan, C. (2013). Tourism and corporate social responsibility. A critical review and research agenda. *Tourism Management Perspectives*, 122–141.

Constanti, P., Farmaki, A., Yiasemi, I., & Karis, P. (2014). Responsible tourism in cyprus: The rhetoric and the reality. *Emerald*, *6*(1),10–26.

Cooper, C., Scott, N., & Baggio, R. (2009). Network position and perceptions of destination stakeholder importance. *Anatolia*, *20*(1), 33–45.

Crosby, B., & Bryson, J. (2005). A leadership framework for cross -sector collaboration. *Public Management Review*, 7(2), 177–201.

Davis, K. (1960). Can business afford to ignore social responsibilities? *California Management Review*, *2*(3), 70–77.

De Jongh, D., & Prinsloo, P. (2005). Why teach corporate citizenship differently? *Journal of Corporate Citizenship*, *18*.

Department of Environmental Affairs (DEAT). (2002). *Responsible tourism guidelines*. South Africa: Department of Environmental Affairs and Tourism (DEAT).

Dhanani, S., & Islam, I. (2002). Poverty, vulnerability and social protection in a period of crisis: The case of Indonesia. *World Development*, 1211–1231.

Ditlev-Simonsen, C., & Midttun, A. (2010). What motvates managers to pursue corporate social responsibility? A survey among key stakeholders. *Corporate Social Responsibility and Environmental Management*, *18*(1),25–38.

Dodds, R., & Kuehnel, J. (2008). CSR among Canadian mass tour operators: Good awareness but little action. *International Journal of Contemporary Hospitality Management*, *22*(2), 221–244.

Federation of Namibian Tourism Associations Transformation (FENATA). (2004). *Charter*. Windhoek: FENATA.

Frederick, W. (1986). *Theories of social corporate performance: Much done, more to do. Working paper*. Oakland: University of Pittsburgh.

Frey, N., & George, R. (2010). Responsible tourism management: The missing link between owner's attitudes and behaviour in the Cape Town Industry. *Tourism Management*, *3*(5). 621–628.

Gartner, C., & Cukier, J. (2012). Is tourism employment a sufficient mechanism for poverty reduction? A case study from Nkhata Bay Malawi. *Current Issues in Tourism*, *15*(6), 545–562.

Government of the Republic of Namibia. (2013). *National development plan 4 (2012/13–2016/17)*. Windhoek, Namibia: National Statistics Agency.

Grant, M. (2004). Innovation in tourism planning processes: Action learning to support a coalition of stakeholders for sustainability. *Tourism Hospitality Planning and Development*, *1*(3), 219–237.

Gurerk, O., Irlenbusch, B., & Rockenbach, B. (2009). Motivating teammates: The leaders choice beween positive and negative incentives. *Journal of Economic Psychology*, *30*, 591–607.

Ho, K., Seoki, L., & Chang, H. (2010). Impacts of positive and negative corporate social responsibility activities on company performance in the hospitality industry. *International Journal of Hospitality Management*, *29*(1),72–82.

Holcomb, J. L., Upchurch, R. S., & Okumus, F. (2007). Corporate social responsibility: What are top hotel companies reporting? *International Journal of Contemporary Hospitality Management*, 461–475.

Kazlauskarite, R., Bucuiniene, I., & Turauskas, L. (2006). Building employee committment in the hospitality industry. *Baltic Journal of Management*, *1*(3), 300–314.

Lapeyre, R. (2011a). Governance structures and the distribution of income in Namibia communal lands: A new institutional framework. *Journal of Economic and Social Geography*, 302–315.

Lapeyre, R. (2011b). The Grootberg lodge partnership in Namibia: Towards poverty alleviation and empowerment for long term sustainability? *Current Issues in Tourism, 14*(3), 221–234.

Makower, J. (1994). *Beyond the bottom line: Putting social responsibility to work for your business and the world*. New York: Simon & Schuster.

Ministry of Environment and Tourism (MET). (2011). *Tourist arrival statistics report*. Windhoek: Ministry of Environment and Tourism.

Namibian CommunityNACSO. (2016, May 29). *What is CBNRM?* Retrieved from NACSO Web site: www.nacso.org.na/what_is_cbnrm.php.

Namibian Statistics Agency (NSA). (2019). *Key highlights of the Namibia labour force survey 2018 report*. Windhoek: NSA.

National Planning Commision. (2017). *Fifth national development plan*. Windhoek: Namibia.

Ndlovu, J., Nyakunu, E., & Kavita, E. (2011). Re-defining the Rural Space in the Kuiseb Delta: Examining the potential tourism benefits for the Topnaars in Namibia. *Journal of Hospitality Management*, 48–56.

Nicanor, N. (2001, May 28). *Practical strategies for pro poor tourism: NACOBTA the case study*. Retrieved from www.propoortourism.org.uk/namibia_cs.pdf: www.odi.org/resources/docs/4022.pdf.

Novelli, M., & Hellwig, A. (2011). The UN millenium development goals, tourism and development: The tour operators perspective. *Current Issues in Tourism*, 205–220.

Republic of Namibia. (2004). *Namibia vision 2030-policy Namibia*. Windhoek: Government of Namibia.

Saarinen, J. (2014). Critical sustainability: Setting the limits to growth and responsibility in tourism. *Sustainability, 6*(11), 1–17.

Saarinen, J. (2017). Enclavic tourism spaces: Territorialization and bordering in tourism destination development and planning. *Tourism Geographies, 19*(3), 425–437.

Saarinen, J. (Ed.). (2019). *Tourism and sustainable development goals: Research on sustainable tourism geographies*. London: Routledge.

Saarinen, J., & Niskala, M. (2009). Selling places and construction local cultures in tourism: The role of the ovahimba in namibian tourism promotion. In P. Hottola (Ed.), *Tourism startegies and local responses in Southern Africa*. Oxon: CABI.

Saarinen, J., & Rogerson, C. (2014). Tourism and the millenium development goals: Perspectives beyond 2015. *Tourism Geographies, 16*(1), 23–30.

Saarinen, J., Rogerson, C., & Hall, C. M. (2017). Geographies of tourism development and planning. *Tourism Geographies*, 307–317.

Sahlins, M. (1972). *Stone age economics*. Chicago: Atherton.

Sharpley, R. (2000). Tourism and sustainabe development: Exploring the theoretical divide. *Journal of Sustainable Tourism*, 1–19.

Sharpley, R. (2009). *Tourism development and the environment: Beyond sustainability?* London: Earthscan.

Spenceley, A., & Goodwin, H. (2007). Nature based tourism and poverty alleviation: Impacts of private sector and parastatal enterprises in and around Kruger National Park, South Africa. *Current Issues in Tourism, 10*(2), 255–277.

Svensson, G. (2015). Exchange and change in northern Norway: On reciprocity in nature-based tourism. Tourism. *Culture & Communication, 14*, 169–182.

Timothy, D., & Ioannides, D. (2002). Tour operator hegemony: Dependence and oligopoly in insular destinations. In Y. Apostopoulus & D. J. Gayle (Eds.), *Island tourism and sustainable development Carribean, Pacific and Mediterranean experiences*. Westport: Praeger.

Timur, S. G. (2009). Sustainable tourism development. How do destinations stakeholders perceive sustainable urban tourism? *Sustainable Development, 17*(4), 220–232.

United Nations. (2015). *Transforming Our world: The 2030 agenda for sustainable development. Resolution adopted by the general assembly.* New York: United Nations.

United Nations. (2017). *Economic development in Africa Report 2017: Tourism for transformative and inclusive growth.* Retrieved from :https://unctad.org/en/PublicationsLibrary/aldafrica2017en.pdf.

Visser, W. (2006). Revisiting Carroll's CSR pyramid an African perspective. *Corporate Citizenship in Developing Countries*, 26–56.

Visser, W. (2008). Corporate Social Responsibility in Developing Countries. In Crane, A., McWilliams, A., Matten, D., Moon, J. and D.S. Siegel (Eds.), *The Oxford Handbook of Corporate Social Responsibility*, 473–499. Oxford: Oxford University Press.

Visser, W. (2011). The age of responsibility: CSR 2.0 and the new DNA of business. West Sussex: John Wiley & Sons.

Weeden, C. (2002). Ethical tourism: An opportunity for competitive advantage? *Journal of Vacation Marketing, 8*(3), 141–153.

World Bank. (2001). *Attacking poverty: World development report 2000/01.* New York: The World Bank.

World Bank Group. (2017). *Tourism for development: 20 reasons sustainable counts for development.* Washington, DC: The World Bank.

World Business Council for Sustainable Development. (2001, July 07). *Making a difference toward the Johannesburg summit 2002 and beyond.* Retrieved from www.iatp.org/files/Busi ness Case for Sustainable Development.

Zhao, W., & Ritchie, B. J. (2007). Tourism and poverty alleviation: An integrative research framework. *Current Issues in Tourism, 10*(2), 119–143.

Zikmund, W. G., & Babin, B. J. (2010). *Essentials of marketing research.* Mason, OH: South Western.

11 Revisiting devolution in community-based natural resources management in Zimbabwe

Towards inclusive governance approaches

Jones Mudimu Muzirambi, Regis Musavengane and Kevin Mearns

Introduction

Community participation in conservation and ecotourism management has been the focus of research discourse for a considerable time (Chambers, 1994; Moser & Felton, 2006; Musavengane, 2019a). Researchers have highlighted a number of challenges facing community-based natural resources management (CBNRM) authorities, including the refusal to devolve power and/or share power with local communities, restricting participation and leading to the alienation of local residents from important decision-making processes (Balint, 2006; Kamphorst, Koopmanschap, & Oudwater, 1997). For the past decade, literature shows that ignoring local people's interests and excluding them from management and decision-making are the main sources of conflicts between stakeholders (Laudati, 2007).

The framework developed by Sacchetti and Campbell (2015) on community development studies highlights the importance of creating an environment of cooperative relations among communities of interest to enhance common welfare. This chapter is influenced by the conceptual framework developed by the common pool resource theory (Ostrom, 1990) and collaborative theory, which emphasise the engagement and empowerment of citizens in the definition and monitoring of rules and identification of solutions to local issues. Colbry, Hurwitz, and Adair (2014) describe cooperative efforts or collaboration as any on-going interpersonal interaction not characterised by significant power imbalances, with the express purpose of achieving common goals. It is with this background that policy-makers recognise the need for local community involvement in the management of natural resources in order to achieve sustainable wildlife management systems, yet, policy-makers seem reluctant to devolve power and decision-making to grassroots levels. For the purpose of this chapter, power is described as the authority for management to make decisions on and benefit from a resource. Devolution of power would therefore refer to conceding decision-making authority and responsibility to the local community level. In Zimbabwe, due to the lack of legislative instruments that provide rural communities and institutions, e.g. wards, ownership rights over wildlife and wildlife revenue, Communal Areas Management Programme for Indigenous Resources' (CAMPFIRE) management principles could not be vigorously enforced but only encouraged. Furthermore, the fragmented policies over the use and control of most natural resources resulted in the failure of devolution of wildlife authority to the grassroots levels. This means that resource control remained in the higher governance levels – the Rural Development Council (RDC).

The United Nations Development Programme (UNDP) and the World Bank emphasise that human development projects cannot succeed without focusing on rights, capacity and governance (UNDP, 2002). Governance refers to interactions among structures, processes and traditions that determine the way in which power and responsibilities are exercised, decisions are made and citizens or other stakeholders have a say (Kauffman, Kraay, & Mastruzzi, 2005; Ruhanen, Scott, Ritchie, & Tkaczynski, 2010). CAMPFIRE in Zimbabwe was initiated to transform the governance of wildlife, seeking to integrate local communities into the wildlife and natural resource management governance system (Harrison, 2015). In the context of this chapter, CAMPFIRE and CBNRM in Zimbabwe shall be used interchangeably throughout the chapter. Measham and Lumbasi (2013) attributed the failure of CBNRM programmes across the world to the following factors: top-down project initiation, lack of economic incentives, lack of autonomy and incompatible livelihoods, and opportunity costs. Kamphorst et al. (1997) concluded that CAMPFIRE has not significantly benefited the poor rural communities as was expected.

In light of the above, this chapter seeks to investigate the challenges that exist in the governance of CBNRM in Zimbabwe, especially relating to the devolution of power and community engagement. The objectives of the study are (i) to critically analyse the nature and challenges of CBNRM governance, (ii) to discuss the political decentralisation and devolution of power in Zimbabwe and its impact on the CAMPFIRE initiatives, (iii) to explore the challenges to the devolution of power to communities through the analysis of case studies in the CAMPFIRE programme, and (iv) to determine the status of devolution of power in the CBNRM in Zimbabwe.

Methodology

This study takes a qualitative approach in which a comparative documentary analysis of several cases is undertaken, presented and discussed. Compared to other methods, a document analysis is less expensive and unobtrusive (Neuman, 2011). The analysis heavily depends on studies that were case study-based with the connotation of obtaining indigenous views from the local people. Critical indigenous inquiry is concerned with issues of representation, legitimacy, accountability and benefits (Bryman & Burgess, 1993; Denzin & Lincoln,

2008). The chapter is limited to case studies because the case study approach tends to answer the "when," "how," "what" and "why" questions which facilitates in scrutinising particular phenomenon (Yin, 2003), in this case, governance of CBNRM in Zimbabwe. Data were collected from journal articles, books, policy and government documents. The criteria for the inclusion of case studies include availability of primary qualitative data obtained from the communities, CAMPFIRE programme documentation and details of governance issues in the particular case study. Studies that relied on secondary source data were excluded from the study. The review process of archival records involved a search for literature using the research search engines – Google Scholar, Scopus and Web of Science. Key words such as "CAMPFIRE," "community-based natural resources" and "common pool resources" in Zimbabwe were used to guide the search. This resulted in the selection of three cases that may be seen as representative of the national CAMPFIRE program, namely, (i) Binga, located in the Matebeleland North, northwest of Zimbabwe, (ii) Mahenye, a ward located in southeastern Zimbabwe, and (iii) Hwange, in Matabeleland north province, Zimbabwe.

The data were analysed qualitatively through a process of coding and theming – thematic analysis. The themes emerged from the obtained data sets from the primary and secondary data. Thematic analysis "is a method for identifying, analysing, organising, describing and reporting themes found within [a] data set" (Nowell, Norris, White, & Moules, 2017, p. 3). The synthesis of the findings facilitated interrogation of the implementation strategies of the pro-poor policies affecting natural resource management. Consequently, constructive criticism and recommendations on pertinent issues on local governance, power relations and sustainability, with respect to Zimbabwean CBNRM, were made.

Findings and discussion

In an attempt to contextualise the findings relating to the CAMPFIRE programme, a sort background is provided on governance within a nature conservation and ecotourism context; thereafter a short introduction is provided of the CAMPFIRE programme. The discussions of the findings focus more on the devolution of power to local levels and the specific lessons learnt from the three case studies investigated. Finally, recommendations are made from the lessons learnt.

Governance of nature conservation and ecotourism

The governance of economic processes, property rights and local political organisations lies at the heart of community-based natural resource management. As Child and Barnes (2010) point out, devolving property rights to communities shifts resource governance, responsibility and benefits appropriately to the local level. Success of the CBNRM initiatives is directly linked to the concept of sustainability.

Three challenges of managing stakeholder relationships include: (i) strengthening the particular sustainability interests; (ii) creating and harmonising mutual sustainability interests and (iii) empowering stakeholders to act as agents for nature conservation and sustainable development (Horisch, Freeman, & Schaltegger, 2014; Pegas & Castley, 2014). Many CBNRM and ecotourism project initiatives struggle to meet their intended goals due to the refusal of authorities to devolve power to lower levels, restrictions on participation by local communities to project implementation only and inequitable sharing of costs and benefits (Gandiwa, Heitkonig, Lokhorst, Prins, & Leeuwis, 2013; Mutandwa & Gadzirayi, 2007; Snyman, 2017). Policy-makers and communities are therefore challenged to create contexts and rules that best serve the development of socio-economic activities, e.g. sustainable conservation without compromising the sustainability of the natural resource system (Campbell & Sacchetti, 2017).

Collaborative governance brings all stakeholders together in collective forums with public agencies to engage in consensus-oriented decision-making (Ansell & Gash, 2008; Emmerson, Nabatchi, & Balogh, 2012). Collaboration is characterised by a joint decision-making approach, where power is shared and stakeholders have collective responsibility for their actions and subsequent outcomes (Bello et al., 2017; Zou, Huang, & Ding, 2014).

Community-based natural resource management in Zimbabwe-The CAMPFIRE programme

CAMPFIRE, a national CBNRM initiative of Zimbabwe, has become one of the most studied CBNRM models worldwide (Roe & Nelson, 2009), from which many developing countries have learnt lessons and improved their natural resource management governance. Child (2003, p. 6) explains clearly that CAMPFIRE's design had at its core,

> the empowerment of community members at village level to control wildlife and its revenue, the internalisation of costs and benefits at this level, and an underlying belief that wildlife was the most suitable land use option in many of these remote areas.

The intended goals of CAMPFIRE were compromised due to the political dynamics of Zimbabwe, which contributed to numerous amendments to the CAMPFIRE Agreement during the signing process. The implemented version of 1989 was very different from the original CAMPFIRE design. Instead of devolving the authority to the local communities in the programme, the 1982 amendment of the Parks and Wildlife Act of 1975 gave the Rural District Councils (RDCs), as the lowest, legally recognised level of government, the Appropriate Authority (AA) status to manage and control the communal lands under their control, including the wildlife (Frost & Bond, 2008; Mandondo, 2000).

One of the main criticisms of CAMPFIRE has been a failure to produce effective community participation, restricting them to implement centrally designed plans and the reluctance of district councils to devolve genuine authority for wildlife management. The top-down approaches have been brutally unsympathetic to the needs of the local residents (Bramwell & Lane, 2012; Nash, 2014; Muzirambi & Mearns, 2018). In agreement with Kamphorst et al. (1997), control and management from above generate social conflict (Muzirambi & Mearns, 2015). The Attorney General's office reversed all democratic elements of the CAMPFIRE Agreement in the Environmental Management Act of 2000 (Mandondo, 2000). The decision of the Attorney General's office was informed by the Constitution of Zimbabwe, which limited decentralisation of authority to the lowest legally recognised level of government – the RDC.

The most salient issues emanating from research literature include the lack of a water-tight policy framework which facilitates genuine devolution of wildlife management authority to the local communities and no political will to empower and capacitate structures below the RDCs to achieve an acceptable level of autonomy to deal with issues. As alluded to by Ntuli and Muchapondwa (2018), the performance of most communities left a lot to be desired with respect to institutional attributes important for conservation.

The challenges of CAMPFIRE have been worsened by political and economic factors that have prevailed since early 2000 (Balint & Mashinya, 2006; Taylor, 2008) and the refusal to transfer benefits directly to local communities (Child, 1996; Gandiwa et al., 2013; Muboko & Murindagomo, 2014). Benefits from wildlife resources included revenue from safari hunting and tourism, employment and game meat (Balint & Mashinya, 2008), infrastructural development e.g. construction of school and clinic blocks, drilling of boreholes and installation of grinding mills (Child, 1993). The main benefit is revenue generation. While not legally binding, it was generally understood in principle that at least 50% of the revenue given to the RDCs should be passed on to local communities involved in CAMPFIRE; 35% goes to wildlife management costs in general and 15% to the RDC as an administrative levy. The RDCs did not adhere to this principle.

Campbell and Sacchetti (2017) argued that, due to lack of an enabling policy, lack of accountability, lack of good governance and corruption, the CAMPFIRE programme has not benefited the poor rural communities as intended. Monitoring and evaluation processes from the inception of CAMPFIRE in 1989, however, did not help to address these challenges. Yet, most natural resource management programmes in the region, (for example in Botswana, Namibia, Tanzania and South Africa) learned from the weaknesses of CAMPFIRE and implemented successful CBNRM initiatives, benefiting their communities and conservation.

With reference to the findings of Mudimba and Tichaawa (2017) and Muzirambi and Mearns (2018) local communities wanted to be involved when nature conservation and tourism policies were being made so that they would be able to influence policy-makers to include their needs and protect their interests. The question arises, then, why was this not possible in CAMPFIRE? This study creates a platform for more robust, discursive and political views which would evaluate CAMPFIRE and suggests implementable recommendations for its improvement.

Political decentralisation and devolution of power to local levels in Zimbabwe and the impact on CAMPFIRE

Harrison (2015) and Kurebwa (2015) provide a deeper insight into the impact of the political-economic processes during decentralisation on the community-based natural resource management. According to Helmsing (1991), at independence in 1980 the government of Zimbabwe introduced a wide range of reforms with the aim to improve local governance into becoming a more democratic and responsive system to address the needs of the people. The proximity of local government to consumers makes them more accountable and compels them to be transparent (Chatiza, 2008; Chigwenya, 2010).

Kurebwa (2015) observed that while the District Councils Act decentralised more power to the districts, this was viewed by critics as an attempt to recentralise power through the Ministry of Local Government and through the District Administrator, a central government employee. This study argues that there are systemic challenges to the whole decentralisation process in Zimbabwe. Briggs et al. (2019) confirmed that social norms that favour collaboration were partially present as national government and the rural district councils were reluctant to decentralise authority to a community level. There are 250 instances in terms of the RDC Act where the minister may interfere with the affairs of the RDC. The minister has substantial control over the finances of the local authorities (Kurebwa, 2015). Derman (1995) explained that one of the major causes of the reluctance by both national government and RDCs to devolve CAMPFIRE to the communities is the high value of wildlife as an economically viable natural resource. By 2001 over 40% of revenue from CAMPFIRE was retained by some government district offices compared to the guidelines of 15% (Balint & Mashinya, 2006). Harrison's (2015) reports on the weaknesses of local governance structures compromising the ability of the community to hold leaders accountable for their actions in the CAMPFIRE programme.

Challenges associated with the devolution of power and community participation in CBNRM

Analysis of case studies of the CAMPFIRE initiative

Ribot (2002a) noted that participation in development committees at provincial, district, ward and village levels was exclusionary, as only ruling party-aligned persons were allowed to participate in the decision-making processes. The discussion so far points to the fact that decentralisation can either occur through deconcentration or devolution. Deconcentration entails the transfer of come entrustments from the central government to the branches of the government, and often the central government remains in control, supervises or may withdraw the entrustments (Mandondo, 2000; Pomeroy, 1999; Ribot, 1999; Ribot & Viet, 2002b). On the other hand, devolution entails the total transfer of entrustments to local communities, and the central government has no control over the governance and management processes (Meinzen-Dick & Knox, 1999; Muboko & Murindagomo, 2014). The following CAMPFIRE case studies provide an opportunity to draw lessons on effective devolution and participatory processes in rural areas where natural resources are communally owned. Emerging themes from these case studies will be analysed to provide a deeper understanding of devolution in CBNRM. The communities included in the synthesis are Binga, Mahenye and Hwange.

Binga case study

Binga is located in the Matebeleland North, the northwest part of Zimbabwe, bordering Zambia, one of the most isolated districts in Zimbabwe. It is predominantly occupied by the Tonga ethnic group, which is an ethnic minority group in Zimbabwe. Before the erection of the Kariba Dam Wall across the Zambezi River, they enjoyed a relative lifestyle of subsistence farming, characterized by cultivation in the fertile flood plain, fishing in the river and hunting wild game. People were forcefully relocated to safe areas to avoid damage from flooding and to pave the way for the development of wildlife reserves and tourism ventures. However, the fishing and tourism industries are heavily dominated by outsiders, and the local people remain poor (Conyers, 2002). Mutana (2013) observed that in Binga local people are involved in menial jobs such as mending boats and park fences, which pay meager remuneration. They do not have the means of production and are often neglected in decision-making (Mutana, 2013). In an interview with senior managers at a tourism organisation in Sikalenge, Binga, Mutana (2013) noted that managers who should promote the participation of local people in tourism decision-making attributed their non-involvement to a lack of skills. Qualifications and skills should not be a deterrent to participate in CBNRM decision-making. Musavengane and Simatele (2016) noted that effective co-management of community natural resources requires a collaborative effort and communities need to be empowered. Training of local people is a positive sign to show the willingness of tourism organisations to empower local communities through education (Musavengane, 2019a, 2019b).

Turning to CAMPFIRE institutions, Binga Rural District Council (BRDC) is an elected body, with twenty-one (21) councilors from the twenty-one (21) wards. The council has a number of committees of which the Conservation Committee is responsible for the CAMPFIRE program (Conyers, 2002). Like other departments in all council committees, the CAMPFIRE department has its own employees (i.e. training officer, game guards and clerical staff) who report to the designated CAMPFIRE manager. The department is externally funded, and most of the funds are earmarked for the training of local people. The funds were initially channeled through a Zimbabwean non-governmental organization, the Zimbabwe Trust, and later through the CAMPFIRE Association (Conyers, 2002). However, there has been a decline in funding for the CAMPFIRE department due to political challenges since 2000. The department now heavily depends on CAMPFIRE revenue, which is not enough to sustain the operations of the department. To ensure equal representation, and to align with CAMPFIRE Guidelines and council policy, each of the district's twenty-one wards has selected CAMPFIRE sub-committees in each of the villages within a ward (Conyers, 2002). However, the effectiveness of village committees is questionable as they seem to exist only on paper. The election of members of the village committees is supposed to be done by all community members; however, only a few participate due to the large spatial distribution of households in Binga. This then poses a threat to fair representation of the people and may lead to a scenario of "representation without participation" of community members (Agyei & Adjei, 2017). The fundamental roles of ward and village committees include taking responsibility for CAMPFIRE's revenue, staffing, educating the general public on wildlife conservation and reporting and resolving animal-people conflicts.

A look at the benefit-sharing arrangements of CAMPFIRE in Binga is key to measuring the success of the projects. Safari hunting is the main source of revenue for the community. However, Conyers (2002) noted that the ward dividends were not systematically distributed to the wards, which contradicts both the CAMPFIRE Guidelines and at least one council resolution. Much of the remitted ward dividends were used to build schools, as well as improve other social services such as health facilities and water reservoirs. The council used some of the proceeds to meet administrative costs and buy assets (some of which are a bit luxurious), which is a concern as the money should have been channeled to wards (Conyers, 2002). Some of the money channeled for administrative purposes was not well recorded, which posed a challenge in accounting for it. Poor accountability may point to the existence of corruption by the political elites and government actors and may undermine the use of natural resources to promote national development (Pigg, 1990). However, the ward dividends helped the Binga community to pursue communal projects that the Government of Zimbabwe (GoZ) could no longer provide due to economic challenges, for example, repairing of boreholes. Nevertheless, the CAMPFIRE gains of individual households remain minimal. It can therefore be deduced that decentralisation of control of communal natural resources is not an absolute to attaining community needs. Second, decentralisation is not a guarantee that CAMPFIRE benefits will accrue to individual households. Further, decentralisation of effective power of controlling natural resources can be resisted by other institutions that may feel like clinging onto power as the only way to access financial gains.

Mahenye case study

The Mahenye case study provides an opportunity to analyse and assess the implications of the post-2000 political and economic crisis involving CAMPFIRE. Balint and Mashinya (2006) established strong links between the current state of CAMPIRE to the episodes of the Zimbabwe land reform era (from 2000) and beyond. Mahenye ward is located in southeastern Zimbabwe which borders with Mozambique; it lies adjacent to Gonarezhou National Park, separated by the Save River. Although community-based natural resources management had commenced in the early 1980s,, the Government of Zimbabwe authorised Chipinge District's CAMPFIRE program in 1991, making Mahenye one of the first officially recognized sites to pursue CBNRM through the CAMPFIRE arrangement (Balint & Mashinya, 2006; Murphree, 2002; Peterson, 1991). Prior to 1991, community members did not benefit from the CAMPFIRE project, as money earned from sport hunting in Mahenye Ward was retained at the district level. This frustrated community members as they expected to benefit from the project (Murphree, 2002). After 1991 well defined structures were established after the launch of the national CAMPFIRE Guidelines. However, the success story did not last as it was impacted by the national economic and political challenges.

In their research on "the decline of a model community-based conservation project," Balint and Mashinya (2006) reported a number of post-2000 challenges in governing Mahenye. These challenges span across devolution, participation and decentralisation issues – critical elements for the success of CAMPFIRE projects. Post 2000, contrary to the main objectives of CAMPFIRE, to empower locals, the majority of the respondents reported that the project was no longer managed to benefit the community, but rather the elite (traditional leaders and political party loyalists). There has been deterioration in the governance system at Mahenye since 2000, as democratic processes were abandoned and replaced by nepotism, favoritism and corruption. For example, instead of electing leaders every two years as stipulated in the by-laws, the chief imposed himself to lead the CAMPFIRE project regardless of the fact that he has no formal authority over the operations of CAMPFIRE. In order to facilitate his intention to control the CAMPFIRE programme and pursue corrupt activities in Mahenye, he appointed his brother to a top leadership position, and since then no elections have been held (Balint & Mashinya, 2006). These circumstances have eroded the social capital within the community and will eventually destroy communal efforts. "If any collaborative CBNRM scheme is to be successful, there is a need for fundamental social and political changes that must be accompanied by a positive alteration in power relations among all of the interested stakeholders" (Musavengane & Simatele, 2016 p. 815).

Since the change in leadership in 2000, election processes were abandoned in Mahenye, and efforts to promote community participation and maintain transparency and accountability have been weakened. No Annual General Meetings have been held, and the auditing of financial records have not been done (Balint & Mashinya, 2006). This compromises trust, social cohesion and collaboration among community members and breeds discontent. It is evident that power relationships are at the core of the disenfranchisement of local people's participation in Mahenye. Despite records of receiving money toward community projects such as building school infrastructure, no remission of such monies was made to the schools. These practices continue to reduce community confidence in leaders and community-based tourism projects. Balint and Mashinya (2006, p. 813) have observed that "that local failures in governance and capacity contributed significantly to the decline in community benefits and the near universal distrust and disillusionment voiced by our respondents." Good governance builds trust and creates goodwill to strengthen relationships, which will further enhance stakeholder satisfaction and commitment and increase legitimacy (Mpinganjira, Roberts-Lombard, & Svensson, 2017; Siakwah, Musavengane, & Leonard, 2019). Cohan and Prusak (2001, p. 46) highlighted that "knowing who people are and what they are doing, builds social connections and trust, just as secrecy builds suspicion."

Hwange case study

Hwange district is located in the Matabeleland North province, Zimbabwe. The communal wards are located in and border national protected wildlife areas and tend to benefit from CAMPFIRE projects. Dube (2019, p. 337) notes that the "Hwange district CAMPFIRE office revealed that the main wildlife producer wards were Mabale (17), Sidinde (8) and Silewu (15)." Villagers in these wards complained of leadership who like to run CAMPFIRE projects but without considering the interests of the community. These leaders and a few elite benefited from CAMPFIRE programmes, a situation which led CAMPFIRE to be regarded as insignificant by local people (Dube, 2019). Furthermore, there seems to be conflicting interests between wildlife producers in the wards and wildlife safari operators. The villagers trade their conservation efforts through the Rural District Council (RDC) with safari operators. The mandatory use of RDC has created perceptions of RDC as riding on the backs of the villagers. It is vital for RDC to communicate effectively with the villagers on the proceeds of wildlife to avoid speculation and suspicion. Musavengane and Simatele (2016) state that communication is important in building strong relations among different actors.

In addition, villagers who encounter conflicts with wildlife receive an equal portion of revenue as those who are less affected by wildlife (Dube, 2019). Villagers have no say on the allocation of revenue; instead decisions are made by the authorities who have allegedly assumed CAMPFIRE decision-making roles without due process. The voices of the local people on wildlife threats tend to be ignored, and if problem animals are reported, authorities will only kill them when they find buyers, meaning that money matters more than the livelihoods of local communities. Hove, Ngwerume, and Muchemwa (2013) noted that genuine grassroots-grown community participation in managing natural resources is vital when making communal decisions. Dube (2019) attributed the tension between villagers and CAMPFIRE authorities to weak ties between the RDC and the village representatives, such as the traditional chiefs and ward councilors. This situation has led CAMPFIRE management to be more representative of, while not being accountable to, the grassroots community. They seem to account to the political leaders and traditional authorities while neglecting those seemingly of a lower class.

Implications from the analysis of the case studies

This section synthesizes the findings and issues that have emerged from the three CAMPFIRE case studies and presents recommendations for future actions. There is evidence that policymakers and conservation authorities recognise the need for community involvement in the management of natural resources in order to achieve sustainable wildlife management systems but are reluctant to devolve power to the grassroots community. It has also been noted that there is a lack of an enabling policy framework which facilitates the genuine democratization of natural resource management in Zimbabwe and the non-existence of the political will to do so. The main challenges affecting CAMPFIRE, from literature, include, among others, the lack of good governance, disrespecting of human rights at the local level, lack of capacity of both the local government (RDCs) and grassroots communities, corruption, lack of transparency and accountability, and gross political interference (Balint, 2006; Balint & Mashinya, 2006; Harrison, 2015). This study exposes the negative perceptions towards top-down approaches and highlights the associated conflicts between conservation authorities and affected communities.

Conclusion

The goal of this chapter was to explore and interrogate the challenges in the governance of CBNRM in Zimbabwe. A critical analysis of the nature and challenges of natural resource management governance in general and case studies in the CAMPFIRE programme in particular were done. The political perspectives of decentralisation and devolution of power in Zimbabwe were explored in order to attempt to understand the context in which the CBNRM was operating and provide possible reasons for the failure of total devolution of wildlife authority to the grassroots levels. The political and economic elites were reluctant to give up access and benefits that accrue from wildlife resources (Gandiwa et al., 2013). Case studies were used to provide practical evidence of issues concerning CBNRM governance in Zimbabwe. Finally, constructive criticism and appropriate recommendations were made e.g. policy reforms for conducive environment for genuine devolution, effective local governance, capacity building, depoliticisation of the CBNRM, increase in autonomy, inculcation of transparency and accountability.

Recommendations

In view of the challenges facing CAMPFIRE, the study urges policy-makers to embark on robust, inclusive and democratic policy reforms which create conditions for genuine devolution of wildlife authority to grassroots communities (Campbell & Sacchetti, 2017). There is a need to institute policies which enable local communities to actively participate in a more open, responsive and effective local governance and benefit-sharing processes of the natural resources within their areas. This study suggests policy reforms that enable active participation of local communities. The creation of effective monitoring and oversight mechanisms by the government to protect the rights of the local communities is recommended. Capacitation of all stakeholders for their roles in the new democratised CBNRM dispensation must be included. Local citizens need to be empowered through appropriate education, skills development and legislative reforms so that they are able to hold their leaders accountable to the mandates of the general community (Nelson, 2010; Harare Workshop, 2014). It is critical for the country leaders to de-politicise natural resource management and community development programmes, increase autonomy and ensure transparency so as to achieve some of the sustainable development goals. It is further recommended that communication strategies among stakeholders need to be improved to enable easy flow of information and build trust (Muzirambi & Mearns, 2018). Facilitation of active community participation will help to develop a sense of ownership and belonging, which is critical for sustainable natural resource management and community development.

References

Agyei F. K., & Adjei, P. O. (2017). Representation without accountability in forestry: Experiences from the Social Responsibility Agreement in Ghana. *Forest Policy and Economics*, *80*, 34–43. doi:10.1016/j.forpol.2017.03.003

Ansell, C., & Gash, A. (2008). Collaborative governance in theory and practice. *Journal of Public Administration Research and Theory*, *18*(4), 543–571. doi:10.1093/jopart/mum032

Balint, P. J. (2006). Improving community-based conservation near protected areas: The importance of development variables. *Environmental Management*, *38*(1), 137–148. doi:10.1007/s00267-005-0100-y

Balint, P. J., & Mashinya, J. (2006). The decline of a model community-based conservation project: Governance, capacity and devolution in Mahenye, Zimbabwe. *Geoforum*, *37*(5), 805–815. doi:10.1016/j.geoforum.2005.01.011

Balint, P. J., & Mashinya, J. (2008). CAMPFIRE during Zimbabwe's national crisis: Local impacts and broader implications for community-based wildlife management. *Society and Natural Resources*, *21*(9), 783–796.

Bello, F. G., Lovelock, B., & Carr, N. (2017). Constraints of community participation in protected area-based tourism planning: The case study of Malawi. *Journal of Ecotourism*, *16*(2), 131–151. doi:10.1080/14724049.2016.1251444

Bramwell, B., & Lane, B. (2012). Towards innovation in sustainable tourism research? *Journal of Sustainable Tourism*, *20*(1), 1–7. doi:10.1080/09669582.2011.641559

Briggs, D., Ban, N. C., Castilla, J. C., Gelach, S., Mills, M., Gandiwa, E., . . . Possingham, H. P. (2019). Insights on fostering the emergence of robust conservation actions from Zimbabwe's campfire program. *Global Ecology and Conservation, 17*, 1–10. doi:10.1016/j.gecco.2019.e00538

Bryman, A., & Burgess, R. G. (Eds.). (1993). *Analysing qualitative data*. New York: Routledge.

Campbell, C., & Sacchetti, S. (2017). Community-based, collaborative solutions to sustainable economic development in and around Biosphere Reserves (Ed.). *Journal of Entrepreneurial and Organisational Diversity, 6*(1), 1–9. doi:10.5947/jeod.2017.001

Chambers, R. (1994). The origins and practice of participatory rural appraisal. *World Development, 22*(7), 953–969. doi:0305750X(94)E0029-W

Chatiza, K. (2008). *Opportunities and challenges in institutionalising participatory development: The case of rural Zimbabwe*. Unpublished Doctoral Thesis. Swansea University.

Chigwenya, A. (2010). Decentralisation without devolution and its impact on service delivery: The case of Masvingo Municipality in Zimbabwe. *Journal of Sustainable Development in Africa, 12*(1), 1–12.

Child, B. (1993). Zimbabwe's CAMPFIRE programme: using the high value of wildlife recreation to revolutionize natural resource management in communal areas. *The Commonwealth Forestry Review, 72*(4), 284–296.

Child, B. (1996). The practice and principles of community-based wildlife management in Zimbabwe: The campfire programme. *Biodiversity and Conservation, 5*(3), 369–398. doi:10.1007/BF00051780

Child, B. (2003). *Origins and efficacy of modern community based natural resources management (CBNRM) practices in the Southern African Region*. Gland, Switzerland: IUCN.

Child, B., & Barnes, G. (2010). The conceptual evolution and practice of community-based natural resource management in Southern Africa: Past, present and future. *Environmental Conservation, 37*(3), 283–295. doi:10.1017/S0376892910000512

Cohan, D., & Prusak, L. (2001). *In good company*. Boston: Harvard Business School Press.

Colbry, S., Hurwitz, M., & Adair, R. (2014). Collaborative theory. *Journal of Leadership Education, 13*(4), 63–75. doi:10.12806/V13/14/C8

Conyers, D. (2002). *Whose elephants are they? Decentralization of control over wildlife management through the campfire program in Binga District, Zimbabwe. Environmental Governance in Africa, Working Paper #4, Institutions and Governance Program*. Washington, DC: World Resources Institute.

Denzin, N. K., & Lincoln, Y. S. (2008). *Collecting and interpreting qualitative materials* (Vol. 3). Thousand Oaks, CA: Sage Publications.

Derman, B. (1995). Environmental NGOs, dispossession, and the state-the ideology and praxis of African nature and development. *Human Ecology, 23*(2), 199–215. doi:10.1007/BF01191649

Dube, N. (2019). Voices from the village on trophy hunting in Hwange district, Zimbabwe. *Ecological Economics, 159*, 335–343. doi:10.1016/j.ecolecon.2019.02.006

Emmerson, K., Nabatchi, T., & Balogh, S. (2012). An integrative framework for collaborative governance. *Journal of Public Administration Research and Theory, 22*(1), 1–29. doi:10.1093/jopart/mur011

Frost, P. G. H., & Bond, I. (2008). The campfire programme in Zimbabwe: Payments of wildlife services. *Ecological Economics, 65*, 776–787. doi:10.1016/j.ecolecon.2007.09.018

Gandiwa, E., Heitkonig, I. M. A., Lokhorst, A. M., Prins, H. H. T., & Leeuwis, C. (2013). CAMPFIRE and human-wildlife conflicts in local communities bordering Northern Gonarezhou National Park, Zimbabwe. *Ecology and Society, 18*(4), 7. doi:10.5751/ES-05817-180407

Government of Zimbabwe (GoZ). (1988). *Rural district councils act* [Chapter 29: 16] Harare: Government Printers.

Harare Workshop. (2014). *Progressing CBNRM in Zimbabwe*. In Harare. Zimbabwe.

Harrison, E. P. (2015). *The governance of natural resource management in Zimbabwe: Unravelling the relationships between conservation and development*. Unpublished Doctoral Thesis. University of Leeds.

Helmsing, A. H. J. (1991). Transforming rural local government: Zimbabwe's post-independence experience. In N. D. Mutizwa & A. H. J. Helmsing (Eds.), *Rural development and planning in Zimbabwe* (pp. 441–481). Aldershot: Gower Publishing Company.

Horisch, J., Freeman, R. E., & Schaltegger, S. (2014). Applying stakeholder theory in sustainability management: Links, similarities dissimilarities and a conceptual framework. *Organisation and Environment, 27*(4), 328–346. doi:10.1177/1086026614535786

Hove, M., Ngwerume, E., & Muchemwa, C. (2013). The urban crisis in Sub-Saharan Africa: A threat to human security and sustainable development. *International Journal of Security and Development, 2*(1), 7–18. doi:10.5334/sta.ap

IUCN (*International Union for Conservation of Nature*). (2010). *Improving protected area governance for livelihood security and rights in Southern Africa*. Unpublished paper, Workshop, 13–16 April, Golden Gate National Park, South Africa.

Kamphorst, D., Koopmanschap, E., & Oudwater, N. (1997). Effective participation in wildlife management in Zimbabwe's CAMPFIRE programme. *European Journal of Agriculture Education and Extension, 4*(3), 173–182. doi:10.1080/13892249785300311

Kauffman, D.; Kraay, A., & Mastruzzi, M. (2005). *Governance matters IV: Governance indicators for 1996–2004 (Working Paper)*. Washington, DC: World Bank Institute.

Kurebwa, J. (2015). A review of rural local government system in Zimbabwe from 1980–2014. *IOSR Journal of Humanities and Social Science (IOSR-JHSS), 20*(2), V: 94–108. doi:10.9790/0837-202594108

Laudati, A. (2007). *The greening of the fortress: Reclaiming the politics of inclusion in a green era*. Unpublished PhD Dissertation, University of Oregon.

Mandondo. (2000). Situating Zimbabwe's natural resource governance system in History. *Center for International Forestry Research*, Occasional Paper No. 32, pp. 1–23.

Matyszak, D. (2011). *Formal structures of power in rural Zimbabwe*. Retrieved from www.researchadvocacyunit.org (accessed 15 March 2015).

Measham, T. G., & Lumbasi, J. A. (2013). Success factors for community-based natural resource management (CBNRM): Lessons from Kenya and Australia. *Environmental Management, 52*(3), 649–659. doi:10.1007/s00267-013-0114-9

Meinzen-Dick, R., & Knox, A. (1999). Collective action, property rights, and devolution of natural resource management, a conceptual framework. Paper to Workshop on Devolution of Natural Resource Management. Puerto Azul, Philippines, 21–25 June 1999.

Moser, C., & Felton, A. (2006). *Intergenerational asset accumulation and poverty reduction in Guayaquil, Ecuador 1978–2004*. Brookings/Ford Workshop, Asset-based Approaches, Conference paper: June 27–28, 2006.

Mpinganjira, M., Roberts-Lombard, M., & Svensson, G. (2017). Validating the relationship between trust, commitment, economic and non-economic satisfaction in South African buyer-supplier relationships. *Journal of Business & Industrial Marketing, 32*(3), 421–431. doi:10.1108/JBIM-04-2015-0073

Muboko, N., & Murindagomo, F. (2014). Wildlife control, access and utilisation: Lessons from legislation, policy, evolution and implementation in Zimbabwe. *Journal for Nature Conservation, 22*(3), 206–211. doi:10.1016/j.jnc.2013.12.002

Mudimba, T., & Tichaawa, T. M. (2017). Voices of local communities regarding their involvement in the tourism development process in Victoria Falls, Zimbabwe. *African Journal of Hospitality, Tourism and Leisure, 6*(4), 1–15.

Murphree, M. (2002). Protected areas and the commons. *Common Property Resource Digest, 60*, 1–3.

Musavengane, R. (2019a). Using the systemic-resilience thinking approach to enhance participatory collaborative management of natural resources in tribal communities: Toward inclusive land reform-led outdoor tourism. *Journal of Outdoor Recreation and Tourism, 25*, 45–56. doi:10.1016/j.jort.2018.12.002

Musavengane, R. (2019b). Land reform and the promotion of collaborative community-based ecotourism at Somkhanda Game Reserve, South Africa. In M. Mkono (Ed.), *Positive tourism in Africa. Contemporary Geographies of Leisure, Tourism and Mobility* (pp. 23–35). London: Routledge.

Musavengane, R., & Simatele, D. M. (2016). Community-based natural resource management: the role of social capital in collaborative environmental of tribal resources in Kwa-Zulu Natal, South Africa. *Development, 33*(6), 806–821. doi:10.1080/0376835X.2016.1231054

Musavengane, R., & Simatele, D. M. (2017). Significance of social capital in collaborative management of natural resources in Sub-Saharan African rural communities: A qualitative meta-analysis, *South African Geographical Journal, 99*(3), 267–282. doi:10.1080/03736245.2016.1231628

Musavengane, R., Tantoh, H. B., & Simatele, D. M. (2019). A comparative analysis of collaborative environmental management of natural resources in sub-saharan Africa: A study of Cameroon and South Africa. *Journal of Asian and African Studies, 54* (4), 512–532. doi:10.1177/0021909618825276

Mutana, S. (2013). Rural tourism for pro-poor development in Zimbabwean rural communities: Prospects in Binga rural district along Lake Kariba. *International Journal of Advanced Research in Management and Social Sciences, 2*(4), 147–164.

Mutandwa, E., & Gadzirayi, C. T. (2007). Impact of community-based approaches to wildlife management: Case study of the campfire Programme in Zimbabwe. *International Journal of Sustainable Development and World Ecology, 14*(4), 336–344. doi:10.1080/13504500709469734

Muzirambi, J. M., & Mearns, K. F. (2015). Active community participation in nature conservation and tourism management: A case study analysis of the state of power relations in South Africa. *Proceedings of BEST-EN Think Tank XV: The Environment People Nexus in Sustainable Tourism – Finding the Balance Conference*, 17–21 June, Kruger National Park, South Africa.

Muzirambi, J. M., & Mearns, K. F. (2018). Charity or Partnership? Striking a relational balance in wildlife conservation and ecotourism development. *Journal of Tourism and Hospitality Management, 6*(2), 80–92. doi:10.15640/jthm.v6n2a7

Nash, R. (2014). *Wilderness and the American mind* (5th ed.). New Haven, CT: Yale University Press.

Neuman, W. L. (2011). *Social research methods: Qualitative and quantitative approaches* (7th ed.). Boston: Allyn and Bacon.

Nowell, L., Norris, J., White, D., & Moules, N. (2017). Thematic analysis: Striving to meet the trustworthiness criteria, *International Journal of Qualitative Methods, 16*, 1–13.

Ntuli, H., & Muchapondwa, E. (2018). The role of institutions in community wildlife conservation in Zimbabwe. *International Journal of Commons, 12*(1), 134–169. doi:10.1016/j.jnc.2017.04.003

Ostrom, E. 1990. *Governing the commons: the evolution of institutions for collective action.* Cambridge University Press, Cambridge, UK.

Pegas, F., & Castley, G. (2014). Ecotourism as a conservation tool and its adoption by private protected areas in Brazil. *Journal of Sustainable Tourism, 22*(4), 604–625. doi:10.1080/09669582.2013.875550

Peterson, J. H. (1991). *CAMPFIRE: A Zimbabwean approach to sustainable development and community empowerment through wildlife utilization.* University of Zimbabwe, Harare: Centre for Applied Social Sciences.

Pigg, K. E. (1990). Accountability in Community Development. *Journal of the Community Development Society, 21*(1), 19–32. doi:10.1080/15575339009489969

Pomeroy, R. S. (1999). *Devolution in fisheries co-management.* Paper to Workshop on Devolution of Natural Resource Management. Puerto Azul, Philippines, 21–25 June 1999.

Ribot, J. C. (1999). Decentralization, participation and accountability in Sahelian forestry: Legal instruments of political-administrative control. *Africa, 69*, 23–65. doi:10.2307/1161076

Ribot, J. C. (2002a). *African decentralization: Local actors, power and accountability.* United Nations Research Institute for Social Development [UNRISD] Programme on Democracy, Governance and Human Rights Paper Number 8, Geneva, Switzerland: UNRISD.

Ribot, J. C. (2002b). *Democratic decentralization of natural resources: institutionalizing popular participation.* Washington DC: World Resources Institute.

Roe, D., & Nelson, F. (2009). The origins and evolution of community-based natural resource management in Africa. In D. Roe, D. R. Nelson, & C. Sandbrook (Eds.), *Community management of natural resources in Africa: Impacts, experiences and future directions* (pp. 5–12). London: IIED.

Ruhanen, L., Scott, N., Ritchie, B., & Tkaczynski, A. (2010). Governance: A review and synthesis of literature. *Tourism Review, 65*(4), 4–16. doi:10.1108/16605371011093836

Sacchetti, S., & Campbell, C. (2015). Creating space for communities: Social enterprise and the bright side of social capital. *Journal of Entrepreneurial and Organisational Diversity, 3*(2), 32–48. doi:10.5947/jeod.2014.012

Siakwah, P., Musavengane, R., & Leonard, L. (2019). Tourism governance and attainment of the Sustainable Development Goals in Africa. *Tourism Planning & Development.* doi:10.1080/21568316.2019.1600160

Snyman, S. (2017). The role of private sector ecotourism in local economic development in Southern Africa. *Journal of Ecotourism, 16*(3), 247–268. doi:10.1080/14724049.2016.1226318

Taylor, J. (2008). Community-based conservation: Trends and issues. In M. A. Rechlin & D. Taylor (Eds.), *Community-based conservation: Is it more effective, efficient and sustainable? Summary of scientific evidence relating to community-based conservation* (pp. 9–19). Future Generations Graduate School of Applied Community Change and Conservation, March, 2008.

United Nations Development Program (UNDP). (2002). *Human development report, 2002: Deepening democracy in a fragmented world.* New York: Oxford University Press.

Yin, R. K. (2003). *Case research: Design and methods* (3rd ed.). Thousand Oaks: Sage Publications.

Zou, T., Huang, S., & Ding, P. (2014). Toward a community-driven development model for rural tourism: The Chinese experience. *International Journal of Tourism, 16*(3), 261–271. doi:10.1002/jtr.1925

Part III

Conservation and tourism development debates in Southern Africa

Origins, narratives and progress

12 The economics and governance of the wildlife economy in drylands in Southern Africa

Brian A. Child

Introduction

The United Nations recently produced a stark warning about the unprecedented loss of nature, including an 82% decline in the biomass of wild mammals since 1970 (Diaz, Settele, & Brondizio, 2019). Protected areas, wildlife, and biodiversity are closely associated with rural people. Their survival is therefore intertwined with their inherent potential to become an important part of the rural economy, largely through hunting and tourism. The whole world will gain if rural people benefit enough from wildlife and tourism that they cannot afford to lose, not only for reasons of conservation but also to maintain viable rural livelihoods against the growing global risks of unprecedented migration and destabilization (Child, 1995b). The importance of Southern Africa to the debate about conservation, community, tourism and development is that it is a global outlier. Like the rest of the world, Southern Africa had lost most of its wildlife by the 1960s, especially on private land. However, unlike the rest of the world, with the exception of North America and small parts of northern Europe, wildlife has begun to recover despite major land use and demographic challenges (Child, 2019b). This chapter describes the policies and practices that led to these recoveries, but also notes that recovery is partial because policy reform has in most places been only partial, especially in rural communities.

Historically, Southern Africa was at the forefront of the loss of wildlife, with hunting, farming and fencing wiping out animals at a very similar scale to that associated with the American Bison. The magnificent but shocking historical account of this destruction of nature by man, compiled by John Pringle (1982), only makes the recovery story described in this chapter more powerful. Wildlife in Namibia declined from 8 million wild animals prior to European settlement to less than half a million animals by the 1960s (Brown, 2019), trends shared by other countries in the region. However, following radical policy reforms in the late 1960s, wildlife recovered to 3 million animals in Namibia (Brown, 2019), 6–18 million animals in South Africa (Bothma, Suich, & Spenceley, 2009; Taylor, Lindsey, & Davies-Mostert, 2016), and quadrupled in Zimbabwe in the 1990s. So far, trends have been less positive in high potential wildlife countries like Mozambique and Zambia where policies have been less bold.

Consequently, as one of the only places in the world where wildlife is recovering, Southern Africa provides a new policy model (Child, 2019b; Jones & Murphree, 2001). This model is based on reintegrating wildlife into the economy as a productive asset. By doubling or even quadrupling the income from land use in drylands, wildlife may well have saved some 10,000 private ranchers from financial and environmental bankruptcy (van Hoven, 2015). This success is now expanding to rural communities, with Southern Africa pioneering community-based natural resource management in the 1980s (Hulme & Murphree, 2001).

This chapter endeavours to clarify the economic basis of this recovery, describing the much-misunderstood differences and complementarities between the so-called consumptive and non-consumptive financial models. Secondly, it will unpack the radical policy reforms that made this recovery possible. And, finally, it will turn to the lessons of CBNRM and the need to ensure that rural communities are active participants in the tourism economy, not passive recipients.

Methods

In responding to criticisms of community conservation, Barrow and Murphree (2001) suggest that scholars are right in criticizing overly simplistic donor projects. However, they suggest that this criticism is unfair if aimed at the scholar-practitioners shaping sustainable use in Southern Africa. Indeed, they suggest that formal scholarship remains ignorant of decades of learning that reside in the oral and grey literature of these rich networks. Consequently, a significant attempt was made by a community of practice centered on the IUCN Southern African Sustainable Use Specialist Group to write down this knowledge (Suich & Child, 2009; Child, 2004; Nelson, 2010; Child & Jones, 2006; Hulme & Murphree, 2001). This chapter continues this tradition. It draws on three decades of interacting with, and being part of, these learning networks (which are still accessed sparingly by scholars), constructing a "literature" review from personal observation, interviews, discussion, workshops and oral and grey materials as much as from the more formal literature.

The wildlife economy: tourism, hunting and wildlife products

The popular press is filled with controversy about trophy hunting and wildlife products, and suggestions that photo-tourism can easily substitute hunting to pay for the recovery of wildlife. This understanding is incorrect and is not shared by land managers or communities. In this section, we dissect the potential of photo-tourism (tourism from now on), hunting and wildlife products for financing wild landscapes and rural development.

Park-based tourism can be a powerful engine for economic growth. Recent studies show that tourism generated a total economic impact of R 6.6 billion in the greater Kruger National Park. In South Luangwa National

Park in Zambia, tourism generated US$38 million in economic activity, over US$6 million in taxes and several thousand jobs. It also doubled the average income of some 9,000 rural people living near the park gate and resulted in rapid commercial development in this community (Chidakel & Child, in review). Tourism is undoubtedly an important mechanism for funding many national parks, but this requires major investments in communications, infrastructure and park management, without which most of the parks in the region are very far from being viable through tourism. Indeed, only about ten of Southern Africa's hundred or so state-owned parks and protected areas are performing at anything like this level, and even within these parks tourism is highly concentrated (Bothma et al., 2009; Child, Musengezi, Parent, & Child, 2012). Like a shopping mall, tourism is a cluster industry that generates a lot of revenue from a highly concentrated portion of the real estate. Tourism is good at generating income and jobs. However, tourism is not a good tool for conserving vast landscapes because, geographically, prime areas become "over-loved" (Simmonds et al., 2018) while tourists shun the rest of the landscape. Hunting is a much more effective tool for rewilding large areas of land, especially park buffer zones and private and community lands. For example, following the hunting ban in Botswana, vast landscapes were abandoned because tourism could not use them. Wildlife and local people suffered without the water, protection from poachers, meat or money provided by hunting outfitters.

The controversy over hunting and elephants in Botswana (Verreynne, 2019) has also contrasted the luxury of international tourism lodges with the poverty of communities sharing these lands. Most tourism strategies were designed decades ago, and there is a need to actively reorient tourism as a tool for financing parks, promoting rural development and, not least, targeting the emerging middle-class African market.

The controversy and misperceptions about the wildlife economy are often rooted in the failure to differentiate between wildlife inside national parks and wildlife outside parks where land has been set aside for economic development, not conservation. The major driver of the loss of wildlife and wild lands has been the tripling of agricultural lands globally (Diaz et al., 2019). This trend has been reversed in Southern Africa. New policies recognise that public wildlife would not survive on private land and reversed colonial policies by devolving the ownership of wildlife to private and community landholders (Suich & Child, 2009). Wildlife markets were encouraged and developed to ensure that wildlife living in intact habitats became more profitable, more so than converting these habitats into spaces for domestic crops and animals. Landholders showed that well-managed hunting earns two to four times as much as using the same habitats to produce meat, even before accounting for ecosystem services like carbon, water and biodiversity (Child et al., 2012). Moreover, the environmental recovery of many wildlife properties suggests that a spectrum of indigenous wild species is ecologically more sustainable than an exotic monoculture, as well as more profitable.

The rapid recovery of wildlife outside parks in Southern Africa has been financed disproportionately or entirely by hunting. In Zimbabwe, for example, private wildlife ranches and CAMPFIRE areas doubled Zimbabwe's conservation estate, but over 95% of the financing for these areas was derived from hunting (Bond, 1994). In South Africa, there are four times the wildlife on private land as in protected areas, with roughly 10,000 wildlife properties conserving some 20 million hectares and millions of wild animals. According to DEA (2015), the private wildlife conservation sector grew at between 6–20% annually between 2005 and 2013. Ecotourism generated only 4% of the income for this R 10 billion sector, with the majority coming from hunting. By comparison, park-based tourism in South Africa earns R 4.1 billion annually, of which just under half comprises in-park revenues like accommodations (23%), park fees (13%), and guiding services (4%), while more than half is generated by off-park activities like food and transport (54%) and travel agents (6%). Likewise, banning hunting can have extremely negative effects on communities and habitats, including villages in Botswana (Mbaiwa, 2018; Onishi, 2015).

Building a wildlife economy requires moving beyond the unsubstantiated and emotional discussions about tourism versus hunting to understand their strengths, weaknesses and synergies. Tourism can generate large revenues and a significant number of jobs, but in surprisingly few tourist destinations such as Kruger National Park, the Okavango Delta, Victoria Falls, and the Luangwa Valley that have a specific set of features – exceptional wildlife and scenery, a significant investment in infrastructure and a well-accepted brand. By contrast, hunting is viable across a broad swath of the remote, uncomfortable locations and harsh landscapes that comprise roughly 80% of Southern Africa's wildlife estate. Hunting also introduces economic activity into extremely remote rural areas with few other options. Private landholders in Botswana are responding to the hunting ban by destocking wildlife and reintroducing livestock, confirming that the emotional cries to ban hunting would put at risk approximately 80% of the region's wildlife. Armchair conservationists in distant capital cities forget that hunters are some of the most passionate and well-informed conservationists who protect many of the buffer zones around parks and tourism areas. Hunting plays a critical role in financing the recovery of wildlife areas, until such time as tourism can pay more (in a few rare cases). Indeed, the massive rewilding of Southern Africa was financed almost entirely by hunting, especially on private and community land, but even in parks like Pilanesburg where carefully zoned hunting provided a substantial contribution to the park budget while the tourism product was still being developed. The beauty of hunting as an economic tool is that selling a small number of high-quality animals and expensive outfitting services generates good profits while also allowing wildlife populations to expand rapidly. Hunting harvests only 1–3% of a herd that can grow at 15–30% annually; for every ten buffalo calves born each year, for instance, only one adult bull is shot.

Thus, hunting is the true high-revenue low-impact use. Meat production is the reverse, compensating for the relatively low value of meat (compared to hunting trophies or tourism) with high offtake levels. As a stand-alone production system, game meat production and indeed meat production generally is unsustainable in drylands (Child et al., 2012): cattle ranching has declined except where it is subsidised; it is difficult to name a single enterprise that depends entirely on game meat, and bushmeat harvesting is notoriously problematic. Nonetheless, game meat has the potential to become an important and healthy by-product of the wildlife economy because the harvesting of excess game animals is an inevitable consequence of expanding hunting and tourism economies. Moreover, there is a need to exploit the ethical, health and land-use benefits of wild meat, especially given the problematic expansion of factory farming (Lymbery, 2017).

The development of effective wildlife economies is, however, threatened by shrill attacks in the media, including banning trade in valuable wildlife products and undermining hunting. This leaves a few data points to understand the value of wildlife products, which are a natural by-product of successful conservation. At the African Wildlife Economy Summit, President Mnangagwa stated Zimbabwe had US$600 million of stored ivory and rhino horns, enough to fund the entire parks and wildlife estate for 20 years, but far less than the value of South African and Namibian rhino horns. Yet these massive numbers only reflect the value of two raw products. Zimbabwe culled 46,775 elephants between 1960 and 1990 but failed to protect the vegetation because elephant numbers increased from 57,000 in 1984 to 76,000 by 1991. Nonetheless, this experience showed that raw hide was worth more than ivory (Child, 1995b), with a luxury leather manufacturing industry multiplying this value many fold through jobs and foreign currency. Like leather, by far the greatest value of ivory lay in ornamental carving, increasingly to the standard of collectable art.

The effects of tourism and political correctness on communities and ecosystem management are also getting far too large to ignore. Tourists favour charismatic species (especially predators and elephants), and detest "management," a euphemism for killing animals. The effect of too many elephants in rural communities in Botswana is illustrated by the recent documentary, *Voices from the Frontline* (https://vimeo.com/343453859). Data from Madikwe Game Reserve hints at the ecological consequences of such anthropomorphic management regimes (Davies, 2013). With a doubling time of about 12 years, elephants tripled from 30% of the biomass in 1995 to 66% in 2012, far too many for the vegetation. Over the same period, predator numbers (lions, wild dogs) more than doubled. Squeezed between elephants (which ate the vegetation) and predators (which ate them), prey biomass crashed from 2 million to 500,000 kilograms, maintained only through regular restocking of "lion food," which unfortunately introduced disease, including tuberculosis, which crashed the value of disease-free buffalo. These problems are widespread in tourism areas where hunting is not allowed, as we see by the substantial loss of vegetation and herbivore species, even in national parks as large as Chobe and Hwange (Valeix et al., 2007; Child, 2019a).

Replacing colonial policies of public ownership with the sustainable governance approach

There is a stark contrast between the rapid expansion of wildlife in Southern Africa and global wildlife decline. Southern Africa recognised that colonial wildlife policies were misaligned and outdated, and changed them. Colonial wildlife policies originated in North America in the early 1900s when President Theodore Roosevelt and his colleagues responded to the demise of wildlife on the colonial frontier by banning market hunting and declaring wildlife (and often fisheries and forests) to be public resources, often on public land. This most capitalist of countries, ironically, turned wildlife into a socialised good by nationalising it. These policies then spread to the European colonies through the London Conventions of 1900 and 1933.

Outside North America's peculiar circumstances, these policies have been catastrophic for wildlife. Private landholders have replaced publicly owned wildlife with privately owned livestock on a global scale. This was not because wildlife was less valuable than livestock; indeed, subsequent research has shown that wild species and intact habitats are often more valuable than domesticated ones (MEA, 2005; TEEB, 2009). Wildlife disappeared because of differential ownership: Farmers looked after domestic species that they owned, and replaced wild species that they did not own. By the 1960s farsighted wildlife administrators, my father included (Child & Child, 2017), recognised this problem. One of the features of Southern Africa is that wildlife managers and administrators have been working together for many decades. Starting in the 1960s, wildlife administrators from Zimbabwe, Namibia, Mozambique, Botswana, South Africa and Swaziland met regularly and came up with a solution to this problem. As circumstances in their countries allowed, they boldly reformed wildlife policies in two major ways. In a strategy that combines proprietorship and price, they devolved the ownership of wildlife to landholders and made wildlife as valuable as possible by encouraging markets and trade, and reducing the burden of permits and government fees.

Legally, the most sophisticated approach was in Zimbabwe. A farmer from near Victoria Falls put the colonial concept of the King's Game out of its misery (Child, 1995b). He was prosecuted for shooting an endangered roan antelope on his land but argued to the High Court that if the state claimed ownership of wildlife, it was also liable for the costs imposed by wildlife, including using his land for grazing (Beadlem & Macdonald, 1969). He won his case, making the old Wildlife Conservation Act of 1960 unworkable, because government was rendered liable for costs associated with wildlife, including death, and wildlife legally included species like quelea, rodents and even locusts.

Waiting for such an opportunity, wildlife administrators wrote the path-breaking Parks and Wildlife Act of 1975. This set up six categories of protected areas nearly 20 years before IUCN came up with very similar definitions. It also radically changed the ownership status of wildlife. Wildlife was defined as *res nullius* (i.e. owned by no-one), but by conferring "appropriate authority" status on private landholders, it devolved to them full proprietorship (Child, 1995b), including the rights to use, manage, sell and exclude others from wildlife on their land (Schlager & Ostrom, 1992).

However, some species of wildlife are mobile or fugitive and move beyond the boundaries of individual properties. South Africa and Namibia solved this problem by linking wildlife ownership to game fencing, with the serious problem that fencing is expensive and causes habitat fragmentation. Determined to avoid fencing, Zimbabwe used collective action. The Natural Resources Act of 1941 had already empowered neighbourhoods of about 50 farmers in a sub-catchment to form democratic associations called Intensive Conservation Areas (ICA). Once an ICA was formed, the government devolved to it strong legal powers of self-regulation. Low-cost mechanisms like local knowledge and peer pressure (Ostrom, 1990), rather than state coercion, became the primary mechanisms for preventing or repairing soil erosion, deforestation, and so on, backed up by the ICA's powers to take legal actions or impose financial sanctions, something that was rarely necessary. Democratic, self-regulating ICAs were so popular and effective that 400 ICAs soon blanketed the country. Government provided assistance in the form of a conservation and extension officer, and oversight through an annual visit and joint inspection overflight with the Lands Inspectorate.

ICAs had clear boundaries. All landholders affected by the rules could participate in making them, so rules were fair, parsimonious and effective. Farmers monitored each other's soil erosion, grazing, forests and wildlife. Graduated sanctions were applied where individuals caused environmental damage and broke agreements; for instance, the ICA could force a member to pay for fixing soil erosion, failing, which it could, to fix the erosion itself and bill the recalcitrant farmer. There were rapid and low-cost mechanisms for conflict resolution including government facilitation (the extension officer) and a Natural Resource Court. Policy and legislation deliberately empowered landholders not only with property rights but also with collective self-regulation (Child & Child, 2015). Fifty years later, these very same design principles for local commons were codified by Elinor Ostrom (1990).

As we seek to design conservation institutions for the 21st century, it is worth noting that the ICAs provided the democratic foundation of the Natural Resources Board, along with civil society. The democratic Natural Resources Board was the primary agency responsible for conservation in Zimbabwe between 1941 and 1977, not a government agency. Indeed, politicians including the prime minster, and government conservation agencies, were legally answerable to it, not the other way around. We have become so used to government leadership that we have forgotten that other mechanisms are available and can be cheaper and more effective.

Governing wildlife as a private good with common pool properties

Wildlife ownership is more complex than the ownership of domestic species. To understand the theoretical elegance of Zimbabwe's solution (designed by farmers and not academics), we need to define private, common pool and public goods. Following Ostrom and Hess (2007), a private good is excludable and subtractable, and the free market is effective at allocating them. By contrast, a public good such as a mountain view, a street light or a justice system cannot be individualized and privatized (it is non-excludable), and its use by one person does not subtract from its use by another (i.e. it is non-subtractable and gets used up). Common pool resources lie somewhere in between, being partly excludable and partly subtractable. Ostrom (1990) pointed out that local collective action was often much better at managing common pool resources than either the private market or the public state.

Roosevelt and the European colonial powers legally defined wildlife as a public asset. However, contemporary wildlife is not a public good. It gets used up (it is subtractable). Its rising value combined with technology increasingly allows an owner to exclude a non-owner from taking it (it is excludable). Wildlife, theoretically, is a private good (it is excludable and subtractable), albeit with some common pool properties, such as when a kudu or an elephant crosses to a neighbour's property. It is not a public good, although it does provide public benefits, including option and existence values (Krutilla, 1967).

Free market economics is based on the idea that bargaining over private resources will allocate resources in the most efficient way by internalizing costs and benefits, provided transaction costs are low (Coase, 1960). This works for simple resources like privately owned plants and animals. However, Zimbabwean legislation recognised that wild species are more complex and came up with an elegant solution:

- Reflecting the private nature of wildlife, the Parks and Wildlife Act of 1975 privatized wildlife at the level of a landholder and, later, a community. This internalised the majority of costs and benefits associated with wildlife.
- Reflecting the common pool properties of wildlife, the Natural Resources Act of 1941 empowered communities of landholders to collectively manage externalities, as proposed by Ostrom (1990).

The purpose of this legislation was not to conserve wildlife per se but to allocate natural resources outside parks in the best way possible by combining markets and collective action to internalizing costs and benefits, including negative externalities like deforestation and soil erosion.

The importance of rights, and especially community rights, for forest and wildlife conservation is emerging in the literature (Veit & Ding, 2016; Pearce, 2016). It is increasingly obvious that rights (i.e. proprietorship) are a necessary condition for sustainability, including incentive-led wildlife conservation.

However, the understanding of markets, trade and the nature of regulation is muddled. With a long history of demonetizing wildlife and public governance, western conservationists often get confused about markets and promote both payments for ecosystem services (which increased the price of wild areas) and demand reduction (which reduces the price of wildlife).

Markets and trade play a critical role in increasing the price of wildlife so that it can compete economically for land. Adam Smith demonstrated that wealth is created through the division of labour, specialization, and diversification and that this requires exchange and trade – Adam Smith's highly productive pin maker cannot eat pins. For these exact reasons, Southern Africa fights hard at CITES to promote legitimate wildlife trade because prices for wildlife products (tourism, hunting, products) are many fold higher in a global market than if restricted to local uses. The competitiveness of wildlife also requires rethinking and reversing conventional bureaucratic practices such as subsidized access to hunting and tourism in parks, administrative pricing, licensing fees on hunting and wildlife regulations that increase the costs of doing business. These all lower the value of wildlife, undermining the objective of 'maximising the value of wildlife to the person living with it.' This is why Zimbabwe did away with all wildlife fees on private and community lands, and most regulations as well (Child, 1995b).

Southern Africa's wildlife policies follow the proprietorship-price hypothesis. If landholders or communities own wildlife (proprietorship), and it is valuable (price), they are likely to increase it; if neither is true, they will replace it with other species, usually domestic. The sustainable governance approach broadens this theory by emphasising subsidiarity and social adaptive management (Child, 2019b). Subsidiarity refers to the principle that higher levels in a hierarchy should always serve lower levels (Meadows, 2008) and in practice is reflected in the devolution of ownership of wildlife to the lowest level. Adaptive management, based on social interactions around quality information and data, is necessary in an uncertain and constantly changing world.

The practice of private wildlife governance in Southern Africa still lags behind these theoretical ideas, and progress has in some cases been reversed (e.g. by Zimbabwe's Parks and Wildlife Authority Act of 2014). This new approach has nonetheless resulted in substantial rewilding of the region and political acceptance of the importance of a wildlife economy that was unthinkable four decades ago. Together, this makes a powerful case that wildlife is a private good with common pool attributes, not a public good that provides a wide range of benefits when its ownership is devolved to landholders and communities.

Community-based natural resource management

By the 1980s, the expanding wildlife economy on private land wildlife drew a stark contrast with the loss of wildlife in community areas. Wildlife administrators partially replicated private conservation policies on communal lands, leading to the emergence of community-based natural resource management, pioneered by CAMPFIRE in Zimbabwe. Once again, the region worked together and learned from each other, with wildlife-based CBNRM emerging in Botswana, Mozambique, Zambia and, most strongly, in Namibia.

CBNRM is conceptually very similar to private conservation, except that the land unit is a community rather than an individual (Child, 2004). In addition to the principles of proprietorship and price, therefore, communities need to be organised as well-governed and productive land units, ideally as a village company with land and resource title where members have legal rights as shareholders.

CBNRM has not yet progressed this far for a number of reasons. Historically, communities were exploited by denying them land or resource rights (Child, 2019b). Colonial regimes blocked the emergence of land and resource rights in communities to enable them to extract resources without paying for them, to force rural people into the labour market in mines and towns and to maintain indirect rule through the chiefs (Acemoglu & Robinson, 2012). Not much has changed. To this day, rural communities own neither their land nor their wild resources. Communal lands are accepted as normal, but this is far from true. They perpetuate historical disempowerment, together with a tragedy of the commons and highly negative economic and environmental outcomes. Moreover, the failure to grant communities the same land and resource rights that are enjoyed by the modern sector, including title, prolongs this injustice and is the underlying cause of extensive rural poverty. There is a widespread reluctance by the modern African state to give rural people the same rights as their urban cousins. Indeed, Nobel Laureate Douglass North argues that a major cause of under-development are extractive and modern elites who empower and enrich themselves by denying rights to ordinary citizens, including rural communities, and then acquiring these rights for themselves (Menard & Shirley, 2014). Similarly, Somerville (2015) blames under-development in a gatekeeper state which acquires wealth through rent seeking and the control of access to people and resources, rather than creating wealth and taxes by framing the conditions for the public good. It is common to witness the dynamics of extractive and authoritarian governance in rural areas. I recently witnessed the efforts, energies and rights of a highly motivated community being undermined by government officials who feared that empowering the community would disempower themselves. This leads to blatant iniquity, with many examples of foreign investors being given land and lodge sites on community land, while denying these very same rights to communities.

CBNRM is designed to rectify this situation by replacing extractive authoritarian systems of governance inherited from the colonial period with inclusive, democratic governance aligned both with the participatory nature of African villages and the liberal, democratic norms that underlie the modern age of prosperity.

The performance of CBNRM across the region mirrors how well the principles of proprietorship, price and inclusive governance are applied. It is no coincidence that the strongest CBNRM programme is in Namibia, where communities retain 100% of the revenues from wildlife just as they do for domestic species like cattle and crops. By contrast, CBNRM is struggling where the benefits are retained centrally, with little or nothing reaching neither the community nor individual community members.

In the past 30 years, the growth in the capacity and commitment of communities to wildlife conservation has been remarkable and often exceeds that of line ministries. This growth in technical, and organizational capital is a natural consequence of the determination of some governments to devolve rights to communities. This passion and capability were on full display at the June 2019 African Wildlife Economy Summit, where communities made a powerful case for owning, conserving and trading wildlife. Their Declaration, "Voices of the Communities: A New Deal for rural communities and wildlife and natural resources," reflects their determination to turn wildlife into a rural economic engine, laying out the requirements for this, including the devolution of rights, good governance, participation, investment and evidence-based management. Unfortunately, in August in Geneva, the parties of CITES will be doing the exact opposite – striving to take away the rights of communities to own, manage and trade the wildlife with which they live.

The second issue is price, and the ability of wildlife and intact habitats to compete economically with users who transform the land. As discussed above, the price of wildlife depends on access to global markets, administrative constraints such as regulation, red tape and license fees, and on product and value chain development. CITES and wildlife trade bans are currently preventing rural people from maximizing the value of their wildlife in the global marketplace, with hundreds of millions of dollars of wildlife products currently lying unused in vaults, safes and on the hoof. The economic potential of CBNRM communities also continues to be undermined by the reluctance to trust rural people to set offtake quotas and negotiate deals. Interestingly, communities in both CAMPFIRE and Namibia quickly negotiated far better deals with the private sector than the government was able to make on their behalf (Child, 1995a; NACSO, 2018). Yet in many places the value of wildlife continues to be reduced where governments insist on negotiating on behalf of communities. It is also reduced by bureaucratic constraints such as excessive regulations and requirements. How successful would cattle ranching be if the government set the quota for offtakes, did the marketing on behalf of the farmer and imposed a plethora of regulations, licenses and fees?

The third challenge concerns within-community governance and the flow of money and decision-making within communities. Far too little consideration is given to the design of the institutions for micro-governance or the differences between representational (committee-based) and participatory (community-based) forms of governance. There is a marked tendency to perpetuate authoritarian norms and top-down committees and managers, compared to face-to-face democratic accountability and participation, which are associated with representational forms of governance and committee-based management and associated with corruption, financial mismanagement and elite capture, with very little benefit sharing or participation by ordinary people occurring in communities of over 300–400 members (Child, 2019b); the rich get richer and the poor get committees (Anderson, Mehta, & Schwarz, 2013).

These problems can be avoided through such simple measures as defining communities at the scale of a village following Dunbar's number (150–250 members) and ensuring that all people participate in all decisions and review them regularly (Child & Wojcik, 2014). It is surprising how often implementers forget Elinor Ostrom's core rule that effective common property regimes require that all people affected by decisions participate in making them (Wall, 2017). In practice, this means that decisions need to be taken and reviewed regularly by the whole community meeting face to face. It defines the role of leaders as bringing together people and information to make informed and participatory choices, not to make decisions on their behalf. Data from Southern Africa shows that in small communities with participatory governance, more than 60% of wildlife revenue benefits communities in the form of cash payments or useful projects, whereas in large communities with committee-based governance less than 10% does (Child, 2019b). These profound differences result from small changes in the design of local institutions, and we need to pay attention to these issues.

Conclusions

This author recently (24–25 June 2019) attended the African Wildlife Economy Summit in Victoria Falls along with four presidents, numerous ministers, park managers, game farmers, tourism operators and, above all, local communities. This author has been in many wildlife conferences in the past, but this was the first one where the narrative was set and dominated by Africans, with community members, especially prominent Africans, who displayed a passion for wildlife conservation and sustainable use, technical competence in how to achieve this and a strong demand that the growth of the wildlife economy be accelerated by devolving rights to communities more completely. Many years ago, an esteemed colleague of this author, James Murombedzi, commented that all the progress we were making in CBNRM at the time would come to naught unless it evolved into a political and social movement. At the African Wildlife Economy Summit, this author personally witnessed the visible emergence of this social movement.

The devolution of proprietorship to landholders and communities over many years has more than quadrupled wildlife populations on private land (Carruthers, 2008), often saving cattle ranches from bankruptcy. In some places, it has slowed or reversed the loss of wildlife on communal lands, often against great odds (Taylor, 2009; NACSO, 2018). Success is predicated on four factors: the genuine devolution of rights and benefits – rather than paying lip service to community conservation – the development of wildlife markets, sound governance and evidence-based learning.

There are increasing buy-ins to wildlife conservation at the level of landholders and communities that is rare elsewhere. There is also, as hinted above, the beginning of the virtuous circle that characterizes inclusive governance (Acemoglu & Robinson, 2012). Communities, benefiting from some rights, are demanding more from their own governments. They are challenging the international community and mechanisms like CITES that are truncating their rights to use and trade in wildlife without informed consent. In standing up for their rights, they are now opposing the globalisation of a wildlife resource that is rightfully theirs. Global conservation benefits will be achieved by strengthening local commons (i.e. private community ownership), not by globalising resources that are rightfully local.

This author proposes that Southern Africa seek to develop an inclusive US$30 billion wildlife economy by 2030, with this ten-fold expansion driving similar gains in wild lands, wildlife numbers and rural livelihoods. This will require investments in infrastructure, substantial improvements in the management of parks and further devolution to private landholders. However, the greatest winners could be rural communities because they own the most land and have the most potential to improve. This New Deal requires that communities are shareholders, not stakeholders, acquiring full rights to own land and wildlife, including rights to use, manage, sell and exclude others from their resources (Schlager & Ostrom, 1992). Communities that combine these rights with high standards of internal governance (possibly formalized as village companies in which members are equal shareholders) are likely to attract high levels of investment, leading to employment and ownership in rapidly expanding tourism, hunting and product businesses. These businesses will depend on rehabilitating the wildlife spectacles and huge landscapes of northern Botswana, the Luangwa Valley, and the Kavango-Zambezi and Limpopo Transfrontier Conservation Areas, with Southern Africa becoming a global tourism destination that combines beaches, wildlife, hospitality and culture.

References

Acemoglu, D., & Robinson, J. (2012). *Why nations fail: The origins of power, prosperity, and poverty*. New York, Crown Business.

Anderson, J., Mehta, S., & Schwarz, J. (2013). The rich get richer and the poor get committees? Institutional arrangements and requirements for local forest management in Africa. *Nature & Faune*, 27, 26–30.

Barrow, E., & Murphree, N. (2001). Community conservation. From concept to practice. In D. Hulme & M. Murphree (Eds.), *African wildlife and livelihoods. The promise and performance of community conservation*. Oxford: James Currey Ltd.

Beadlem, C. J., & Macdonald, J. A. (1969). R. v. Moresby-White. Criminal law – Wild Life Conservation Act [Chapter 199] – s. 41 – construction of – right of private landowner to kill animals on property discussed. *Rhodesian Law Reports*, 484–490.

Bond, I. (1994). *Importance of elephant hunting to CAMPFIRE revenue in Zimbabwe*. Harare: WWF- Southern African Regional Programme Office.

Bothma, J. D. P., Suich, H., & Spenceley, A. (2009). Extensive Wildlife Production on Private Land in South Africa. In H. Suich & B. Child (Eds.), *Evolution & innovation in wildilfe conservation*. London: Earthscan.

Brown, C. (2019). Natural resource competitiveness as a driver for sustainable development. *Southern African Wildlife Thought Leaders Meeting*. Namibia.

Carruthers, J., & Arruthers, J. (2008). 'Wilding the farm or farming the wild'? The evolution of scientific game ranching in South Africa from the 1960s to the present. *Transactions of the Royal Society of South Africa, 63*, 160–181.

Child, B. (1995a). *A summary of the marketing of trophy quotas in CAMPFIRE areas 1990–1993*. Harare: Department of national parks and Wildlife Management, Zimbabwe.

Child, G. (1995b). *Wildlife and People: The Zimbabwean success. How the conflict between animals and people became progress for both*. Harare: Wisdom Foundation.

Child, B. (2004). *Parks in transition. Biodiversity, rural development and the bottom line*. London: Earthscan.

Child, B. (2019a). Botswana – A crisis to too many, not too few. *Mmegionline (Botswana)*. Retrieved from https://africasustainableconservation.com/tag/brian-child/. Dikgang Publishing Company.

Child, B. (2019b). *Sustainable governance of wildlife and community-based natural resource management*. London: Routledge.

Child, B., & Child, I. (2017). Graham Child, and the emergence of a new paradigm of wildlife conservation. *Pachyderm, 58*, 173–174.

Child, B., & Jones, B. (Eds.). (2006). *Practical tools for community conservation in Southern Africa: special edition*. London: International Institute for Environment and Development (IIED).

Child, B., Musengezi, J., Parent, G., & Child, G. (2012). The economics and institutional economics of wildlife on private land in Africa. *Pastoralism Journal, 2*, 1–32.

Child, B., & Wojcik, D. (2014). *Developing capacity for community governance of natural resources: Theory & practice*. Bloomington: AuthorHouse.

Child, G., & Child, B. (2015). The conservation movement in Zimbabwe: An early experiment in devolved community based regulation. *Journal of Wildlife Research, 45*, 1–16.

Coase, R. H. (1960). The problem of social cost. *The Journal of Law and Economics, 3*, 1–44.

Davies, R. (2013). *Madikwe game reserve management plan*. North West Parks and Tourism Board.

Department of Environmental Affairs (DEA). (2015). *Situation analysis of four selected subsectors of the biodiversity and conservation sector in South Africa, and transformation framework*. Pretoria: Department of Environmental Affairs, South Africa.

Diaz, S., Settele, J., & Brondizio, E. S. (2019). *Summary for policymakers of the global assessment report on biodiversity and ecosystem services of the Intergovernmental Science-Policy Platform on Biodiversity and Ecosystem Services*. IPBES.

Hulme, D., & Murphree, M. (2001). *African wildlife & livelihoods. The promise and performance of community conservation*. Oxford: James Currey.

Jones, B., & Murphree, M. (2001). The evolution of policy on community conservation in Namibia and Zimbabwe. In D. Hulme & M. Murphree (Eds.), *African wildlife & livelihoods. The promise and performance of community conservation*. Oxford: James Curry.

Krutilla, J.V. (1967). Conservation reconsidered. *The American Economic Review, 57*, 777–786.

Lymbery, P. (2017). *Dead zone: Where the wild things were*. London, Oxford, New York, New Dehli, and Sydney: Bloomsbury Publishing.

Mbaiwa, J. E. (2018). Effects of the safari hunting tourism ban on rural livelihoods and wildlife conservation in Northern Botswana. *South African Geographical Journal, 100*, 41–61.

Mea. (2005). *Millennium ecosystem assessment. Ecosystems and their services*. Washington, DC: Island Press.

Meadows, D. H. (2008). *Thinking is systems. A primer*. London: Earthscan.

Menard, C., & Shirley, M. M. (2014). The contribution of Douglass North to new institutional economics. In S. Galani & I. Sened (Eds.), *Institutions, property rights and economic growth: The legacy of Douglass North*. Cambridge: Cambridge University Press.

NACSO. (2018). The state of community conservation in Namibia. *Annual Report 2017*. Windhoek: Namibian Association of CBNRM Support Providers.

Nelson, F. (Ed.). (2010). *Community rights, conservation and contested land: The politics of natural resource governance in Africa*. London: Earthscan.

Onishi, N. (2015). A Hunting Ban Saps a Village's Livelihood. *New York Times*, September 12 2015.

Ostrom, E. (1990). *Governing the commons: The evolution of institutions for collective action*. Cambridge: Cambridge University Press.

Ostrom, E., & HESS, C. (2007). *Private and common property rights*. Bloomington: Workshop in Political Theory and Policy Analysis, Indiana University.

Pearce, F. (2016). *Common ground: Securing land rights and safeguarding the earth*. Oxfam, International Land Coalition, Rights and Resources Initiative.

Pringle, J. (1982). *The conservationists and the killers. The story of game protection and the Wildlife Society of South Africa*. Cape Town: Books of Africa.

Schlager, E., & Ostrom, E. (1992). Property-rights regimes and natural resources: A conceptual analysis. *Land Economics, 68*, 249–162.

Simmonds, C., Mcgivney, A., Reilly, P., Maffly, B., Wilkinson, T., Canon, G., . . . Whaley, M. (2018). Crisis in our national parks: How tourists are loving nature to death. *The Guardian*.

Somerville, K. (2015). *Africa's long road since independence*. Hurst and Co.

Suich, H., & Child, B. (Eds.). (2009). *Evolution & innovation in wildlife conservation. Parks and game ranches to transfrontier conservation areas*. London: Earthscan.

Taylor, A., Lindsey, P., & Davies-Mostert, H. (2016). *An assessment of the economic, social and conservation value of the wildlife ranching industry and its potential to support the green economy in South Africa*. Johannesburg: The Endangered Wildlife Trust.

Taylor, R. (2009). Community based natural resource management in Zimbabwe: The experience of CAMPFIRE. *Biodiversity and Conservation, 18*, 2563–2583.

Teeb. (2009). The economics of ecosystems and biodiversity: Climate issues update. Retrieved from www.teebweb.org/media/2009/09/. The Economics of Ecosystems and Biodiversity.

Valeix, M., Fritz, H., Dubois, S., Kanengoni, K., Alleaume, S., & Saïd, S. (2007). Vegetation structure and ungulate abundance over a period of increasing elephant abundance in Hwange National Park, Zimbabwe. *Journal of Tropical Ecology, 23*, 87–93.

Van hoven, W. (2015). Private game reserves in Southern Africa. In R. Van Der Duim, M. Lamers, & J. Van Wijk (Eds.), *Institutional arrangements for conservation, development and tourism in eastern and Southern Africa. A dynamic perspective*. London: Springer.

Veit, P., & Ding, H. (2016). *Protecting indigenous land rights makes good economic sense*. Washington, DC: World Resources Institute.

Verreynne, F. J. (2019). *Botswana and lifting the hunting ban. Your Botswana*. Gabarone, Botswana: Your Botswana.

Wall, D. (2017). *Elinor Ostrom's rules for radicals. Cooperative alternatives beyond markets and states*. London: Pluto Press.

13 Role of forest resources in local community livelihoods

Implications for conservation of Chobe Forest Reserve, Botswana

Joyce Lepetu and Hesekia Garekae

Introduction

For centuries, the majority of people from developing countries have been immensely dependent on environmental resources, particularly forests (Mukul, Rashid, Uddin, & Khan, 2016). On that account, forests are considered a vital resource for human well-being, more especially for communities living adjacent to and around forested areas (Córdova, Wunder, Smith-Hall, & Börner, 2013; McElwee, 2010; Uberhuaga, Smith-Hall, & Helles, 2012). Forests provide a vast array of ecosystem goods and services to humans and the natural environment (Howe, Suich, Vira, & Mace, 2014). Ecosystem services are classified into four typologies: provisioning, regulating, cultural and supporting services (Millennium Ecosystem Assessment, 2005). Contrary to the "fortress" conservation approaches of establishment of protected areas, which were mainly skewed towards pristine preservation of resources and characterised by alienation of local people (Wells, Brandon, & Hannah, 1992), scholars argued for alternative integrative management approaches which recognise that forests and human needs are inextricably intertwined (Ros-Tonen, 2000). On that note, non-timber-oriented forest management approaches were promoted as a viable strategy for conservation because of their arguably low detrimental effects on the forest structure (Ros-Tonen, 2000). These approaches recognise the contribution of forests in human's sustenance juxtaposed with their role in conservation of the same resource. According to the World Bank (2004), forests provide support for nearly half of the 2.8 billion people living on US$2 or less a day. In that light, a growing body of empirical evidence has demonstrated the functional role played by forests in rural livelihoods.

Globally, numerous studies have quantified the contribution of forests to rural household economies. A meta-analysis of 51 case studies from 17 developing countries revealed that forest-derived income on average accounted for 22% of the total household income (Vedeld, Angelsen, Bojö, Sjaastad, & Berg, 2007). In North and South America, studies conducted revealed that annual forest income ranged from 14% to 20% of the total household income (Córdova et al., 2013; Uberhuaga et al., 2012). Studies conducted in Asia found that communities were strongly dependent on forests for their livelihood sustenance and overall well-being. Gautam (2006) stated that over 95% of the Nepalese population depends on forests for both timber and non-timber forest products (NTFPs). In India, 400 million people were estimated to be reliant on non-timber forest products (NTFP's) (MoEF Report 1999 cited in Bhavannarayana, Saritha, Usha, & Rao, 2012), while in Bangladesh, NTFPs cash benefits varied from 10% – 20% of the total household income (Mukul et al., 2016). Likewise in sub-Saharan Africa, findings from Mamo, Sjaastad, and Vedeld (2007) in Dendi District, Ethiopia, revealed that forest income resulted in a share of 39% of the total household income. In Ghana, forest income accounted for 38% of the total household income (Appiah et al., 2009). In Southern Africa, forest income ranged from 30% – 45% of the total household income (Kalaba, Quinn, & Dougill, 2013).

On the foregoing, it is evident that forest products are integral to local livelihoods and contribute substantially towards household economies. Forest resources form part of livelihood diversification and may serve as the sole source of livelihood, especially to the poorest households. Households harvest forest products for both subsistence and commercial consumption. Uberhuaga et al. (2012) noted that forests provide an important source of subsistence and cash income to rural households. Through forests, households generate cash income, which in turn contributes towards their household economies. Moreover, in some studies, the contribution of forests to household economies often has been observed to be higher than other income-generating options such as agricultural activities (see McElwee, 2010).

Research has explored the contribution of forests to rural livelihoods (Appiah et al., 2009; Garekae, Lepetu, & Thakadu, 2019). However, most studies conducted on the forest-livelihood nexus have been undertaken in tropical rainforest regions (Ezebilo & Mattsson, 2010; Shackleton, Shackleton, Buiten, & Bird, 2007). Of the few studies conducted in Africa, a limited number of them placed focus on the dry forests of Southern Africa. Furthermore, the knowledge and understanding on the patterns of forest reliance among households in terms of the amount collected as well as the medium of collection is still inadequate, despite this being critical to resource management and planning. In Botswana, there is little knowledge on the intricate relations of forests and people, more especially on the level of reliance on forest resources, their uses and value in rural livelihoods. Taylor, Mateke, and Butterworth (1996) contend that very few governments in Africa are aware of the extent of use or the value of NTFPs in the informal sector. The low recognition of the forestry sector in Botswana may downplay the potential role of forests in both local and national economies. Against this backdrop, the chapter assesses the extent of households' reliance on non-timber forest products and their contribution towards livelihoods of Chobe enclave communities. Specifically, the study determines the level of households' reliance on forests and establishes and quantifies the amount of forest products harvested from the Chobe Forest Reserve (CFR).

Conceptual framework

This study is informed by the Sustainable Livelihood Framework (SLF). According to Ellis (2000, p. 19), a "livelihood comprises the assets (natural, physical, human, financial and social capital), the activities, and the access to these (mediated by institutions and social relations) that together determine the living gained by an individual or household." Assets are stocks of capital that households draw on for their well-being sustenance (Ellis, 2000). Institutions are the "humanly devised constraints that shape human interaction" (North, 1990, p. 3). Social relations are relationships among people within the society, stratified according to factors such as gender, caste, class, age, ethnicity and religion (Ellis, 2000). Following Ellis (2000), "social relations, institutions and organisations are critical mediating factors for livelihoods because they encompass the agencies that inhibit or facilitate the exercise of capabilities and choices by individuals or households" (p. 39). A sustainable livelihood is a function of its ability to cope with and recover from stresses and shocks, without diminishing the resource base (Chambers & Conway, 1992). In the context of this study, SLF is used in understanding the contribution of forest resources (assets) in rural livelihoods (strategies/activities) of the Chobe enclave communities.

Materials and methods

Description of the study setting

The study was conducted in Chobe District, northern Botswana (Figure 13.1). Chobe District borders three countries: Namibia on the north and west, Zambia on the north, and Zimbabwe on the east (CDDP, 2003). The district is comprised of protected areas: national parks, wildlife management areas and forest reserves. The Chobe River delineates the Chobe region from the Namibian boundary. The Chobe District is endowed with the only six gazetted forest reserves in Botswana (see Garekae, Thakadu & Lepetu, 2016, for detailed description). The Chobe forest reserves and Chobe River are integral to livelihood sustenance in the Chobe district. The forest reserves provide various resources for domestic use, household energy needs, building materials and agricultural inputs while the Chobe River is a source of riverine resources for food, building and crafting materials. Moreover, the district abounds with varied wildlife populations, especially elephants and buffaloes. The Chobe District is made up of ten settlements, with the township of Kasane serving as the district administrative centre. The rest of the settlements are categorised into two distinct clusters: east (Kazungula, Lesoma, Pandamatenga) and west (Mabele, Kavimba, Kachikau, Satau, Parakarungu) cluster communities. The western cluster is commonly known as the Chobe enclave. The major economic activities in the Chobe District include crop production (rain-fed and flood recession), livestock production, tourism-based industries, wage employment, small and medium enterprises and commercial arable farming (CDDP, 2003).

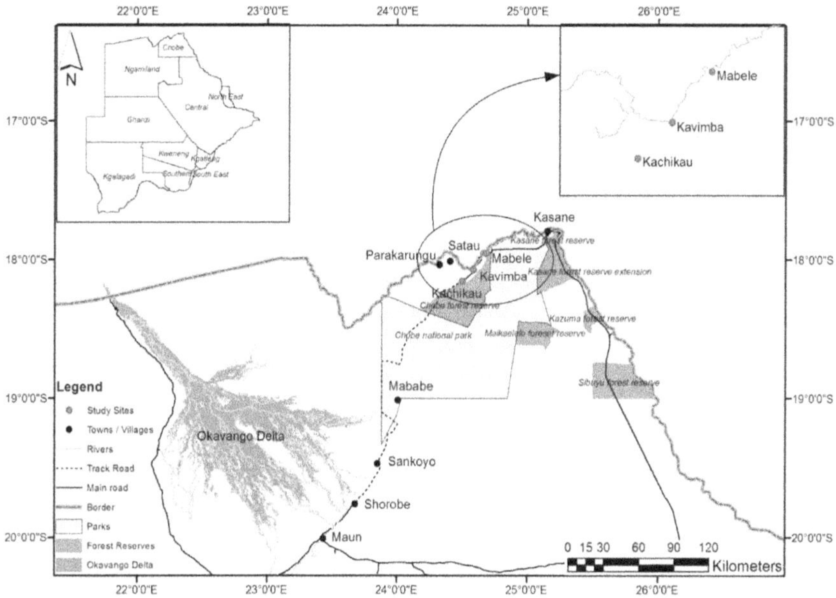

Figure 13.1 Map of Chobe District depicting the study areas

Source: (Authors, 2019)

Since the Chobe area is endowed with varied natural resources, partnership between the government and the communities has been in existence for a while through the Community-Based Natural Resource Management (CBNRM) initiative. Community Based Organisations (CBOs) were formed for implementing the CBNRM programmes in the area. These are Kalepa, Seboba and Chobe Enclave Conservation Trust (CECT). Kalepa and CECT CBOs are constituted by the eastern and western cluster communities respectively while Seboba comprises the Kasane community. Chobe District is multi-ethnic, with the BaSubiya being the largest group and found in almost all Chobe enclave settlements but dominant in Satau and Parakarungu (CDDP, 2003). The BaTawana and Basarwa live in Kachikau and Pandamatenga respectively. There is also notable presence of the BaNabjwa, BaTebele, BaHambukushu and Bakalaka in smaller proportions within Chobe's ethnical composition. The study participants were drawn from the Chobe enclave community. The enclave is comprised of a stream of five villages spanning the Chobe River basin (Figure 13.1). The total population size of the enclave was 4,182 (Statistics Botswana, 2011b). Three villages: Mabele, Kavimba and Kachikau were purposively selected to form part of the study area (Figure 13.1). The villages were mainly chosen based on their proximity to CFR. The key economic activities in these villages are agriculture (crop and livestock production) and wage employment, supplemented by small-scale businesses (beer making and natural resources products selling) (Jones, 2002).

Collection and analysis of data and material

Cross-sectional design was employed where observations on the sampled population were made at a single point in time (Babbie, 2016). The study largely draws from a combination of household surveys and in-depth interviews. The duos were complemented by reviewing of secondary data sources (published and unpublished documents) and observation. A standardised semi-structured questionnaire was used to solicit data from the household survey participants. The questionnaires were administered to a sample of 183 households randomly selected from the villages of Mabele, Kavimba and Kachikau (Table 13.1). This

Table 13.1 Background characteristics of the respondents

Variable	*Items*	*N*	*M (SD)*	*%*	*n*
Gender	Male			38.8	71
	Female			61.2	112
Age (years)		182	49.33 (17.28)		
Ethnicity	Basubiya			53	97
	Bayei			13.7	25
	Batawana			10.9	20
	Hambukushu			6.6	12
	Basarwa			7.1	13
	Banabjwa			.5	1
	Others			8.2	15
Education level	None			13.1	24
	Primary			44.8	82
	Secondary			36.1	66
	Tertiary			5.5	10
Employment status	Full-time employed			12.6	23
	Part-time employed			6.6	12
	Self-employed			14.2	26
	Unemployed			54.1	99
	Others			12.6	23
Average monthly income	< 500			42.8	77
	> 500 < 1000			32.2	58
	> 1000 < 1500			7.8	14
	> 1500 < 2000			3.9	7
	> 2000 < 2500			6.1	11
	> 2500 < 3000			3.3	6
	> 3000			3.9	7
Household size (years)		183	4.85 (2.64)		
Length of residency (years)		175	40.26 (20.73)		
Place of origin	Other settlements in Chobe District			29.4	53
	Outside Chobe District			10.0	18
	Always lived here			60.6	109

sample accounts for about 34% of the total households from the study villages. Household heads were targeted for interviews or, in exceptional cases, any household members who were 18 years old and above were considered to represent the household head. The questionnaires were administered to the respondents through face-to-face interviews, conducted by the researchers at the interviewees' homes and at agreed convenient times. All ethical protocols were observed.

The survey questionnaire solicited information on the respondent's profile and forest resource utilisation. The amount of forest products harvested was assessed through various media, including: head-load, wheelbarrow, cart-load, sledge load, re-used bags, bundles and logs. With regard to the in-depth interviews, six key informants were purposively sampled. The informants were chosen based on their position, expertise and knowledge on issues of natural resources utilisation, management and conservation (Marshall, 1998). The informants comprised three local chiefs (being traditional or tribal authorities), one community trust member and two officials from the Department of Forestry and Range Resources (DFRR). These informants were responsible for management and conservation of natural resources both at the community (trust member) and district and/or country level (DFRR officials). The informants provided information on communities' reliance on forest resources and, most importantly, regulatory instruments governing access to the resources. The interviews lasted between 30 and 60 minutes.

A validity and reliability of data collection instrument was ensured. The formulation of the instrument items was guided by the existing literature, and then a team of experts from the author's respective institutions reviewed the instrument for content validity. Lastly, the Statistical Package for Social Sciences (SPSS) version 23 was used for managing data. Descriptive statistics such as frequencies, proportions, measures of central tendency and dispersion were used to analyse quantitative data while thematic analysis was used to analyse qualitative data. In this study, a thematic analysis procedure outlined by Braun and Clarke (2006) was followed. Furthermore, some excerpts from the transcribed data items were used as a thick description to demonstrate the respondents' viewpoints.

Results

Respondents' profile

Out of the total sampled population, 61.2% (n=112) were females. Most of the respondents were on their mid-elderly ages (M=49.33, SD = 17.28). About 44.8% (n = 82) of the respondents had some primary education, a few (5.5%, n=10) attained a tertiary education level, while 13.1% (n=24) of the respondents reported that they have never received any form of education. The majority of the respondents were unemployed (54%, n=99), followed by those employed on a full-time basis (12.6%, n=23) and those who considered themselves to be self-employed (14.2%, n=26). Household monthly income varied between two extremes, with the majority reporting an average income of less than US$45[1] (42.8%, n=77), while a few indicated US$270 or more (3.9%, n=7). The mean household size was 4.85 (SD=2.64), ranging from 1 to 16 household members.

Amount and extent of use of specific forest products

Forest income was derived from the direct use values of selected products such as firewood, poles, thatching grass and wild fruits (see Table 13.2). Despite fewer households having reported collecting kraal and fence poles, they contributed to the highest proportion of forest income respectively (M = 2 067.88, SD = 2524.56 and M = 1 858.78, SD = 2631.39), followed by firewood (M = 1 713.28, SD = 1285.28). The mean annual forest income per year per household was BWP3 858.34 (SD = 3155.5), ranging from BWP319.4 to BWP18 636.00.

FIREWOOD

Firewood was the most collected and used forest product across the study villages. Firewood was still an important source of energy among many households. It was mainly used for heating, especially during winter season and for cooking. Although some households maintained that they use other alternative energy sources, they resorted to firewood usage for cooking some meals that tend to take a longer time to cook. Therefore, firewood was still a vital energy source for many, if not all, of the households in the study villages. About 86% (n = 157) of the respondents reported collecting and using firewood. The preferred firewood species across the study villages were *Baikiaea plurijuga* (Mokusi), *Colophospermum mopane* (Mophane), *Combretum elaeagonoides* (Moswaapeba) and *Combretum apiculatum* (Mohihiri). *Baikiaea plurijuga* was mostly used across the

Table 13.2 Mean annual value (in pula currency) of forest products per year per household

SOURCE OF FOREST INCOME	*MABELE (SD)*	*KAVIMBA (SD)*	*KACHIKAU (SD)*	*MEAN TOTAL (SD)*
Firewood	1 896.86 (1224.31)	2 089.00 (1539.65)	1 476.67 (1177.00)	1 713.28 (1285.28)
Fencing poles	3 145.71 (4468.90)	1 814.29 (712.81)	1 226.43 (1143.30)	1 858.78 (2631.39)
Building poles	326.67 (234.69)	466.67 (115.47)	334.00 (242.64)	339.64 (234.93)
Kraal poles	2 262.67 (2642.19)	1 462.50 (1910.08)	2 260.00 (2921.26)	2 067.88 (2524.56)
Thatching grass	746.80 (521.84)	1 026.85 (161.69)	1 247.35 (2284.69)	1 173.54 (2091.32)
Motsentsela	1 277.60 (1962.43)	1 103.38 (1118.94)	867.62 (941.61)	939.84 (1097.13)

three study villages. *Colophospermum mopane* and *Combretum apiculatum* were predominantly used in Mabele while *Combretum elaeagonoides* was predominantly used in Kachikau.

Households used various methods for collecting firewood: either through head-loads, wheelbarrows, cartloads or sledge loads. The mean annual direct-use value of firewood per year per household was BWP1 713.28 (SD = 1285.28), ranging from BWP240.00 to BWP6 720.00.

WILD FRUITS

Wild fruits were the second most collected and used forest product after firewood. Wild fruits form part of the household dietary needs. A total of 126 (68.9%) of the respondents indicated that they collected wild fruits. The mean level of dependency on wild fruits was 2.49 (SD = 2.01), demonstrating low dependency. Furthermore, 32.5% (n = 51) indicated that wild fruits were extremely important to their livelihoods while 3.8% (n = 6) maintained neutral. Across the study sites, wild fruits were mainly collected from *Berchemia discolor* (Motsentsela) and *Adonsonia digitate* (Mowana). Motsentsela fruits were mainly collected between the months of January and April while Mowana were collected between the months of July and September. Mowana fruit was mainly collected for subsistence consumption while Motsentsela was collected for both subsistence and commercial usage. The mean annual direct-use value of Motsentsela per season per household was BWP939.84 (SD = 1097.13), ranging from BWP319.40 to BWP6 388.00.

THATCHING GRASS

Thatching grass was the third most harvested and used forest product. The preferred grass species was *Cymbopogon excavates* (Mokamakama). Thatching grass was mainly harvested between the months of July, August and September. The mean level of dependency on thatching grass was 1.81 (SD = 2.10), indicating low dependency. Although households were somewhat dependent on thatching grass, their response on the importance of thatching grass to their livelihoods was in contrary (M = 2.20, SD = 2.29). Grass was harvested in the form of bundles. The mean annual thatching grass bundles harvested per season per household was 62.86 (SD = 112.02), ranging from 2 to 800 bundles, while the mean annual direct-use value per season per household was BWP1 173.54 (SD = 2091.32), ranging from BWP37.34 to BWP14 936.00.

FENCING POLES

Fencing poles were primarily used for paling yards and fields. In the study areas, most of the homestead fences were constructed with poles extracted from the forest reserve. Less than half of the respondents (26.8%, n = 42) were highly dependent on poles extracted from CFR while 2.5% (n = 4) reported very low dependency. Similarly, 35% (n = 55) of the respondents noted that fencing poles were extremely important to their livelihoods while a few (0.6%, n = 1) were undecided. The preferred species for fencing poles were *Baikiaea plurijuga, Colophospermum mopane, Combretum elaeagonoides* and *Combretum apiculatum. Colophospermum mopane* and *Combretum apiculatum* were mostly used in Mabele while *Baikiaea plurijuga* and *Combretum elaeagonoides* were mostly used in Kachikau.

The amount of fencing poles used by various households was assessed in terms of the number of cartloads, logs and sledge loads collected per year per household. This assessment was applied to the other pole types. Generally, the most frequent medium of fencing poles collection was through "number of logs" followed by sledge loads and cartloads. The mean number of fencing logs used per year per household was 92.94 (SD = 131.57), ranging from 2 to 800 logs, while for cartloads it was 1.31 (SD = .75), ranging from 1 to 3. The mean annual direct-use value of fencing logs used per year per household was BWP1 858.78 (SD = 2631.39), ranging from BWP40.00 to BWP16 000.00.

BUILDING POLES

Building poles were the fifth most frequently used forest products. The poles were primarily used for roofing mud shelters and also for constructing open outdoor structures. The majority of the households had at least one structure roofed with poles and thatching grass in their yards. The most preferred species were *Baikiaea plurijuga, Colophospermum mopane, Combretum apiculatum, Terminalia sericea* (Mogonono), *Combretum elaeagonoides* and *Croton gratissimus* (Moologa). *Colophospermum mopane* was predominately used in Mabele while *Baikiaea plurijuga, Terminalia sericea, Croton gratissimus* and *Combretum elaeagonoides* were mostly used in Kachikau. About 27% (n = 43) of the respondents reported very high dependency on building poles while 4.5% (n = 7) reported low dependency. Likewise, 38% (n = 59) of the respondents indicated that building poles were extremely important to their livelihoods whereas 1.9% (n = 3) were undecided.

The majority of the households collected building poles in terms of "number of logs." The mean "number of logs" used per year per household was 16.98 (SD = 11.75), ranging from 3 to 50. The mean annual direct-use value of building logs used per year per household was BWP339.64 (SD = 234.93), with a range of BWP60.00 to BWP1 000.00.

KRAAL POLES

Kraal poles were mainly used for constructing cattle enclosures. The most preferred species were *Baikiaea plurijuga, Combretum apiculatum* and *Colophospermum mopane. Combretum apiculatum* and *Colophospermum mopane* were predominately used in Mabele while *Baikiaea plurijuga* was mostly used in Kachikau. A total of 22.9% (n = 36) of the respondents reported very high dependency on kraal poles whereas 1.3% (n = 2) stated very low dependency. Similarly, 30% (n = 47) of the respondents expressed that kraal poles were extremely important to their livelihoods while 10.8% (n = 17) indicated kraal poles to be important. The predominate medium of kraal poles collection was through "number of logs." The mean number of logs used per year per household was 103.39 (SD = 126.23), ranging from 2 to 500 logs, whereas the mean number of sledge loads used per year per household was 43.67 (SD = 50.90), ranging from 1 to 100 loads. The mean annual direct-use value of the number of logs used per year per household was BWP2 067.88 (SD = 2524.56), ranging from BWP40.00 to BWP10 000.00.

OTHER RESOURCES

Some of the resources collected from CFR included handicrafts materials, medicinal plants, fodder and wild vegetables. The amount collected for these resources was not quantified in this study because they were extracted in fewer quantities and at irregular times. In addition, some respondents could not account for their collection.

Discussion

The purpose of this chapter was to assess the extent of households' reliance on non-timber forest products together with its contribution to livelihoods. The findings demonstrated that Chobe enclave residents were dependent on CFR for their various livelihood needs. Generally, households were highly dependent on forest resources for their livelihood sustenance (M=4.24, SD=0.95).

The finding resembles Mukul et al. (2016) observations in Bangladesh. In their study, the level of forest dependency varied with households' area of location, with those living within the national park (Tiprapara residents) being significantly highly dependent on forests for their livelihoods. Forest resources served various livelihoods needs among the study communities, including: domestic use, household energy needs, building materials, agricultural inputs and dietary needs. Based on the foregoing, forest products for household energy includes firewood and poles for agricultural inputs, fencing fields and construction of livestock enclosures. For building and construction material, thatching grass and poles as roofing rafters were commonly used. On one end, wild fruits contributed to dietary needs. Therefore, forest products were integral to local livelihoods in the study area. This finding is consistent with the ascertained substantial value of forest products' contribution to livelihoods (Negash, 2007), particularly in providing food, fuel, medicine and construction materials.

In this study, firewood and poles were continuously collected throughout the year while thatching grass and wild fruits were only harvested during particular seasons. On that account, firewood was the most collected and used forest product across the study sites while fodder was the least utilised. Like in many other developing countries where provision and access to energy in rural areas are still a constraint (Babulo et al., 2009), firewood remained an important source of energy among many households in the study area. It was mainly used for heating, especially during the winter season and for cooking. Although some households maintained that they used other alternative energy sources for either cooking or heating, they resorted to firewood for cooking some meals that tend to take a longer time to cook.

Other energy sources such as electricity, coal, gas and kerosene may not be affordable to the majority of households compared to firewood, which is generally free. Given the extent of households' reliance on fuelwood in this study and elsewhere, it is important to encourage communities to share information on firewood collection with young people so that the current and possibly sustainable practices are transferred across generations.

These findings are consistent with other studies elsewhere. Studies conducted in South Africa (Mtati, 2014; Thondhlana & Muchapondwa, 2014) and Nepal (Panta, Kim, & Lee, 2009) indicated that more than 80% of the households were reliant on firewood as a primary source of energy. As observed from this study and elsewhere, even those households with access to other alternative energy sources were greatly dependent on firewood to meet their energy needs. However, the abject poverty conditions in Chobe enclave will not accord everyone access to alternative energy sources such as electricity and gas because they involve monetary transactions. Hence, people will continue to be reliant on firewood since it is a cheap and/or free commodity. On that note, availability and access to alternative energy sources such as electricity do not seem to gravitate households away from firewood usage. Subsequently, in some areas, firewood has emerged as an important income-generating activity during times of shock; for example, during seasonal food shortages, some households rely on the cash income accrued from firewood (Mamo, Sjaastad, & Vedeld, 2007). Therefore, households may resort to selling firewood in order to complement their household cash income.

Wild fruits formed an integral component of households' dietary needs. Wild fruits serve as a safety net during times of shock (drought, famine and low agricultural productivity) because of their tolerance to adverse conditions compared to staple crops and domesticated ones (Makombe, 1994; Stadlmayr, Charrondière, Eisenwagen, Jamnadass, & Kehlenbeck, 2013). For example, households that realy heavily on agriculture may resort to harvesting wild fruits during the agricultural off-season.

Ecotourism as an alternative for sustainable forest management for CFR

Ecotourism is a rapidly growing industry that could address the conflict between conservation efforts and the needs of local populations.

For ecotourism to be successful within a protected area such as forest reserves, it must focus on three concepts: it must be nature-based, educational, and economically and socially sustainable.

The Department of Forestry and Range Resources is proposing opening up forest reserves for ecotourism purposes as income-generating activities for communities. This proposal sis manifested due to the tourism pressure in Chobe National Park, hence CFR and other forest reserves, such as Kasane Forest Reserves which acts as a buffer zone to CFR, are seen as alternative routes for tourists operators. As a whole, the policy interventions for the sustainable balance between conservation, forest resource use, socioeconomic development and expansion of ecotourism in the buffer zone area can increase both the legitimacy and efficiency of conservation efforts in Chobe National Park.

In a bid to promote sustainable NTFPs utilisation, this study suggests that the Chobe enclave locals and communities look elsewhere to explore other key alternative strategies and technologies such as domestication of preferred species and a shift towards emerging technologies such as energy-efficient firewood stoves. Species domestication will reduce the pressure exerted on forests as communities will now have other resources to harvest from within their vicinity. In consistency with the Sustainable Livelihood Framework and forest literature, this study demonstrates that forests and livelihoods are intertwined. SLF outlines that a livelihood comprises assets, activities and access, which together shape the living of an individual or households. Therefore, forest products provide assets which individuals and households can draw on for their livelihood sustenance. On that note, management, planning and strategies such as ecotourism should take into account the functional role of forests in sustaining livelihoods of adjacent communities.

Lastly, the study expanded the forest-livelihood literature, particularly in Botswana, which was scanty and biased towards biophysical researches. Future research is needed to quantify the economic value of Botswana's forest resources so as to inform polices and planning.

Note

1 BWP1 = USD0.09 (2016)

References

Appiah, M., Blay, D., Damnyag, L., Dwomoh, F. K., Pappinen, A., & Luukkanen, O. (2009). Dependence on forest resources and tropical deforestation in Ghana. *Environment, Development and Sustainability, 11*, 471–487.

Babbie, E. R. (2016). *The practice of social research* (14th ed.). Boston: Cengage Learning.

Babulo, B., Muys, B., Nega, F., Tollens, E., Nyssen, J., Deckers, J., & Mathijs, E. (2009). The economic contribution of forest resource use to rural livelihoods in Tigray, Northern Ethiopia. *Forest Policy and Economics, 11*(2), 109–117.

Bhavannarayana, C., Saritha, V., Usha, P., & Rao, B. P. (2012). Dependency and usage pattern of forest-dwellers on non timber forest products. *Erudite Journal of Ecology and Environment, 1*(1), 1–5.

Braun, V., & Clarke, V. (2006). Using thematic analysis in psychology. *Qualitative Research in Psychology, 3*(2), 77–101.

Chambers, R., & Conway, G. (1992). *Sustainable rural livelihoods: Practical concepts for the 21st century* (No. IDS Discussion Paper Number 296). Institute of Development Studies, Brighton.

Chobe District Development Plan (CDDP). (2003). *Chobe district development plan 6: 2003–2009*. Kasane, Botswana, Botswana.

Cocks, M. L., & Wiersum, K. F. (2003). The significance of plant diversity to rural households in Eastern Cape Province of South Africa. *Forests, Trees and Livelihoods*, *13*(1), 39–58.

Córdova, J. P. P., Wunder, S., Smith-Hall, C., & Börner, J. (2013). Rural income and forest reliance in highland guatemala. *Environmental Management*, *51*, 1034–1043.

Ellis, F. (2000). *Rural livelihoods and diversity in developing countries*. Oxford: Oxford University Press.

Ezebilo, E. E., & Mattsson, L. (2010). Contribution of non-timber forest products to livelihoods of communities in southeast Nigeria. *International Journal of Sustainable Development & World Ecology*, *17*(3), 231–235.

Garekae, H., Lepetu, J., & Thakadu, O. T. (2019). Forest resource utilisation and rural livelihoods: insights from Chobe enclave, Botswana. *South African Geographical Journal*, 1–19. DOI: 10.1080/03736245.2019.1606730.

Garekae, H., Thakadu, O. T., & Lepetu, J. (2016). Attitudes of local communities towards forest conservation in Botswana: A case study of Chobe Forest Reserve. *International Forestry Review*, *18*(2), 180–191.

Gautam, K. H. (2006). Forestry, politicians and power – perspectives from Nepal's forest policy. *Forest Policy and Economics*, *8*, 175–182.

Ghimire, K. B., & Pimbert, M. P. (1997). Social change and conservation: An overview of issues and concepts. In K. B. Ghimire & M. P. Pimbert (Eds.), *Social change and conservation: Environmental politics and impacts of national parks and protected area* (pp. 1–45). London: Earthscan Publications.

Howe, C., Suich, H., Vira, B., & Mace, G. M. (2014). Creating win-wins from trade-offs? Ecosystem services for human well-being: A meta-analysis of ecosystem service trade-offs and synergies in the real world. *Global Environmental Change*, *28*, 263–275.

Jones, B. T. B. (2002). *Chobe enclave, Botswana – lessons learnt from a CBNRM project 1993–2002*. Gaborone: IUCN/SNV CBNRM Support Programme.

Kalaba, F. K., Quinn, C. H., & Dougill, A. J. (2013). Contribution of forest provisioning ecosystem services to rural livelihoods in the Miombo woodlands of Zambia. *Population and Environment*, *35*(2), 159–182.

Makombe, K. (Ed.). (1994). *Sharing the land: Wildlife, people and development in Africa*. IUCN/ROSA Environmental Issues Series No. 1. IUCN/ROSA, Harare, Zimbabwe and IUCN/SUWP, Washington, DC.

Mamo, G., Sjaastad, E., & Vedeld, P. (2007). Economic dependence on forest resources: A case from Dendi District, Ethiopia. *Forest Policy and Economics*, *9*, 916–927.

Marshall, M. N. (1998). Key informant technique. *Family Practice*, *13*(1), 92–97.

Masozera, M. K., & Alavalapati, J. R. R. (2004). Forest dependency and its implications for Protected Areas Management: A case study from the Nyungwe Forest Reserve, Rwanda. *Scandinavian Journal of Forest Research*, *19*(4), 85–92.

McElwee, P. D. (2010). Resource use among rural agricultural households near protected areas in Vietnam: The social costs of conservation and implications for enforcement. *Environmental Management*, *45*, 113–131.

Millennium Ecosystem Assessment. (2005). *Ecosystems and human well-being: Synthesis*. Washington, DC: Island Press.

Mtati, N. (2014). *The relative contribution of non-timber forest products, agriculture and off-farm sources of inome to rural households in Koloni and Guquka, Eastern Cape.* Masters thesis. Retrieved from http://hdl.handle.net/10962/d1018193.

Mukul, S. A., Rashid, A. Z. M. M., Uddin, M. B., & Khan, N. A. (2016). Role of non-timber forest products in sustaining forest-based livelihoods and rural households' resilience capacity in and around protected area: a Bangladesh study. *Journal of Environmental Planning and Management, 59*(4), 628–642.

Negash, M. (2007). Trees management and livelihoods in Gedeo's agroforests, Ethiopia. *Forests, Trees and Livelihoods, 17*(2), 157–168.

North, D. C. (1990). *Institutions, institutional change, and economic performance.* Cambridge: Cambridge University Press.

Panta, M., Kim, K., & Lee, C. (2009). Households' characteristics, forest resources dependency and forest availability in Central Terai of Nepal. *Journal of Korean Forest Society, 98*(5), 548–557.

Ros-Tonen, F. M. A. (2000). The role of non-timber forest products in sustainable tropical forest management. *Holz Als Roh- Und Werkstoff, 58*(3), 196–201.

Shackleton, C. M., Shackleton, S. E., Buiten, E., & Bird, N. (2007). The importance of dry woodlands and forests in rural livelihoods and poverty alleviation in South Africa. *Forest Policy and Economics, 9*, 558–577.

Stadlmayr, B., Charrondière, U. R., Eisenwagen, S., Jamnadass, R., & Kehlenbeck, K. (2013). Nutrient composition of selected indigenous fruits from sub-Saharan Africa. *Journal of the Science of Food and Agriculture, 93*, 2627–2636.

Statistics Botswana. (2011a). *Chobe sub-district population and housing census 2011: Selected indicators for villages and localities.* Gaborone: Statistics Botswana.

Statistics Botswana. (2011b). *Population of town, villages and associated localities.* Gaborone: Statistics Botswana.

Taylor, F., Mateke, S. M., & Butterworth, K. J. (1996). *A holistic approach to the domestication and commercialization of non-timber forest products.* Food and Agriculture Organisation of the Unitied Nations (pp. 1–288). Rome: FAO.

Thondhlana, G., & Muchapondwa, E. (2014). Dependence on environmental resources and implications for household welfare: Evidence from the Kalahari drylands, South Africa. *Ecological Economics, 108*, 59–67.

Uberhuaga, P., Smith-Hall, C., & Helles, F. (2012). Forest income and dependency in lowland Bolivia. *Environment, Development and Sustainability, 14*, 3–23.

Vedeld, P., Angelsen, A., Bojö, J., Sjaastad, E., & Berg, K. G. (2007). Forest environmental incomes and the rural poor. *Forest Policy and Economics, 9*(7), 869–879.

Venter, S. M., & Witkowski, E. T. F. (2011). Baobab (Adansonia digitata L.) fruit production in communal and conservation land-use types in Southern Africa. *Forest Ecology and Management, 261*, 630–639.

Wells, M., Brandon, K., & Hannah, L. J. (1992). *People and parks: Linking protected area management with local communities.* Washington, DC: World Bank; World Wildlife Fund, U.S. Agency for International Development.

World Bank. (2004). *Sustaining forests: A development strategy.* Retrieved from https://books.google.co.bw/ [accessed March 1, 2016]

14 An assessment of supply-side factors and ecotourism in Mauritius

Seetanah Boopen and Sannassee R.V

Introduction

The contribution of the tourism sector towards the economic development of nations has been well documented in the literature. From direct benefits which include creation of employment, generation of added value and tax revenue, boosting of inward foreign direct investment as well as generation of much-needed foreign exchange to wider spillovers through skills diffusion, as well as other related spill-over positive impact, the potential benefits which can be engendered by the tourism sector have been clearly spelt out in the literature (e.g. see Sinclair, 1998; Seetanah, 2011; Nunkoo, Seetanah, Jaffur, Moraghen, & Sannassee, 2019).This is particularly true for the case of Mauritius as the prevalence of the tourism sector towards the economic development of the island is undeniable. Today, the tourism sector is considered as the second contributor pillar to the economy after sugar cane, approximating 18% of GDP and in excess of 60 billion rupees in 2018 (Statistics Mauritius). Tourist arrivals for the year 2018 were around 1,400,000 and total passenger arrivals were essentially by air (99%) and 1% by sea. Various factors have served to generate such an increase in revenue and tourist numbers. Some of these include trust as an element inducing first-time visitors and repeat tourists (Sannassee & Seetanah, 2014; infrastructure (Khadaroo & Seetanah, 2007, 2008); marketing promotion expenditure (Seetanah & Sannassee, 2014) and air access policy (Seetanah, 2019).

Tourism has also had some negative impacts on the physical environment, economic, and socio-cultural landscapes of Mauritius. Indeed, over the past 20 years or so, a number of issues have become topical, including inefficient use of resources (water and energy among others); generation of waste; coastal and marine degradation; damage to the physical environment (land, sea and beaches); as well as loss of biodiversity. In this regard, one of the biggest challenges for a Small Island Development State (SIDS) such as Mauritius relates to its capacity to ensure development of a sustainable and competitive tourism industry. Managing tourism in a sustainable way involves formulating, implementing and instilling appropriate policies and measures to optimise benefits while mitigating negative impacts on the society, culture and environment. The aim of sustainable tourism[1] therefore is to ensure that development brings a positive experience for local people, tourism businesses and tourists themselves.

The current tourism trend and policy focus is presently geared towards beach tourism, which is regarded unsustainable in the long term; hence, the government has, since the turn of the new millennium, embarked on a series of policy measures aimed at fostering green initiatives with the ultimate objective of greater sector sustainability. These include amongst others the National Tourism Development Plan (2000), the Environment Protection Act of (2002) and the Tourism Authority Act (2006) which were enacted in 2008; the Hotel Development Strategy (2008) and finally the Tourism Sector Strategy Plan (2009–2015). Interestingly, the above Acts and Plans have led to the fostering of the eco-tourism concept in Mauritius, which was further reinforced by the Mauritius Ile Durable (MID) concept which was launched in 2008 under the long-term vision for sustainable development. The latter was principally geared towards making Mauritius a world model of sustainable development. In a similar vein, additional measures which were also initiated included, among others, the Blue Flag Programme and an Eco-Label scheme.

One can argue that, despite their very best intentions, successive governments have not been able to propound an integrated eco-tourism plan with the result that the above-mentioned measures, are, at best, piecemeal in nature. To make matters worse, one can even contend that there is no record of any assessment of existing eco-tourism sites on the island, let alone a study reviewing the existing regulations (if any) that govern the conservation/protection of same.

Given the above, the objective of the chapter is to assess the relevance and prevalence of ecotourism sites towards fostering ecotourism. This should also permit to uncover any supply-side constraints which could have hampered potential realisation of benefits from exploiting these sites and propose remedial measures accordingly.

The rest of the chapter is structured as follows: section 2 provides some key information with respect to the tourism industry in Mauritius, as well as a review of the existing regulations and policy frameworks on eco-tourism. Section 3 discusses the methodology and findings, while Section 4 concludes and provides the resulting policy implications.

Ecotourism policies and regulations

The tourism industry in Mauritius: facts and figures

Since its independence, the Mauritian economy has undergone several remarkable economic transformations, with the island successfully diversifying from an agro-centric base into other sectors such as textile, tourism, financial services, seafood industry and the ICT. Diversification of the economy emerged as a result of the deterioration of the economic situation in the late 1970s, with rising petroleum prices and the end of the sugar sector boom. As a result, there was a pressing need to seek alternative sources of export income, and various

Table 14.1 Key tourism indicators

	1980	*1990*	*2000*	*2010*	*2013*	*2016*	*2017*
Population	1060000	1080000	1186140	1195433	1259838	1,275,227	1,264,613
No. Hotels	43	75	95	97	107	111	111
Hotel Rooms	2101	4603	8255	12275	12376	13,547	13,511
Tourist arrivals	115080	291550	656543	702018	993106	1,275, 227	1,341,860
Tourism Receipts (% of GDP)	6%	10%	14%	17%	18%	20%	24%

Source: Compiled by authors from Statistics Mauritius (2018)

measures and policies were devised which fostered investment expansion, both at the local and international level. Today, it can be safely argued that the tourism industry is one of the most important sectors contributing significantly to the country's GDP. The government has been able to take advantage of the tropical island appeal, beautiful beaches, security and absence of tropical disease to promote Mauritius as an attractive destination. Table 14.1 shows tourism key indicators.

In addition, the number of tourist arrivals since 1974 has increased by more than ten-fold with numbers rising to nearly 1,4000,000 in 2018. Similarly, tourism receipts have substantially increased from as low as Rs 135 million in 1974 to in excess of 60 billion rupees in 2018 (Statistics Mauritius).[2] Based on the latest available data on tourist arrivals and information gathered from stakeholders, the forecasted figure of 1,450,000 for tourist arrivals for the year 2019 is maintained, representing an increase of 3.6% over the figure of 1,399,408 in 2018. The Bank of Mauritius is also maintaining its forecasted tourism earnings for the year 2019 at Rs 67.5 billion (Bank of Mauritius, 2018).

While traditionally France, the UK, South Africa, Germany and India remained the major markets, it is noteworthy that tourist arrivals from Saudi Arabia, the Netherlands, China and the Asian regions have been on an increasing trend during the last decade. Employment data for the tourism sector, related to employment in food service, hotels, travel and other service establishments with 10 or more workers, indicates that employment in these establishments increased continuously to reach around 31,000 with the contribution of tourism to employment at around 10% of total employment (Statistics Mauritius, 2018).

Challenges facing the ecotourism industry

Despite the increase in tourist arrivals and the ever-growing contribution to GDP, one could argue that there are a number of challenges, both current and forecasted, which the industry is presently facing and/or will face in the foreseeable future (see Table 15.2).

- **Climate Change**: impacts as identified by the UNWTO (2002) related to i) include warmer temperatures, sea surface temperatures and sea level rise; increased frequency and intensity of storms and heavy precipitation; and more frequent and larger forest fires, among others.
- **Infrastructure and Utilities:** some of the issues identified are as follows: severe water shortages arise during periods of drought (Sept –Dec), with a water shortage foreseen in the future; weaknesses in waste disposal and treatment that could undermine the country's attractiveness as a top-notch tourism destination; constraints in road capacity and traffic congestion, particularly in the vicinity of Port Louis; limits to internet access, which is sometimes unreliable, or lack of connectivity; adequate but comparatively high-cost power supply.
- **Water Supply:** tourism growth will result in increased water demand given the target of two million tourists and the development of the IRS segment. This could pose a challenge to the water sector in terms of the additional capacity required, given a scarcity of current resources, notably during dry seasons or during some peak periods. However, the problem of network losses is a well-publicised issue, although quantitatively, the apportionment of water to tourism is of a smaller order of magnitude than the actual network losses. The importance of reducing these losses has been recognised with a pilot project to reduce lost water involving around 40,000 customers. The aim is to extend these efforts to the whole island with the aim of reducing lost water from 48% as of 2008 to 25% by 2019. In addition, although there has been an increasing use of desalination plants for hotels and resorts, experience elsewhere suggests that this is not insuperable at coastal locations in terms of its impact on hotel pricing structures. Given the relatively higher cost of electricity in Mauritius, it may not be the capital cost of the desalination plant that is necessarily the constraint, but the operational costs.
- **Waste Disposal:** wastewater treatment is particularly an issue for smaller hotels where inadequate standards of operation of treatment via septic tanks can result in environmental degradation. This suggests a need for improved levels of training in relation to the operation of these systems. There is also a need to ensure an invariably high level of operation of sewage treatment plants at the larger hotels. In addition, solid waste is a significant issue, with the Mare Chicose landfill close to saturation.
- **Land Constraints:** given the island's relatively small size (2040kms^2), allocation of land to hotels and other tourism-related businesses are becoming more and more of an issue for the developing authorities.
- **Environmental Issues:** vulnerability due to the effects of climate change, including tidal waves and surges; deterioration of the coral reefs through global warming and tourism activities such as diving and water sports;

increasing pollution and pressure on environmental assets (deforestation, depletion of non-renewable resources, pressure on marine eco-systems, vehicular traffic); beach erosion; and coral bleaching.

Current tourism policies

Prior to 2000, tourism development in Mauritius was done on an ad hoc basis with no formal long-term planning for the destination. However, this was to be changed with the implementation of the National Tourism Development Plan (NTDP) in 2000, which reinforced the island's previous and current position as a high-quality and upmarket tourist destination (Deloitte & Touche, 2002). Furthermore, the Tourism Authority Act, which was enacted in 2006 and amended in 2008, made better provisions for regulating the operation of tourist enterprises and pleasure craft (boats used for fishing, water sports and so on). In addition, insofar as hotel development projects were concerned, the Act urged hotel developers to install eco-friendly and energy-saving practices such as desalination and recycling plants. Since 2000 there has been no other national tourism planning document for the island. In 2008, the government revised the Hotel Development Strategy (HDS), requiring hoteliers, among others, to strictly follow the recommendations of the Planning Policy Guidance (PPG) for coastal development. The document offers guidelines on land management, architectural design and eco-friendly practices, amongst others. Other areas of intervention include tourism-related development control through EIA mechanism, control of recreational activities in the lagoon and the promotion of energy-efficient and environmentally friendly technology in hotels, amongst others. Finally, the Tourism Sector Strategy Plan (2009–2015) recommended ways and means of achieving an environmentally sound, socially acceptable and economically viable tourism development.

In addition, other measures which have been taken include: revamping of the tourism policy with a focus on the gradual liberalization of air access; strengthening of the marketing strategy by positioning Mauritius as a cruising destination; the Mauritius Brand Strategy was launched in October 2009 to strengthen and enhance the image of the Mauritius destination and to ensure greater visibility of the Mauritian brand worldwide; provision of professional assistance has been given to small and medium enterprises under the Empowerment Programme to improve the quality of their products so that they can act as reliable suppliers for the tourism industry; measures were taken to protect the environment (pollution control, sound environmental management, protection of natural resources, landscaping and so on); promoters of new hotel projects are encouraged to use renewable energy and adopt eco-friendly practices and use, as far as practicable; energy-saving devices; use of eco-friendly outboard engines for crafts at sea are now being used to protect the marine environment; encouraging low-rise and low-density hotel development to avoid the disfigurement of the coastal landscapes; and the installation of permanent mooring buoys have been installed at dive sites to protect the coral from damage caused by anchors.

Methods and analysis

Face-to-face in-depth interviews were carried out with the different stakeholders of the tourism industry, including representatives of the private sector and business associations (Association des Hôteliers et Restaurateurs de l'île Maurice (AHRIM)) and the Association of Tourism Promoters), the ministries (Ministry of Tourism and Leisure and the Ministry of Environment and Sustainable Development), relevant authorities (Beach Authority, the Natural Disaster Risk Reduction and Management Centre and the Mauritius Tourism Authority) and a couple of researchers in the field from the University of Mauritius. The respondents were interviewed independently after having been contacted through mail and sent a background of the research together with some leading questions. Questions were geared towards assessing the current eco-tourism products on offer, the pertinence and relevance of regulations and regulatory frameworks pertaining to eco-tourism, if any, and identifying any constraints or problems which are hindering the development of a successful eco-tourist sub-sector in Mauritius. Follow-up questions were asked depending on the interviewees' responses. The responses were subsequently synthesised under a few dimensions which are reported and discussed in the next section.

Lack of an overarching policy framework

Unfortunately, the discussions with the stakeholders tend to point to the fact that there is no holistic approach to tackling the various issues related to the tourism sector. There is a dire need for a strategic plan which would delineate the various strategies to be adopted for the various sub-sectors within the tourism industry. Although the Tourism Act (2006) briefly mentions "the promotion of sustainable development of the Tourism sector" as one of the objects of the Tourism Authority which was created under the same Act, there was no specific reference to eco-tourism per se. Furthermore, it is obvious that any holistic approach to fostering and promulgating the tourism sector is scanty and piecemeal, let alone for the eco-tourism sub-sector.

In this regard, the report commissioned by the Mauritius Ministry of Agro-Industry and Fisheries and carried out by the US Forest Service in 2008 entitled, "A New Vision for Sustainable Ecotourism in Mauritius" is the only study carried out with a specific focus on ecotourism. Firstly, the report highlighted the need for an interactive process amongst institutions working towards attracting ecotourists and the conservation of biodiversity. The report also noted the need to enhance the geographical character of the island to foster sustainable ecotourism, and it was crucial to preserve the environment, culture, heritage and the well-being of the residents in the promulgation of same. Secondly, there were already a number of good examples of sustainable ecotourism in Mauritius, including the likes of Blue Bay, Isle Aux Aigrettes, the SSR Botanical Garden and Valley Ferney. But most importantly, the findings clearly argued that it was fundamental to prioritize sustainable ecotourism if the island was to achieve its target of two million visitors and it was to move beyond the image of the 3Ss. As such, it was indeed necessary for the government to balance the tourist influx with the island's conservation efforts. Interestingly, the report also catered to funding proposals, and to that end, entrance fees to conservation sites, private investment, sponsorships, foundations and the establishment of a Conservation Fund were some of the funding strategies which the government could propose to ensure the sustainability of its ecotourist sites. Unfortunately, however, most of the above propositions, if not all, are yet to be implemented, and the underlying reasons as to why this is the case will be discussed at length later on in the report. Furthermore, criticisms also abounded (as gathered from the interviews) as regards the sensitization of the outcomes of such a report. Most stakeholders were unaware that such a study was commissioned. Similarly, what also transpired from the various meetings is the argument that it is common practice to commission and conduct studies, but the biggest issue remains one of implementation and monitoring. To that end, a more cynical view was given by one interviewee in his endeavour to provide a plausible explanation to the painstakingly slow growth of our eco-tourism sub-sector. He argued that one of the supply-side constraints inhibiting the growth of eco-tourism sites and the sustainability of the island's core assets, including beaches and coral reefs, is not the result of a lack of regulations pertaining to the sector, but rather due to a lack of proper monitoring of the same. For instance, although the Beach Authority was created to oversee the proper management of the island's beaches, yet, no beach management plan has been devised. Additionally, there is still no accepted view as to how to combat beach erosion. At some point in time, the MID project was meant to cater to the same; however, it has been disbanded since the beginning of this year.

Given the above, the stakeholders argued that it is crucial that the forthcoming promulgation of a new set of inland eco-tourism guidelines be widely diffused to all stakeholders to ensure proper implementation and monitoring of the same. The proposed inland eco-tourism guidelines present both a potential to cater to the new demand as well as an avenue for tourism product diversification. It proposes new avenues with the potential for exploitation, including land leased for shooting and fishing purposes, which may prove crucial and even more so in view of the scarcity of beach sites and the growing trend for eco-tourism worldwide. These guidelines promote three main criteria with respect to eco-tourism projects. They should: be well planned, financed, managed and marketed to meet the stringent environmental and recreational requirement of ecotourism development; effectively promote the preservation of the entire local ecosystems; and be economically viable to attract financing and be sustainable.

In addition to the above, however, some stakeholders have argued that despite the absence of any overarching policy framework with respect to the sector and that any sectoral policy prescriptions are done in a piecemeal manner, there have been nevertheless a number of measures which have been adopted in an attempt to "green" the tourism industry, some of which could be related to eco-tourism practices.

Firstly, the Environment Protection Act (2002) consolidated and reinforced the institutional and legal framework for the protection of environmental assets. As a result, hotel developers were required to submit Environmental Impact Assessment results with their application for hotel development. In addition, the Tourism Fund Act of 2006 was a first step towards social sustainability, which meant that hotel developers have had to contribute to a tourism fund to enable infrastructural development in those local communities that will be affected by such development. Hoteliers are also implementing Corporate Social Responsibility initiatives such as building schools, sponsoring employees' children for tertiary education, embellishing the environment and promoting local arts and crafts so that local communities will continue to support hotel development (Beachcomber, 2009). Government is also working to support efforts to raise the level of recycling, which have been supported by the tourism industry including via the composting of all green wastes.

As regards energy supply, both the government and the private sector have been implementing policies and measures alike to encourage the adoption of energy-saving devices and eco-friendly practices. This includes the use of renewable energy and more particularly solar power generation when designing new hotel and bungalow projects. A proposal for these renewable energy measures is now one of the conditions of the EIA license

In addition, a number of desalinations have been installed which has helped reduce the pressure exerted by the hotels on an already-stressed public water system. Furthermore, the major hotels are already operating waste treatment plants although there are still some small hotels, campement (bungalow) sites and commercial premises linked to the tourism industry where the wastewater disposal systems are inadequate and unregulated.

Also, a greater number of hotels are moving towards the Green Globe certification, which can only serve to enhance the island's image as a green destination. Similarly, the government has already embarked on seeking Blue Flag status for its beaches, with the Albion Beach being the first to be earmarked. Blue Flag status is awarded to those beaches that achieve excellence in water quality, environmental education and information, and adhere to international safety standards. The main objective of the government in that endeavour is to promote inter-alia the sustainable use of the coastal resources and sound national policies on lagoon water quality, reefs, protection of the beaches and safety. Interestingly, such a process has seen the involvement of the local community with residents of Albion being major stakeholders in the same.

Spatial planning of the lagoons has also become of prime importance, and the government has already embarked on the zoning and sustainable management of the lagoon. In addition, to move towards the "greening" of the tourism industry, the government of Mauritius is in the process of introducing an eco-label scheme for the environment and sustainability of the sector. The eco-labelling will be conducted through a certification programme. In this respect, the Ministry of Tourism and Leisure, in collaboration with the Mauritius Standards Bureau (MSB) and other key stakeholders of the Tourism

Industry, have developed eco-label standards for the tourism industry. The eco-label standards will be applicable to the following tourism businesses: accommodations (hotels, guesthouses, tourist residences); restaurants; tour operators; tourist attractions (heritage, natural and cultural); and pleasure craft and related activities such as boating, scuba diving, helmet diving and parasailing.

Finally, other actions which have been undertaken include, amongst others:

- vigorous efforts to address beach erosion in sensitive areas, especially along coastlines (e.g. Flic en Flac);
- maintenance of touristic sites and the rehabilitation of some attractions with an embellishment programme at key touristic locations. Measures to achieve a greener destination via buffer zones, limits on access and constraints on the circulation of vehicles in sensitive areas are being implemented;
- Achieve green productivity – (e.g. capture solar energy, bio gas, reduce wastages).

Constraints related to the island's topography

The majority of interviewees were quick to point out that there are a number of constraints limiting the expansion of an eco-tourism sub-sector in Mauritius. Firstly, the island's topography does not favour the expansion of inland eco-tourism. In this regard, there were calls for further fostering and promoting the "Iles Vanilles" concept given that islands such as the Seychelles and Mauritius were predominantly resort locations. The topography of islands such as Reunion and Madagascar was more conducive to the fostering of an eco-tourism concept, and as such tourists could be offered an experience mixing a resort with an eco-tourism option. Unfortunately, the outcome so far has delineated an opposite trend with tourists visiting the island favouring a package which includes Mauritius and Seychelles only. Such a trend clearly points to the intrinsic specificity of the island's visitors who are predominantly favouring the 3Ss concept.

Overwhelming perception of Mauritius as a resort destination

Whilst there was an overwhelming consensus that there was a dire need to move beyond the 3Ss, the same stakeholders were realistic in their assessment of the potential impact of the eco-tourism sector in that they argued that the island will always be recognized as a resort destination and that the latter will remain our core product. However, there was also the general belief that any initiative to promote an eco-tourism sub-sector will fail unless there is a change of mindset amongst the various stakeholders, including promoters, policy makers and the population at large. There needs to be the realization of the need for sustainable growth which includes the-long-term sustainability of the tourism sector. With the aforementioned problems and challenges, it is obvious that an over-reliance on beach tourism is not sustainable in the long term and as such, an amalgamation with other types of tourism is crucial. Thus, the stakeholders argue that there is a dire need for more frequent sensitization and educational campaigns with the main aim of highlighting the importance of green and sustainable tourism concepts. Similarly, there are also claims that the sole focus of any business model in the tourism sector is a return on investment. There are a number of projects including eco-lodges that have failed to take off given the very meagre return on investment. However, the interviewees argue that the prevalence of sustaining the sector through the establishment and promulgation of green and eco-projects, albeit the returns on investment on the same are low at the onset, needs to be heightened and this calls for a change in mindset amongst investors and entrepreneurs.

Lack of stakeholders' collaboration

More importantly though, the discussions also point to a need for greater inter-institutional communication, collaboration and coordination amongst the various tourism stakeholders to foster a greater harmonisation of policy prescriptions and implementation. For instance, one interviewee argued that any proposal for a holistic plan with respect to the tourism sector calls for greater interaction at the highest level from which would emanate clear guidelines and responsibilities which will then be cascaded for implementation purposes. Similarly, there are arguments that the realization of sustainable eco-tourism can sometimes be challenging due to the multi-jurisdictional management issues at eco-tourism sites.

Financial constraints

The main deterrent to fostering eco-tourism on the island is the lack of funding or the willingness by various stakeholders, including the government, to invest in the protection and/or conservation and refurbishing of the eco-sites. At some point in time, a Tourism Fund was established to finance tourism and tourism-related projects, but unfortunately, such a fund has now been integrated into the consolidated fund which entails that availing of the required funding to foster new and to sustain existing eco-sites has become even more painstaking. This is why there is the general belief that entrance fees to nature parks as well as the use of CSR funds through the establishment of a foundation dedicated solely to the protection and conservation of such sites is a viable option.

Other supply-side constraints

Finally, other supply-side constraints which have been identified include, amongst others: the limited area under protection and inadequate active conservation management of native ecosystems; inadequate protection for biodiversity, especially on private land; incomplete inventory with respect to eco-sites; limited research or monitoring to support adaptive management; limited awareness of the population at large; limited development of conservation as a profitable venture; commitment to conservation not necessarily a priority for government; lack of data and information; lack of consistent guidelines and procedures; and lack of public awareness and sensitization.

Conclusion and policy recommendation

The aim of this chapter was to identify the various supply-side factors hindering the expansion of an eco-tourism sub-sector and propose remedial measures accordingly. This was undertaken through a review of the existing regulations and legal frameworks and through discussions with various sector stakeholders. Discussions and in-depth interviews with stakeholders reveal that the main constraints to promoting an eco-tourism sub-sector include: lack of finances with respect to the fostering and conservation of eco-sites; lack of a holistic approach for the strategic orientation of the sector; lack of educational and sensitisation programmes and the prioritisation of return on investment when making investment decisions in the sector; the island's topography is not conducive to fostering eco-tourism; and finally, a lack of inter-institutional communication, collaboration and coordination amongst the various tourism stakeholders.

The above clearly points to the need for wholesale changes with respect to the current structure and policy measures presently underpinning the eco-tourism sub-sector. Given the supply side constraints which the sector is facing currently, the following measures and courses of actions are proposed.

1 There is a clear need for a holistic approach to tourism in Mauritius which would also cater to the requirement of the eco-tourism sub-sector. For instance, with respect to eco-tourism projects, it is crucial that the Ministries overseeing the tourism and the environment sectors co-ordinate their actions and combine their expertise, clearly delineating each ministry's role to avoid any duplication of initiatives and to ensure policy coherence. In addition, such a plan should foster a participatory approach which would include all stakeholders, including the private sector and the local community. As such, during the implementation of eco-projects, a comprehensive strategy needs to be conceived, and this should cover the entire range of necessary eco-tourism aspects from participatory planning and product development to marketing and monitoring.
2 Closely related to the importance of integrating the local community into any eco-tourism project formulation and design is the need to educate all of the tourism stakeholders as to the importance of fostering the long-term sustainability of tourism as a core product. This calls for the sensitization of all parties through educational programmes, frequent discussions and a sensitization campaign which can only serve to bolster a change in mindset from one which is overtly favouring return on investment to one

which also considers the element of sustainability. Additionally, training programmes should become a major component of any ecotourism project with local communities, NGOs and protected areas' agencies being the main target groups.

3 Similarly, there needs to be a greater synergy with the private sector and fostering private public partnerships between private companies on the one hand, and the government, local communities and NGOs on the other. This is particularly important for generating the required funding for developing new, and maintaining existing, eco-sites and to ensure proper monitoring. This is even more crucial in view of the ever-contracting budgetary allocation being given to these types of projects. One should remember that the Tourism Fund no longer exits, and finding alternative sources of funding is crucial. In this regard, the setting up of foundations using CSR funds and private sponsorships may prove useful. Additionally, government should be able to take bold measures with respect to charging entrance fees to eco-sites, as is the case presently for the SSR Botanical Garden. For instance, charging fees for accessing Albion Beach, which has been earmarked for Blue Flag status may be the only way forward to ensure sustainability of the same. Nonetheless, one should not underscore the interplay between the strategies proposed thus far. As an example, to ensure the availing of funds, it is crucial that there is a change in mindset amongst the various stakeholders, including the private sector, which can only be achieved through educational programmes and through their integration at the onset in the design and implementation of any eco-tourism projects.

4 Certification has a central role to play in promoting environmental and sustainability performance and satisfying consumer demand for responsible travelling. Initiatives such as Blue Flag, Fair Trade and Green Globe certifications should be reinforced, and tourism enterprises need to be continually sensitized as to the benefits thereto, especially in view of the predicaments linked to environmental degradation and non-sustainability of the island's current tourism trend. In this regard, there is a need for policies geared towards incentivizing hotels and other tourism businesses to become more sustainable. Comprehensive tourism development strategies need to be developed in partnership with the community and other industry stakeholders, including tour operators. To that end, incentives may help leverage sustainability initiatives in resorts to develop niche travel products which may allow local communities to preserve their cultural heritage while at the same time potentially increasing market share and profitability.

5 Discussions with stakeholders also revealed the absence of any branding exercise with respect to the island offering opportunities for eco-tourism experience. Branding of the same is critical for informing prospective visitors and for meeting their expectations once they arrive. In this regard, the relevant authorities need to embark on a branding and marketing exercise of its ecotourism opportunities. One should not forget that the island

does possess some unique selling points which include amongst others the Dodo, its mixed cultural heritage and its natural parks. To that end, delivering a single brand message for the entire island is fundamental in creating communication efficiencies, solidifying a good target market and enhancing overall visitor expectations; in a nutshell, developing a branding message that introduces and entices visitors and authenticates the Mauritian brand.

Notes

1 Freedman (1995) defines "ecotourism as a segment of the travel industry which appeals to the environmentally conscious and which has a low impact on the surrounding area while contributing to the local economy".
2 Statistics Mauritius (2018) http://statsmauritius.govmu.org/English/StatsbySubj/Pages/INTERNATIONAL-TRAVEL-and-TOURISM.aspx

Bibliography

Ahmed, N. (2015). *Trade in services: Bangladesh context. Multi-year expert meeting on trade, services and development.* Geneva, 11–13 May 2015. UNCTAD.

Bank of Mauritius. (2018). Bank of Mauritius Annual Report 2018. BoM, Port-Louis. Retrieved from www.bom.mu/publications-statistics/publications/annual-report/annual-report-year-ended-30-june-2018.

Beachcomber. (2009). "Going green and leading the way in the development of sustainable tourism: Helping the environment and aiding local communities". In *Beachcomber Newsletter June.* Available at www.beachcomber.com.au/news/2009/jun/going_green [accessed 29 August, 2019]

Davis, P. B. (1999). Beyond guidelines: A model for Antarctic tourism. *Annals of Tourism Research, 26*(3), 516–533.

Deloitte & Touche. (2002). Tourism Development Plan for Mauritius, Mauritius.

De Vicente, J. (2004). *Demand for certification: Tourist industry and marketing experts.* Center on Ecotourism and Sustainable Development. The International Ecotourism Society.

De Witt, L., Merwe, P., & Saayman, M. (2011). *An ecotourism model for South African National Parks.* International conference on tourism & management studies – Algarve 2011.

Dickey, A., & Higham, J. (2005). A spatial analysis of commercial ecotourism businesses in New Zealand: A *c* 1999 benchmarking exercise using GIS. *Tourism Geographies*, 7, 373–388.

Dodds, R., & Joppe, M. (2003). The application of ecotourism to urban environments. *Tourism, 51*, 157–164.

Edgall, D. L. (2006). *Managing sustainable tourism: A legacy for the future.* Haworth Press, New York.

Freedman, A. J. (1995). Ecotopia. *National Review, 47.*

Hawkins, D. (2004). A protected areas ecotourism competitive cluster approach to catalyse biodiversity conservation and economic growth in Bulgaria. *Journal of Sustainable Tourism, 12*, 219–244.

Higham, J., & Lück, M. (2002). Urban ecotourism: A contradiction in terms? *Journal of Ecotourism, 1*, 36–51.

Honey, M. (2008). *Ecotourism and sustainable development: Who owns paradise?* (2nd ed.). Washington, DC: Island Press.

Jenkins, R.-G. (2014). *Agriculture and tourism development: The case of St Kitts*. Masters in Environmental Studies, University of Waterloo, Canada.

Khadaroo, J., & Seetanah, B. (2007). Transport infrastructure and tourism development. *Annals of Tourism Research, 34*(4), 1021–1032.

Lawton, L., & Weaver, D. (2001). Ecotourism in modified spaces. In D. Weaver (Ed.), *Encyclopedia of ecotourism* (pp. 315–326). Wallingford: CAB International.

Mackoy, R., & Osland, G. (2004). Lodge selection and satisfaction: Attributes valued by ecotourists. *Journal of Tourism Studies, 15*, 13–25.

Mao, N., Grunfeld, H., DeLacy, T., & Chandler, D. (2014). Agriculture and Tourism Linkage Constraints in the Siem Reap-Angkor region of Cambodia. *Tourism Geographies*, doi.org/10.1080.14616688.2014.915878.

Meyer, D. (2007). Pro-poor tourism: From leakages to linkages: A conceptual framework for creating linkages between the accommodation sector and poor neighbouring communities. *Current Issues in Tourism, 10*, 558–583.

Nunkoo R., Seetanah, B., Jaffur, Z. K., Moraghen, W., & Sannassee, R. V. (2019). Tourism and economic growth: A meta regression analysis, forthcoming in *Journal of Travel Research*.

Osland, G., & Mackoy, R. (2004). Ecolodge performance goals and evaluation. *Journal of Ecotourism, 3*, 109–128.

Pope. (2005). A study of the nature-based tourism supply side in Zambia. *Whydah Consulting Ltd.* Development and Natural Resource Consultants.

Rogerson, C. M. (2012). Tourism-agriculture linkages in rural South Africa: Evidence from the accommodation sector. *Journal of Sustainable Tourism, 20*, 477–495.

Rogerson, Jayne M. (2014). Maximising the local development potential of Nature Tourism accommodation establishments in South Africa. *African Journal of Hospitality, Tourism and Leisure, 3*(1), 2223–814X.

Sannassee, R.V., & Seetanah, B. (2014). The influence of trust on repeat tourism: The Mauritian case study. *Journal of Hospitality Marketing & Management* 140819121514009.

Seetanah, B. (2011). Assessing the dynamic economic impact of tourism for island economies. *Annals of Tourism Research, 38*, 291–308.

Seetanah, B., & Khadaroo, J. (2008). Transport infrastructure and tourism development. *Annals of Tourism Research, 34*(4), 1021–1032.

Seetanah, B., & Sannassee, R. V. (2014). Marketing promotion financing and tourism development: The case of Mauritius. *Journal of Hospitality Marketing and Management*. http://dx.doi.org/10.1080/19368623.2014.914359.

Seetanah, B., Sannassee, R. V., Teeroovengadum, V., & Nunkoo, R. (2019). Air access liberalization, marketing promotion and tourism development. *International Journal of Tourism Research, 21*(1), 76–86.

Silva, G., & McDill, M. (2004). Barriers to ecotourism supplier success: A comparison of agency and business perspectives. *Journal of Sustainable Tourism*, 12, 289–305.

Sinclair, M.T. (1998). Tourism and economic development: A survey. *The Journal of Development Studies, 34*, 1–51.

Smeral, E. (1998). The impact of globalization on small and medium enterprises: new challenges for tourism policies in European countries. *Tourism management, 19*(4), 371–380.

Torres, R., & Momsen, J. (2004). Challenges and potential for linking tourism and agriculture to achieve pro-poor tourism objectives. *Progress in Development Studies, 4*, 294–318.

United Nations Climate Conference Copenhagen. (2010). Event sustainability report. *Copenhagen Sustainable Meetings Coalition*, 1–76.

United Nations World Tourism Organisation (UNWTO) and United Nations Environmental Policy (UNEP) (2002). Final Report, World Ecotourism Summit, UNWTO and UNEP, pp. 2–137.

Weaver, D. B. (2001). Ecotourism as mass tourism: Contradiction or reality? *Cornell Hotel Restaurant Administrative Quarterly*, *42*(2), 104–113.

Weaver, D. B. (2005). Mass and urban ecotourism: New manifestations of an old concept. *Tourism Recreation Research*, *30*, 19–26.

Weaver, D. B., & Lawton, L. J. (2002). Overnight ecotourist market segmentation in the Gold Coast hinterland of Australia. *Journal of Travel Research*, *40*, 270–280.

Weaver, D. B., & Lawton, L. J. (2007). Twenty years on: The state of contemporary ecotourism research. *Tourism Management*, *28*(5), 1168–1179.

World Trade Organization (WTO). (2010). Trade policy review report by MALAWI. WT/TPR/G/231, 5 May 2010.

15 Investigating sustainable development goals, livelihoods and tourism development in a rural protected area

The iSimangaliso Wetland Park, South Africa

Felicite A Fairer-Wessels

Introduction

Coastal rural communities, especially in and adjacent to protected areas, are faced with basic survival challenges (iSimangaliso Annual report, 2016) that relate directly to Sustainable Development Goals (SDGs): 1 (no poverty), 2 (zero hunger), 3 (good health and well-being) and 5 (gender equality). The sustainable development goals: 4 (quality education), 8 (decent work and economic growth), and 11 (sustainable communities) are also imperative, with goal 13 (climate action) impacting overall. The 2030 Agenda for Sustainable Development adopted by all United Nations Member States in 2015 provides a shared blueprint for peace and prosperity for people and the planet, now and into the future. At its heart are the 17 Sustainable Development Goals (SDGs), which are an urgent call for action by all countries, developed and developing, in a global partnership. They recognize that ending poverty and other deprivations must go hand in hand with strategies that improve health and education, reduce inequality and spur economic growth – all while tackling climate change and working to preserve our oceans and forests.

The most appropriate SDG goals within the context of this research appear to be: 1: no poverty; 2: zero hunger; 3: good health and well-being; 4 quality education; 5: gender equality; 6: clean water and sanitation; 7: affordable and clean energy; 8: decent work and economic growth; 11: sustainable (cities and) communities; 13: climate action; 14: life below water; 15: life on land; 17: partnerships for the goals.

A qualitative method was employed and supplements existing research in the area (Gumede, 2009; Fairer-Wessels, 2017; Jiyane & Fairer-Wessels, 2012; Nzama, 2008;) with the aim to determine, within the context of the sustainable livelihoods approach and sustainable tourism development, whether current primary activities of communities; climate change (refer to SDG 13) and preservation of the Wetland (SDGs 14 and 15); and the importance of skills training for (tourism) employment (SDGs 4 and 8) have an economic impact (SDG 8, decent work and economic growth) on the livelihoods of the selected rural communities in a protected area. The SDGs are seen as the context within which any sustainable tourism development and activity can take place. The specific objectives are to determine:

- the primary activities undertaken by the selected communities in order to sustain their livelihoods, taking climate change into consideration;
- the level of awareness on the importance of preserving the wetland park as a natural resource for sustainable livelihoods and sustainable tourism development within a climate change context among the selected communities;
- the level of awareness on the importance of skills training for employment opportunities as a tool to improve livelihoods through sustainable tourism development among selected communities.

The iSimangaliso Wetland Park World Heritage with its four Ramsar sites (UNESCO, 1999) in Kwa-Zulu Natal, South Africa is used as a case study. This protected area is selected because of its sensitive natural heritage and exceptional biodiversity on which the climate and humankind is impacting. Three communities, geographically spaced and within the iSimangaliso Wetland Park and four Ramsar sites are selected.

Sustainable tourism development and community livelihoods

The most commonly used definition of "sustainable development" is based on the Brundtland report which maintains that sustainable development is development that meets the needs of the present without compromising the ability of future generations to meet their own needs (United Nations, 1987). This definition is unbalanced as it focuses too heavily on socio-economic concerns (the people) and only hints at environmental concerns, while sustainable development must promote the idea that social, environmental and economic development is achievable while protecting the earth's natural resources. The United Nations states that sustainable tourism should aim to: make optimal use of environmental resources that constitute a key element in tourism development, maintaining essential ecological processes and helping to conserve natural heritage and biodiversity; respect the socio-cultural authenticity of host communities, conserve their built and living cultural heritage and traditional values and contribute to inter-cultural understanding and tolerance to ensure viable, long-term economic operations, providing socio-economic benefits to all stakeholders that are fairly distributed, including stable employment, income-earning opportunities and social services to host communities and contributing to poverty alleviation (UN, 1987; UNEP & UNWTO, 2005, p. 11).

According to the UNWTO (2005, p. 12), sustainable tourism is "tourism that takes full account of its current and future economic, social and environmental impacts, addressing the needs of visitors, the industry, and the environment and host communities." Sustainable tourism development expects the informed participation of all relevant stakeholders and strong political leadership if it is to ensure wide participation and consensus building (SDG 17) (UNWTO, 2000). The World Travel and Tourism Council maintains that impacts on natural and cultural environments (SDGs 14 and15) should be positive, that sustainable growth (SDGs 11 and 12) should provide benefits for all sectors of society, including young children, women (SDG 5 and 10) and indigenous people, that a skilled workforce (SDGs 4 and 8) should be developed, that demand for sustainable products (SDG 12) should be stimulated and that technology should be used for innovative solutions (SDG 9) (WTTC, 2013). [SDGs added by the author].

Research on sustainability and livelihoods remains incomplete with Chambers and Conway (1992, p. 7) defining a sustainable livelihood as one that "comprises the capabilities, assets (including both material and social resources) and activities required for a means of living" and states that a livelihood is sustainable "when it can cope with and recover from stress and shocks, maintain or enhance its capabilities and assets, while not undermining the natural resource base." Scoones (2009) again investigates sustainability as an integral element of the livelihoods framework that captures the complex interactions between people and their natural resource base. Other research (Ellis, 2006) on sustainable livelihoods reflects on rural households with a mix of agricultural and non-agricultural activities for livelihoods diversification, while another perspective emphasizes the need to gain a deeper understanding of rural livelihoods within the sustainable rural livelihoods approach and the obstacles of access to education and information for the rural poor (Fairer-Wessels, 2017; Jacobs & Makaudze, 2012).

Tao and Wall (2009, p. 97) propose "sustainable livelihoods" as a more practical approach in contrast to sustainable tourism development, as it has the potential to alleviate poverty, increase rural development and sustain the environment (Chambers & Conway, 1992; Ellis, 2000; Jacobs & Makaudze, 2012; Manwa, 2014; Spenceley & Meyer, 2012).

A study by Nzama (2008) on the ecological management of iSimangaliso Wetland Park indicates that respondents, including Kwa-Zulu Natal Ezemvelo officials and business owners, were positive about tourism development involving the restriction of four-by-four vehicles at Sodwana Bay that had compromised the protection of turtles, and they also felt that new forms of tourism growth should be encouraged, although local community respondents working and residing in the area felt that such restrictions were robbing them of their livelihood. Nzama is of the opinion that World Heritage Site authorities should encourage local communities working in the park to participate fully in activities and educational programs to enhance their understanding of nature and ecosystems.

Skills training in the basic know-how of the tourism businesses could help rural communities to find jobs in tourism and to access the tourist market, and bring benefits to such communities (Ashley & Jones, 2001; Botha, 2006; Okazaki, 2008). Among the obstacles that impede the success of such initiatives in South Africa are rural communities' lack of skills regarding involvement in tourism development. Compounding the problem is that rural communities in developing countries are struggling to meet their basic needs, with day-to-day survival taking precedence over higher needs, and participation in the tourism development process is by no means a priority for rural communities (Apleni, 2013; Tao & Wall, 2009; Tosun, 2000) with "peasant farmers and villagers ill-disposed to ideas of sustainable tourism" (Yasarata, Altinay, Burns, & Okumus, 2010, p. 354).

Climate change and tourism

Climate change is a central concern that influences the survival of humankind at a global level. This includes the natural and cultural resource base. In an attempt to address this global phenomenon all spheres of life have to make a concerted effort to mitigate and reduce their contribution to this potentially catastrophic occurrence (Mearns, 2016). Tourism plays an important role when looking at job creation and aiding impoverished communities worldwide. The need to ensure the sustainability of this sector cannot be overstated (Petrovic & Markovic, 2013). The impact that climate change will have on sustainable tourism is a complicated point of research because climate change and tourism are both dynamic and complex in their own right. Future trends that should be looked out for include the following: a general rise in the global climate at least in part due to greenhouse gas emissions as well as climatic surprises due to the inability of current models to predict with absolute certainty the extent and timeframe in which such extreme events may occur (Dubois & Ceron, 2006). The most vulnerable areas to climate change in terms of their tourism industries are developing nations in the Caribbean, Small Island developing States, Southeast Asia and Africa (Grosling, Hall, & Scott, 2009). With the dependence of rural livelihoods and many tourism businesses on specific climatic features, predicting local scenarios is not only preferable but necessary for their survival and sustainability.

The case study: iSimangaliso World Heritage Site

The Greater St Lucia Wetland Park was proclaimed in 1999 under the World Heritage Convention Act due to its exceptional biodiversity and ecological processes. Sixteen parcels of land were consolidated in 1895 to create an integrated protected area of over 358,534ha stretching 280 km along South Africa's coast fronting the Indian Ocean from Maphelane nature reserve in the south to Kosi bay in the north, including the successful settlement of nine major land claims that sketches the political background of the destination (Briggs, 2006; Gumede, 2009; Hansen, 2014). In 2008 the new name, *iSimangaliso* ('miracle' in isiZulu), was gazetted.

The park consists of two marine protected areas, the St Lucia and Maputaland marine reserves, 11 non-marine areas and four Ramsar sites (St Lucia Lake system, the turtle beaches/coral reefs of Tongaland, Kosi Bay Lake system and Lake Sibaya). Its five interlinked ecosystems, with rich microhabitats of mangrove, coral reefs and raffia-palm forests, include the coastal and marine system, the eastern shore dune forests, the estuary and lake systems, the Mkhuze River papyrus swamps and the acacia savannah on the western shore (Briggs, 2006; Nzama, 2008).

The wetland has been under threat for decades with the health status of Lake St Lucia compromised due to reduced water from the diverted Umfolozi River placing stress on the ecosystem (Briggs, 2006). In addition the upper Umfolozi swamps were drained to make way for sugarcane in the 1930s. Climate change (SDG 13), such as unusual droughts, has increased Lake St Lucia's salinity and killed off the shoreline vegetation, causing bank erosion and silting of the lake. Damage by tourism and overuse of resources (SDG 12) through unsustainable fishing; land claim issues by local communities in the area; infestation by alien invasive plants; and the grounding of an oil tanker near the park in 2002 are threats that have led to irresponsible and unsustainable development (Briggs, 2006; Gumede, 2009; Hansen, 2014; Nzama, 2008).

The iSimangaliso Wetland Park has not only environmental but also recreational and tourism value, for local people and tourists alike, especially at Sodwana Bay, a popular scuba diving and deep-sea fishing destination that annually attracts thousands of enthusiasts from South Africa and abroad. Local communities are also involved in these activities at Sodwana Bay, not only for jobs but also for recreational purposes (Nzama, 2008). Issues of sustainability revolve around conflicts of interest between developers and investors with poor locals wanting to develop agriculture and alternative styles of development (Yasarata et al., 2010).

The daily management of the iSimangaliso Wetland Park is undertaken by Kwa Zulu Natal Ezemvelo Wildlife, a provincial agency mandated to carry out biodiversity conservation in the province of KwaZulu-Natal. The management of the park, however, remains a complex situation that necessitates the need for good governance and leadership.

In order to put sustainable tourism development into practice always involves a political dimension (Yasarata et al., 2010). Several noteworthy government policies on sustainable tourism development have been gazetted since 1996. The first policy was the White Paper of 1996 that focused on sustainable tourism development; followed by the 2002 UN Conference of Environment and Development World Sustainability Forum (hosted in Johannesburg) that set sustainability targets (DEAT, 2002); the National Tourism Sector Strategy (Republic of South Africa, 2010) that focussed on tourism growth for communities and poverty alleviation; the Protected Areas Act of 2011 that regulated community access to natural resources in coastal forestry areas (EKZNW, 2011, 2013); and the Industrial Action Plan (DTI) that earmarked tourism as a growth sector for community benefits and job creation. To date these policies have not yielded substantial benefits.

The study sites

Nzama (2008) recommends that world heritage site authorities should identify activities such as capacity building and empowerment to enhance tourism development and local communities could be encouraged to understand the sensitivity of a natural site. For this investigation three coastal sites with tourism activity and development within the iSimangaliso Wetland World Heritage Site were selected to represent a geographic spread of communities. The sites were the municipality of Mtubatuba (including St Lucia town and Maphelane nature reserve) in the southern region of the park; the municipality of Sodwana Bay in the middle region; and the district municipality of uMhlabuyalingana (including the rural communities of Kosi Bay) in the northern region. These coastal sites are justified geographically along the coast as especially Lake St Lucia (Mtubatuba) and Sodwana have well-developed tourism activity with Kosi Bay to a lesser extent. Most other coastal sites, with the exception of Cape Vidal, are quite underdeveloped from a tourism perspective.

As the total population of the iSimangaliso Wetland Park and its immediate surroundings is in the range of 500,000 people, this research is a snapshot of available unemployed male and female community members at the three sites taken over three consecutive days. This research is limited in participant numbers due to financial and time constraints and should rather be regarded as a pilot study.

Research methods

A qualitative analysis based on focus group discussions (FGDs) with community members at grassroots level, one FGD at each site was utilised. In-depth interviews were held with local experts and tourism operators at each site and with stakeholders at the regional and government level. Interview schedules were used as a research instrument for both groups, and the data from the FDGs and in-depth interviews were transcribed and the content analysed according to the objectives of the study.

The population of the three FDGs consisted of a total of 28 available unemployed male and female (roughly equally represented) adult community members using non-probability convenience sampling (Cater & Low, 2012; Du Plooy, 2009), with 12 participants from Mtubatuba, 6 from Sodwana Bay, and 10 from Kosi Bay. Due to the high levels of illiteracy, the use of FGDs was deemed suitable to allow for oral communication with the participants, with an experienced local fieldworker interpreting the discussions from isiZulu to English.

In-depth personal interviews using non-probability purposive sampling was used for the small population with special knowledge in terms of the objectives, i.e. senior teachers and representatives from: iSimangliso Wetland Park Authority and KZN Ezemvelo Wildlife, Department of Education, Department of

Agriculture, Forestry and Fisheries (DAFF) and local private tour operators from Sodwana Bay, Mtubatuba and Kosi Bay.

Findings and discussion

Primary livelihood activities of the selected rural communities

At the three sites the FDGs and interviews reveal that the communities are primarily engaged in subsistence fishing using traditional fish-traps, mussel harvesting, hunting and subsistence farming. Fishing and farming are severely affected by the drop in water levels (SDG 13). Regarding mussel harvesting and farming, the interviewees said that the communities are not aware (SDG 4,12) of the best seasons for mussel harvesting and farming. Accordingly, mussels are prematurely harvested and crops fail due to planting in incorrect seasons (Jiyane & Fairer-Wessels, 2012). Livestock, their economic base, is perishing from lack of water. This affects both the livelihoods and cultural traditions of the communities, since cattle and goats are needed to perform traditional ceremonies. The interviewees said that the communities need to be skilled in basic farming and other occupations in order to raise them from the level of basic day-to-day subsistence survival and to help them towards a more sustainable livelihoods approach (SDG 12). Community elders are made aware of climatic conditions though the Department of Agriculture, Forestry and Fisheries (DAFF) and Ezemvelo KZN Wildlife but with limited response.

In terms of tourism, the communities of St Lucia and Sodwana are involved in certain tourism activities such as hippo and crocodile watching, horseback riding, snorkelling, scuba diving and deep-sea fishing, with entrepreneurs selling snacks on the beach (SDG 8). At Kosi Bay the community is involved in tourist guiding for birdwatching, turtle hatching (November –February) hiking and fishing.

The iSimangaliso Authority's commercial policy regards the transforming of the tourism sector as central to include equity partnerships between private and mandatory community partners (SDG 17), and job creation and employment equity (SDG 8,10) in tourism facilities and activities, and in the procurement of goods and services from local entrepreneurs and small businesses (SDG 12) that is commendable.

Awareness of climate change and the preservation of the wetland park

FDG participants agree that the wetland is important and must be preserved (SDG 14,15) but express concern about climate change (SDG 13) and how the park is to be protected. They do not feel they have the authority or that it is their duty to protect the park and consider this the responsibility of the iSimangaliso Authority and Ezemvelo KZN Wildlife, since politics has a strong influence on sustainable tourism in developing countries (Farmaki, Altinay, Botterill, & Hilke, 2015).

The FGD men's main concern is the dry rivers and lakes resulting in a severe decrease in fish. They express disappointment with the authorities who are not preserving the natural site by controlling over-fishing through legal licenses and limiting fishing and hunting to certain seasons since jobless people are resorting to fishing for survival. The men also state that the historical planting of pine and eucalyptus plantations caused groundwater depletion but that since the granting of World Heritage status to the park in 1999, these plantations on the lake had been removed to increase water seepage.

The FGD women said they are not responsible to preserve the site, as this is the responsibility of the ward councillors with political clout. This is substantiated by the expert interviewees. Water shortage due to severe drought has previously been their most pressing problem, although water purification from the Jozini and Hluhluwe Dams for local community consumption has alleviated this concern (SDG 6). With regard to climate change, the women maintained that they are not provided with any help from the ward councillors and relevant Departments of Agriculture, Forestry and Fisheries and the Department of Environmental Affairs (DEA) and that only men are privileged to receive pertinent information as this is a patriarchal society (SDG 5,10).

Interviewees maintain that protection of the wetland is important since communities have harvested most of the indigenous trees for firewood and are illegally fishing and hunting for survival. Men in the FGDs indicate that indigenous trees are depleted by wood artisans who carve sculptures as tourism souvenirs, and for medicinal purposes. They state that people need to earn a living, with local elders not able to control over-harvesting of trees.

The iSimangaliso Authority notes the importance of fostering conservation awareness amongst the youth through environmental programmes through local schools substantiated by Nzama (2008). A public awareness programme on all levels is also strongly promoted by Yasarata et al. (2010, p. 355).

Skills and employment opportunities in tourism for sustainable livelihood development and sustainable tourism development

The coastal communities of Mtubatuba, Sodwana and Kosi Bay have unique tourism offerings and a variety of employment opportunities that require skilled labour. Mtubatuba (St Lucia) hosts the largest town with various water and land-based recreational activities with many restaurants and accommodation establishments. Sodwana is regarded as a prime scuba diving and deep-sea fishing location, with both camping and chalet facilities managed by Ezemvelo Wildlife. Kosi Bay with its unique lakes and estuary offers extraordinary snorkelling and fly-fishing experience, but it is not as well-developed a tourist destination as Sodwana Bay and Lake St Lucia. The FGD men specifically mentioned the turtle nesting tourist activity that locally trained Thonga guides oversee with proceeds going directly to the local community.

Vocational skills training can be implemented at local schools in subjects such as tour guiding, basic project management, computer skills, housekeeping, needlework, craft making, pottery, basket and mat weaving, local cuisine, artisans in carpentry, bricklaying, plumbing and more. The interviews indicated that collaboration between the Department of Education and the Department of Environmental Affairs in educating and training of pupils from the primary level, teaching them about the benefits of tourism and the importance of the wetland as a World Heritage Site would be advantageous. During the UN Climate Change Conference in Durban 2011, the media (such as TV, newspapers and radio) made information available on climate issues in the province (UN, 2011), but nearly a decade has passed and the situation remains dire.

Recommendations for sustainable livelihoods and tourism development

Climate change and the environment

Although all of the participants are aware of climate change and its extreme effects on the environment and communities' livelihoods, they are merely surviving within this condition. The crux seems to be that sustainable livelihoods are more critical to attain than sustainable tourism development because communities first need to survive sustainably before they can engage in sustainable tourism activities. First and foremost communities need shelter and food and water before considering tourism activities. Tao and Wall (2009) maintain that by using a practical analogy it may be possible to develop a sustainable livelihoods mindset. For example, if the Kosi Bay women harvesters can be convinced not to harvest the immature mussels and trample the mussel beds, but rather allow them to mature, this will result in a sustainable practice to harvest fully grown mussels that can be sold to local restaurants and become part of the value chain that results in responsible consumption and production (SDG 8,12) (Jiyane & Fairer-Wessels, 2012); or the controlled seasonal (May) harvesting of reeds for the annual Reed Festival (iNcema) and for weaving baskets – because the reeds need to gain maturity before harvesting. A sustainable livelihood therefore comprises the skills such as agricultural skills, assets (including both material and social resources) and activities required for a means of living and which is sustainable when people can cope and recover from misfortune and maintain (or improve, such as annual mussels to harvest) their skills and assets, while not depleting the natural resource base (Chambers & Conway, 1992).

Skills development, employment opportunities and tourism developments

Local authorities should consider undertaking community workshops to empower unemployed adults with hands-on livelihood skills to survive in the context of climate change to make them less reliant on their natural resource base for survival. Rural communities need to be taught skills to alleviate the stress on the remaining indigenous forests. This will assist towards achieving SDGs 1 and 2 in relieving hunger and poverty; and SDGs 8, 11, 12, 13, 14 and 15.

Tourist developments locally are usually small in scale and only contribute in part to a community's socio-economic situation. Communities should be made aware that they are not the primary solution to the socio-economic uplift of an entire community. The findings indicate that communication and collaboration between stakeholders (i.e. communities and the authorities) is often lacking in terms of tourism development and training of skills, which refers to one of the objectives mentioned at the outset. Farmaki et al. (2015, p. 188) maintain that an impenetrable jungle "exists of interpersonal relationships between politicians and powerful business elites who govern the tourism industry, creating undemocratic networks that raise questions over tourism governance structures, destinations management and social learning." All of the participants feel it is the responsibility of the various authorities (i.e. municipal, provincial and national level) to implement training and skills transfer programs for the communities, focusing on the sustainable livelihoods approach, but also supporting sustainable tourism to sustain the natural and cultural resource base.

Conclusion

To date limited research has been undertaken on a broad spread of communities within the protected areas of the iSimangaliso Wetland Park, with authors focusing on certain areas within and on the protected area's buffer zone (Gumede, 2009; Fairer-Wessels, 2017; Hansen, 2014; Mhlongo, 2011; Nzama, 2008). The small number of FGD participants and interviewees of this research is limiting, given that the 2011 population census of the three selected municipalities indicated a total population of nearly 400,000 (Statistics South Africa, 2011); while a Community Survey by StatsSA in 2016 indicated a population growth of over 600,000 in the same municipalities. A survey of the whole area is recommended in 2021 in tandem with the fourth democratic census to identify a wider spread of local communities focusing specifically on unemployed community members, since South Africa currently (2019) has a rate of 27% unemployment countrywide, a dire situation that can only be addressed though skills, training and employment opportunities.

The awareness of communities to preserve the protected area from climate change and their realisation that access to skills development to enhance their sustainable livelihoods is important; also to empower them through further skills in (sustainable) tourism product and service development (refer to SDGs 4, 5, 8, 10, 11, 12, 13, 15). Economic prosperity can only be gained once relevant skills development has taken place, as the lack of trained human resources results in many lost opportunities (SDG 4 and 8). Authorities in protected areas must take ownership in addressing poverty of rural communities through managerial policies and by implementing training programmes.

Considering South Africa's current political context, the livelihood strategy as initially proposed by Tao and Wall (2009) still remains the most feasible approach for survivalist communities in protected areas with the sustainable tourism development approach only attainable once such communities have the necessary skills and have surpassed the survivalist stage. In view of the title of the book, the iSimangaliso Wetland World Heritage Site as a natural resource and protected area will only be conserved for future generations if the challenges of community livelihoods are met and the sustainable development goals are implemented following the principles of good governance.

References

Apleni, L. (2013). *A missed opportunity: Community participation in tourism in South Africa.* A Policy Brief: Sustainable Development Programme, Africa Institute of South Africa.

Ashley, C., & Jones, B. (2001). Joint ventures between communities and tourism investors: Experience in Southern Africa. *International Journal of Tourism Research, 3*(2).

Botha, M. (2006). Measuring the effectiveness of the women entrepreneurship programme, as a training intervention, on potential, start-up and established women entrepreneurs in South Africa. DCom thesis, University of Pretoria, South Africa.

Briggs, P. (2006). *Greater St Lucia Wetland Park.* Johannesburg: 30 Degree South Publishers.

Cater, C., & Low, T. (2012). Focus groups, in handbook of research methods. In L. Dwyer, A. Gill, & N. Saatarem (Eds.), *Tourism: Quantitative and qualitative approaches* (pp. 352–364). Cheltenham: Edward Elgar.

Chambers, R., & Conway, G. R. (1992). *Sustainable livelihood: Practical concepts for the 21st century.* IDS discussion paper 296. Institute of Development Studies, University of Sussex, Brighton.

DAFF: Department of Agriculture, Forestry and Fisheries (DAFF). Retrieved from www.daff.gov.za.

Department of Environmental Affairs and Tourism (DEAT) (1996). The development and promotion of tourism in South Africa (White paper) Government of South Africa.

Department of Environmental Affairs and Tourism. (2002). National responsible tourism development guidelines for South Africa.

Department of Trade and Industry (DTI). (2013). Industrial policy action plan. Economic sectors and employment cluster.

DuBois, G., & Ceron, J. P. (2006). Tourism and climate change: proposals for a research agenda. *Journal of Sustainable Tourism, 14*(4), 399–415.

Du Plooy, G. M. (2009). *Communication research: Techniques, methods and applications.* Lansdowne: Juta.

Ellis, F. (2000). *Rural livelihoods and diversity in developing countries.* Oxford: Oxford University Press.

Ellis, F. (2006). Agrarian change and rising vulnerability in rural sub-Saharan Africa. *New Political Economy, 11*(3), 387–397.

Ezemvelo KZN Wildlife (EKZNW). (2011). *Managing our biodiversity report.* Pietermaritzburg: Ezemvelo KZN Wildlife.

Fairer-Wessels, F. A. (2017). Determining the impact of information on rural livelihoods and sustainable tourism development near protected areas in Kwa-Zulu Natal, South Africa. *Journal of Sustainable Tourism, 25*(1), 10–25.

Farmaki, A., Altinay, L., Botterill, D., & Hilke, S. (2015). Politics and sustainable tourism; The case of Cyprus. *Tourism Management, 47*, 178–190.

Grosling, S., Hall, C. M., & Scott, D. (2009). The challenges of tourism as a development strategy in an era of global climate change. From: Palesuo, E. Rethinking development in a carbon-constrained world: Development of Cooperation and Climate Change.

Gumede, Z. S. (2009). Tourism participation in the iSimangaliso Wetland Park: Perceptions, practices and prospects. Master's thesis, Department of Recreation and Tourism, University of Zululand.

Hansen, M. (2014). *Struggles over conservation space: Social justice in the iSimangaliso Wetland Park, South Africa*. PhD, Lund University, Norway.

iSimangaliso. (2016). iSimangaliso Wetland Park Annual Report 2016. [Retrieved from: https://isimangaliso.com/product/annual-report-2016].

Jacobs, P., & Makaudze, E. (2012). Understanding rural livelihoods in the West Coast District, South Africa. *Development Southern Africa, 29*(4), 574–587.

James, B. (2012). Personal communication with Bronwyn James, Senior manager, Policy, Planning, Research and Development, iSimangaliso Wetland Park Authority. St Lucia.

Jiyane, G.V., & Fairer-Wessels, F. A. (2012). Dissemination of information on climate change: A case study of women mussel harvesters at KwaNgwanase in KwaZulu-Natal. *Mousaion, 30*(1), 19–38.

Manwa, H. (2014). Impacts of Lesotho Highlands water project on sustainable livelihoods. *Mediterranean Journal of Social Sciences, 5*(15), 640–647.

Mazibuko, N. P. (2000). *Community participation in tourism development at KwaNgcolosi, KwaZulu-Natal*. Master's thesis, University of Zululand, South Africa.

Mearns, K. (2016). Climate change and tourism: Some industry responses to mitigate tourism's contribution to climate change. *African Journal of Hospitality, Tourism and Leisure, 5*(2), 1–11.

Mhlongo, S. K. (2011). *Environmental management plan as a tool for tourism development within*. Master's thesis, Department of Recreation and Tourism, University of Zululand, South Africa.

Nzama, A. T. (2008). The nexus between sustainable livelihoods and ecological management of the World Heritage Sites: Lessons from iSimangaliso World Heritage Park, South Africa. *Inkanyiso, Journal of Human and Social Sciences, 1*(1), 34–41.

Okazaki, E. (2008). A community-based tourism model: Its conception and use. *Journal of Sustainable Tourism, 16*(5), 511–529.

Petrovic, M. D., & Markovic, J. J. (2013). Sustainable tourism as a part of comprehensive environmental monitoring: A study of Serbia. *Journal of Environmental and Tourism Analyses, 1*(1), 30–42.

Republic of South Africa. (2010). Department of Tourism: National Growth Sector Strategy. [Retrieved from https://www.tourism.gov.za].

Scoones, I. (2009). Livelihoods perspectives and rural development. *Journal of Peasant Studies, 32*(1), 171–196.

Spenceley, A., & Meyer, D. (2012). Tourism and poverty reduction: Theory and practice in less economically developed countries. *Journal of Sustainable Tourism, 20*(3), 297–317.

Statistics South Africa. (2011). Provincial Profile Kwa-Zulu-Natal, Census 2011. Report no 03-01-74(2011). [Retrieved from www.statssa.gov.za/publications].

Statistics South Africa. (2019). Provincial profile 2019, KwaZulu-Natal. Retrieved from www.statssa.gov.za/publications/Report-00-91-05/Report-00-91-052004.pdf [accessed 5 April 2019].

Sustainable Development Goals (SDG). Retrieved from https://sustainabledevelopment.un.org/?menus=1300.

Tao, T. C. H., & Wall, G. (2009). Tourism as a sustainable livelihood strategy. *Tourism Management, 30*, 90–98.

Tosun, C. (2000). Limits to community participation in the tourism development process in developing countries. *Tourism Management, 21*, 613–633.

United Nations (UN). (1987). Brundtland report. Report of the World Commission on Environment and Development: Our common future. Retrieved 10 April 2019.

United Nations (UN). (2011). Climate Change Conference, Durban: United Nations. [Retrieved from https://unfccc.int/process-and-meetings/conferences/past-conferences/durban-climate-change-conference-november-2011].

United Nations Development Programme (UNDP). (1997). New York. Retrieved from Hdr.undp.org/sites/default/files/reports/258/hdr_1997_en_complete_nostats.pdf.

United Nations Environment Programme (UNEP) & United Nations World Tourism Organisation (UNWTO). (2005). *Making Tourism More Sustainable – A Guide for Policy Makers*. United Nations Environmental Programme (UNEP) and World Tourism Organization (WTO), Paris.

United Nations Scientific and Cultural Organisation (UNESCO). (1999). Retrieved from www.whc/unesco.org/en/list/914 [accessed 5 April 2019].

United Nations World Tourism Organisation (UNWTO). (2000). Retrieved from http://www2.unto.org/ [accessed 6 August 2014].

WTTC (World Travel and Tourism Council). (2013). Tourism for tomorrow: The WTTC perspective. Retrieved from www.wttc.org/our-mission/strategic-priorities/tourism-tomorrow/ [accessed 10 July 2016].

Yasarata, M., Altinay, L., Burns, P., & Okumus, F. (2010). Politics and sustainable tourism development – can they co-exist? Voices form North Cyprus. *Tourism Management, 31*, 345–365.

Part IV

Southern Africa in the world

Cases and lessons

16 Natural resources, sustainable tourism development and community livelihoods relationships

A comparison between Botswana and the USA

Moren T. Stone, Gyan P. Nyaupane, Dallen J. Timothy and Lesego S. Stone

Introduction

Natural resources use and community livelihoods are inseparable and interlinked entities that have drawn global attention in the twenty-first century (Stone & Nyaupane, 2016). The relationship between natural resources and community livelihoods is complex (Nyaupane & Poudel, 2011). In their endeavour to enhance livelihoods, natural resources-dependent communities (e.g. forests, land, soils, water) produce unintended outcomes that may not be consistent with conservation objectives. Due to these relationships, conflicts between natural resources utilisation and community livelihoods have been the focal point of many discussions over the last two decades (Stone & Rogerson, 2011). These conflicts arise due to the non-existence of agreed and conclusive interventions in which relationships between natural resources and community livelihoods can be reconciled without contestations. In Southern Africa, natural resources are mainly hosted by protected areas. In developing countries, most protected areas were established in remote, peripheral regions largely occupied by marginalized and poverty-stricken populations (Sanderson, 2005). Their establishment ignored the needs of local people, thus marginalizing them through access restrictions to resources, resulting in a widespread lack of community support for conservation (Matiku, 2008). Initially, the paramount benefits of protected areas for their existence values and legacy were largely shared by global citizens (Balmford & Whitten, 2003), whereas the costs incurred were absorbed by adjacent local communities (Matiku, 2008). The skewed benefits-costs of living with protected areas necessitated the call for the best possible approach for adjacent communities to benefit. Tourism has been found to be a possible remedy for adjacent communities, taking advantage of the abundant wild resources hosted by protected areas. Research shows that there is a market for nature-based tourism. In 2015, research estimated that globally protected areas receive eight billion visits annually, generating about US$600 billion annually (Balmford et al., 2015). One study shows that 66% of all international travellers are now motivated to travel to experience nature and beautiful scenery (American Express Travel, 2015). In a survey of international travellers from key travel markets, wildlife tourism and nature were consistently ranked in the top five motivating factors for travellers selecting their last vacations (Brand USA, 2015). These figures attest that the market exists.

The natural resources literature, particularly that dealing with conflicts between protected areas and surrounding communities mostly focuses on the Global South, and there is a lack of comparative studies between the Global North and the Global South to explain the issues and the role tourism plays. The purpose of this chapter, therefore, is to explore relationships among natural resources, conservation, community livelihoods and sustainable tourism development in and around protected areas, with specific reference to communities around the Chobe National Park (CNP), Botswana and the Grand Staircase-Escalante National Monument (GSENM), USA. The aim is to compare and contrast the two study sites on how they fair in terms of natural resources, tourism development and conservation. The chapter aims to answer the following questions: what is the relationship between natural resources, tourism and community livelihoods? What impact does tourism have on the sustainability of natural resources?

Background to the study

Nature-based tourism has become a salient component of domestic and international tourism. In the United States, participation in nature-based activities is one of the most important purposes for travel (US Travel Association, 2012). This sub-sector has also registered growth rates generally higher than the tourism industry average. Nature-based activities and many cultural pursuits in the United States take place primarily in protected areas and on other public lands (Nyaupane & Timothy, 2013). It is understood that nature-based tourism is an important ecosystem service that is capable of generating substantial resources for both conservation and local economic development (Balmford et al., 2009). Furthermore, nature and heritage-based tourism can be a tool to create linkages between community development and conservation (Timothy, 2011).

Tourism can help rural communities near protected areas directly and indirectly in several ways. First, local residents may be employed in and operate tourism enterprises. Second, tourism provides a market for local goods and services (Scheyvens, 2007), thereby helping to maximize supply/demand balances and minimize economic leakage (Ollenburg & Buckley, 2007). Finally, tourism tax revenues can benefit residents and communities through investments in local infrastructure and services such as roads, water systems, electricity, and education and health services (Hall, 2007).

Indirectly, tourism can empower communities, enhance a local sense of place, foster community pride and solidarity, and build social capacity which, in turn, improves livelihoods. Empowerment is a multi-dimensional concept that includes economic, social, political and psychological elements (Scheyvens, 1999). Regular economic gains from employment and business opportunities help empower the community economically and improve living standards.

However, to maximize the benefits of tourism in destination communities while simultaneously maintaining the management goals of public lands, strong linkages among conservation, livelihood and tourism are necessary. These cultivate long-term benefits that translate into both conservation and community development goals. This then would transform into sustainable tourism development that calls for ecological, economic and social sustainability.

Principles of sustainable tourism development

The Brundtland report (WCED, 1987) put sustainable tourism development on the world agenda. The World Tourism Organization (UNWTO) defines sustainable tourism as tourism "that meets the needs of tourists and host regions; while at the same time it protects and improves opportunities for the future" (World Tourism Organization, 1993). Sustainable tourism recognises that tourism can support regeneration and economic development, conservation and enhance the quality of life of both visitors and host communities (Cornell, Page, & Bentley, 2009). In the tourism industry, sustainability became a dominant subject due to an increased consciousness of the negative impacts of mass tourism, leading to a quest for alternative forms of management and development options. Sustainability arose as a substitute to neo-classical models of economic development (Choi & Sirikaya, 2006).

Sustainable tourism is guided by the following six principles; it should:

1 contribute to the satisfaction of basic and felt needs of those hitherto excluded in local tourist destinations;
2 reduce inequality and absolute poverty in local tourist destinations;
3 contribute to the emergence of necessary conditions in tourist destinations which will lead local people to gain self-esteem and to feel free from the evils of want, ignorance and squalor;
4 help host communities to be free or emancipated from alienating material conditions of life and from social servitude to nature, ignorance, other people, misery, institution and dogmatic beliefs;
5 accelerate not only national economic growth, but also regional and local economic growth. This growth must be shared fairly across the social spectrum;
6 achieve the above principles indefinitely without compromising the ability of future generations to meet their own need (Tosun, 2001, pp. 290–291).

This chapter adopts sustainable tourism development as the theoretical underpinnings guiding its execution.

Study sites and methods

Description of study sites

Grand Staircase-Escalante National Monument

The chapter focuses on communities adjacent to the GSENM in Utah, USA and CNP, Chobe, Botswana. The GSENM was established on September 18th, 1996, by United States President Bill Clinton using the authority of the 1906 Antiquities Act to protect 1.87 million acres of land for the protection of historic and scientific resources (Bureau of Land Management, 1999). The monument is located in southern Utah. Two-thirds (68%) of the monument lies within Kane County and one-third (32%) lies within Garfield County (Nyaupane & Timothy, 2013).

The monument is primarily surrounded by other federally protected lands, including Dixie National Forest, Capital Reef National Park, Glen Canyon National Recreation Area, and Bryce Canyon National Park. Although the monument shares only 8% of the boundary with private lands, there are many small and large communities adjacent to the monument (Bureau of Land Management, 1999).

On the basis of geographical location and mutual dependence, 16 communities surrounding the monument were selected for the study. The communities are located in the states of Utah (Big Water, Kanab, Mt. Carmel, Orderville, Henrieville, Cannonville, Glendale, Hatch, Alton, Tropic, Panguitch, Bryce Canyon City, Escalante and Boulder) and Arizona (Fredonia and Page), and are

Table 16.1 Population by State and County surrounding GSENM

Community	*State*	*County*	*Population*
Big Water	Utah	Kane	475
Kanab	Utah	Kane	4312
Mt. Carmel	Utah	Kane	112
Orderville	Utah	Kane	577
Glendale	Utah	Kane	381
Alton	Utah	Kane	119
Henrieville	Utah	Garfield	230
Cannonville	Utah	Garfield	167
Hatch	Utah	Garfield	133
Tropic	Utah	Garfield	530
Panguitch	Utah	Garfield	1520
Bryce Canyon	Utah	Garfield	198
Escalante	Utah	Garfield	797
Boulder	Utah	Garfield	226
Fredonia	Arizona	Coconino	1314
Page	Arizona	Coconino	7247
Total			**18338**

Source: (Nyaupane & Timothy, 2013)

diverse in terms of geography, socio-demography and economic characteristics. The total population of the study area is 18,338 (Nyaupane & Timothy, 2013). Each township's population is summarized in Table 16.1.

Trends in economic conditions within the study area follow the national trend; the service sector is a leading source of employment (Bureau of Land Management, 2015). Service sector industries include: utilities; wholesale trade; retail trade; transportation and warehousing; information technology and information services; finance and insurance; real estate, rental and leasing services; professional and technical services; management of companies and enterprises; administrative and waste services; educational services; health care and social assistance; arts, entertainment and recreation; accommodation and food services (Nyaupane & Timothy, 2013; BLM, 2015).

Chobe Enclave Conservation Trust (CECT)

The CECT is a community-based organization (CBO) comprising five villages: Mabele, Kavimba, Kachikau, Satau and Parakarungu. CECT is located adjacent to the Chobe National Park (CNP) where natural resources management through community-based tourism is practiced. The villages are located on a belt that runs along the Chobe-Linyanti River Basin, forming an enclave (see Figure 16.1).

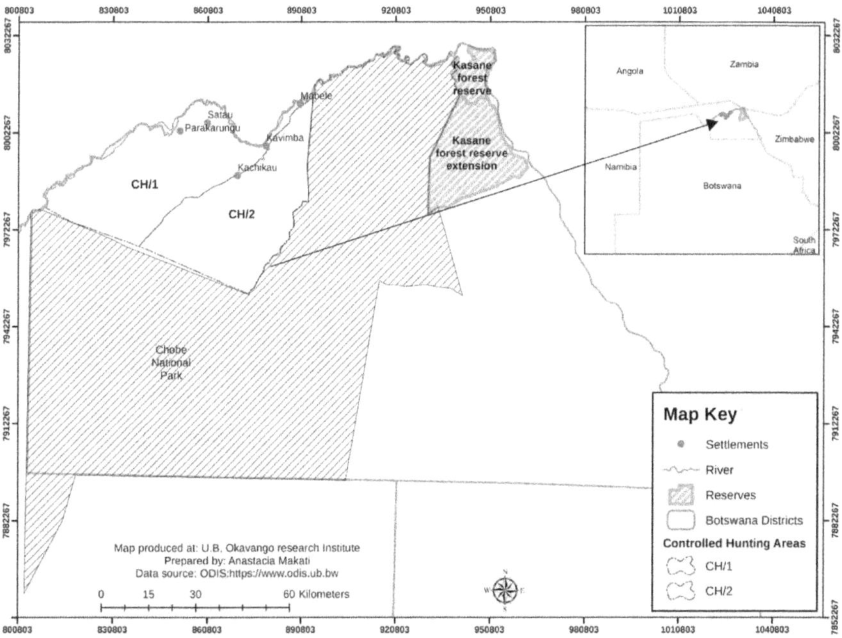

Figure 16.1 CECT villages

Source: Courtesy of Makati (2019) for this book project

Table 16.2 Chobe Enclave Conservation Trust (CECT) study area population

Community	*District*	*Population*
Kachikau	Chobe	1356
Kavimba	Chobe	549
Mabele	Chobe	773
Parakarungu	Chobe	845
Satau	Chobe	605
Total		4128

The estimated population of the Enclave community is 4,128 (Government of Botswana, 2011). Table 16.2 summarizes the CECT area populations. CECT inhabitants are predominately Basubiya, agro-pastoralists dependent on arable farming complemented by sales from livestock. CNP has one of the largest concentrations of biodiversity in Africa (BTO, 2012).

The park was declared a non-hunting area in 1933 and was officially established as a national park in 1968 (BTO, 2012). Before becoming a national park, the land was mainly used as tribal lands for the Basubiya, whose territory spread into neighboring Caprivi in Namibia and Zambia (Jones, 2002). The CNP became prominent in the mid-19th century with colonial hunters, who gathered in the area for big game shooting expeditions (Eyes on Africa, 2002). Before this, wildlife was only hunted customarily by native people. Colonial hunters introduced guns to the native population, resulting in increased killings of wildlife (Eyes on Africa, 2002) Hunting continued until 2014 when it was banned. In June 2019, the hunting ban was lifted. The park creation led to the eviction of the Basubiya tribe from their ancestral land (Jones, 2002). As depicted in Figure 16.1, the CECT community is enclosed by three (3) protected areas – CNP, Kasane Forest Reserve and the Chobe River (Ecosurv, 1996). Before participation in tourism, the CECT was isolated from tourism activities (Nchunga, 1999), a situation which was intensified by effects of the PAs that border the enclave (Jones, 2002).

Methods

Data collection and analysis

The chapter adopted an Appreciative Inquiry (AI) approach to understand the linkages between natural resources, tourism and community livelihoods. AI is an asset-based participatory approach. It advocates for affirmative approaches or positive lenses focusing on the success stories of people and organizations (Nyaupane et al., 2013). This type of research puts theory and practice, action and reflection together to find practical solutions to real-world problems (Reason & Bradbury, 2008). AI is unlike traditional problem-solving or deficiency-based methods that ask people to look backward to identify problems and their causes and then design programs to solve the problems (Nyaupane et al., 2013). AI seeks out the best qualities in individuals and organisations, rather than focusing on problems and deficiencies (Cooperrider & Whitney, 2005). The technique has been shown to be an effective and appropriate research method to study tourism in rural communities (Nyaupane & Poudel, 2012).

Data for both study sites were collected through primary and secondary sources. The data collection process for the GSENM took seven months, from October 2012 to April 2013, while that for the CECT site took one month – August 2012. Data were updated in 2017 and 2018 for the GSENM and CECT sites respectively. Secondary data were gathered alongside primary data throughout the studies. The field data collection process through the AI process involved five steps in a sequence as highlighted in Figure 16.2.

Appreciative inquiry process

The appreciate inquiry process is summarized by Figure 16.2, following a five-step process.

Results

Natural resources, tourism and community livelihoods dynamics

The AI assets audit of the GSENM yielded a diversity of community assets and activities directly linked to natural resources and tourism. Activities include hiking, scenic driving, fishing, hunting, biking and horseback riding, viewing wildlife and beautiful scenery provided by natural areas, including the parks, forests, Hole-in-the-Rock and Grand Staircase-Escalante National Monument. The diverse scenery of the mountains, lakes, streams and desert gives people opportunities for photography and painting. The area has a unique geography that can be enjoyed and studied. Canyoneering and the exploration of nearby forests, parks, lakes and the monument allow tourists to wander through nature. Local markets offer products from the area, and festivals showcase local products, such as the 24th of July pioneer celebrations, marathons, Escalante Art Festival, Heritage Festival, Christmas Craft Fair and heritage and music targeting tourists visiting the GSENM. In response to these GSENM offerings, peripheral communities responded by offering tourism services. There are small local food and lodging facilities, including camping areas owned by local residents. Tourists are able to learn about the local rural lifestyle and geology. There is a deliberate objective in the area to keep tourism small to limit the number of visitors.

At the CECT in Botswana it was found that tourism activities and individuals' engagement in tourism business activities were minimal to non-existent. Instead, a Community-Based Organisation (CBO) approach has been established in the form of a Trust that has been bestowed with running tourism

Figure 16.2 Appreciative inquiry process

Step 1 – Preliminary Field Visit

A five-day (October 13th – October 17) preliminary field visit was conducted to the GSENM study area, while CECT was on August 2nd – 3rd, 2012. The visits were to familiarize the researchers with study sites and create rapport with potential respondents and/or information gate keepers.

⬇

Step 2 – Asset Mapping

The second step in the data collection process was asset mapping. An asset-based mapping tool was designed to audit community assets. The community assets inventory process involved document review and field verifications. The asset-based mapping exercise provided a big picture of the social, cultural and economic environment of the communities in relation to their natural resources use.

⬇

Step 3 – Appreciative Interviews

Interviews with a wide range of stakeholders were executed. A total of 70 appreciative interviews were conducted at the GSENM and 31 at CECT. The respondents interviewed belonged to many stakeholder groups, including community members, village chiefs, tourism entrepreneurs, farmers and ranchers, environmental groups representatives, elected officials, chamber of commerce members and relevant government department agencies representatives.

⬇

Step 4 – Appreciative Inquiry Mini Sessions

The mini-AI sessions were conducted in eight different locations at GSENM and at all five villages at CECT and were open to everyone in the community. To achieve as much public participation as possible, more effort was put into publicizing the mini-AI sessions via personal invitations (emails, in-person invitations), word of mouth, village chiefs and Kgotlas, newspaper announcements, media interviews and flyers at post offices, libraries and town halls. The mini-AI sessions were guided by a semi-structured guide which was developed in advance to steer the discussion in the correct direction, keeping in mind the positive goal orientation of the project. The researchers played the role of both facilitators and moderators.

⬇

Step 5 – Appreciative Inquiry Summits

This was the final step in the data collection process. The purpose of the summits at both study sites was to create regional-level consensus about the use of protected areas' natural resources to improve local livelihoods through tourism development. Participants in the mini-sessions were invited to participate in the main summit with additional stakeholders, including elected officials, federal agencies, tourism offices, environmental groups, tourism entrepreneurs and safari companies. In common with the mini-sessions, the role of the research team was that of facilitators and moderators. The group discussion guide used in the mini-AI sessions was adapted to fit the summit audience.

Source: (Authors, 2019)

activities on behalf of the five villages. Due to non-existent tourism skills, the CBO rented out its concession areas and sells its hunting quotas (pre 2014 hunting ban). Representatives from the five villages are elected every two years to sit on the board of trustees. The main tourism activity at CECT is wildlife-based tourism. In evaluating traditional livelihood undertakings practiced at the CECT area before the introduction of nature-based tourism, key traditional livelihoods were crops and cattle farming, fishing, hunting and wild resources gathering, wholly reliant on natural capital. Agriculture was the central livelihood activity, augmented by hunting and gathering. Subsequent to the adoption of tourism, livelihood undertakings changed to comprise the sale of wildlife quotas, sub-leasing of the community's photographic concession area, employment in tourism establishments and the supply of game meat from trophy hunting. Due to its proximity to the CNP, CECT villages have very high densities of natural capital in the form of wild animals, which significantly increased the value for trophy hunting before the ban on photographic safaris. At GSENM money from tourism trickles directly to households as a number of them provide tourism services directly to tourists. At CECT all money gained from tourism reaches the community through the CBO; the CBO, in consultation with the CECT community decides how to distribute proceeds to beneficiaries. While at GSENM residents participate in tourism individually; at CECT collectivism is promoted.

Active versus passive community tourism participation: sustainability issues

At GSENM tourism helps to support the community financially through employment, entrepreneurial opportunities and fostering local businesses. Tourism helps build global friendships and brings new residents to the community. The community provides services (guides, food and lodging), cultural resource opportunities, information, knowledge and skills required for desert survival and organises events for tourists to enjoy. Natural resources (public lands and resources) provide attractions and educational opportunities to visitors. Visitors return or stay longer because of the love for the area, the monument, state parks and local culture. The abundant scenic beauty also helps promote tourism. Tourism supports public lands and resources through revenue generation and helps provide an understanding about how to minimize impacts and maximise appreciation of public lands and resources. Tourism educates people about respecting the land, archaeological resources, and inaccessible areas in and around the monument. Community members have been proactive in adopting tourism by being hospitable to visitors, providing information and exceptional customer service. The community promotes tourism through travel shows, magazines and newspapers, websites and personal communication to encourage people to visit. Brochures have been created and distributed to promote the area. Community members are also active in the Travel Council, Scenic Byway 12 meetings, the Chamber of Commerce and others to initiate town events. Community members also volunteer at events or help create and display signs and banners to promote the community as a destination. All of these activities have diversified tourism products and activities. In return, tourism helps support the community financially, helps build community pride and revitalises culture.

At CECT a majority of the community does not have direct contact with tourists, neither do they provide tourists services. However, the CBO spearheads tourism activities by leasing out community concession areas and selling wildlife quotas from the Department of Wildlife and National Parks. The community receives a once-off payment for these two activities. The CECT community has not done much to tap into the tourism market by developing tourism products or services further. Comparatively, the GSENM community is actively involved while CECT is passively involved, hence the sustainability of tourism at CECT is in doubt and/or debatable.

Natural resources and tourism: positive and negative outlooks

While not many are involved in the tourism business as entrepreneurs at CECT, many are employed in the tourism industry as waiters/waitresses, receptionists, camps attendants, cooks, cleaners, tour guides, drivers, safari hunting helping hands, guards, and so on. Groups of women and men have responded by producing handicrafts to sell to tourists who visit and/or pass through the villages. Community asset mapping and interview results indicate that employment in tourism and handicrafts businesses have natural resources use implications. Employment creation has been linked to a reduction in dependency on certain natural resources while handicrafts development has heightened the use of certain natural resources. Tourism has therefore reinforced the use of certain plant species used mainly for basket weaving and carving and reduced the reliance on certain species previously used as sources of food. Table 16.3 summarises wild plants that are frequently used to meet the demands of tourists who buy souvenirs from the CECT community. These plant resources uses are on the rise because of the influence of tourism.

Table 16.3 Highly exploited wild plant species

Species	*Family*	*Indigenous name*	*Use*
Ficus sycomorus L.	Moraceae	Mochaba	Furniture /carving
Garcinia livingstonei T. Anderson	Cluisiaceae	Motsaodi	Furniture /carving
Diospyros mespiliformis Hochst. ex A.DC.	Ebenaceae	Mokutsumo	Furniture /carving
Hyphaene petersiana Klotzsch ex Mart.	Aecaceae	Mokololwane	Basket weaving & carving /furniture
Gewia flavescens Juss	Malvaceae	Mokgomphatha	Basket weaving
Berchemia discolor (Klotzsch) Hemsl.	Rhamnaceae	Motsentsela	Dye/colorant

Source: Authors' fieldwork

While tourism development has increased the dependency on certain plant species, it has also contributed to the decrease on the use of other species. The decrease in the use of plants gathered for food is linked to the availability of better diets due to cash obtained through employment in tourism establishments. Table 16.4, below summarises wild plants that are reported to be less frequently used.

In addition, increased human – wildlife conflicts have been reported due to large wildlife populations. Conflicts come about due to the wandering of problem animals (elephants, buffalos, zebras, wild dogs, lions, hyenas and baboons) beyond park boundaries, destroying community crops and livestock. They are also exacerbated by increases in the land area ploughed. Wildlife is also killed and injured due to fencing entanglements and the blocking of wildlife corridors through fencing to protect community field crops.

At GSENM tourism has contributed to broad-based small-scale tourism entrepreneurship; many local residents have converted their residences into guesthouses, opened eateries, sell hiking gears, offer horse rides, bike hiring, festivals, and so on. Negative tourism-related activities have also been reported such as littering, vandalism of canyons, off-track driving, increased crime rates, soil trampling, defacing of landscapes of importance, etc. Residents are also affected by unused tourism facilities and services during the off-peak season. Conflicts are also caused by the community being deprived the opportunity to carry out mining activities in the area, community's perception that their paleontological "finds" are moved to other areas, historical pilgrimages/treks, recreational activities and visual arts permit denials and not being heard because the focus is on "green groups." The community also expressed the lack of communication between them and GSENM compared to other federal public land agencies; this is said to affect travelers/visitors' experience of the GSENM. All of these conditions present a threat to the sustainability of tourism development in the GSENM area.

Table 16.4 Wild plants species less frequently used

Species	*Family*	*Indigenous name*	*Use*
Azanza garckeana (F. Hoffm.) Exell & Hillc.	Malvaceae	Morojwa	Food/fruit
Nymphaea nouchali Burm.f.	Nymphaceae	Tswii	Medicine & food
Ximenia americana L.	Ximeniaceae	Moretologa	Food/fruit & medicine
Hyphaene petersiana Klotzsch ex Mart.	Arecaceae	Mokolwane	Alcohol beverage
Sclerocarya birrea (A, Rich.) Hochst.	Anacardiaceae	Morula	Food/fruit & alcohol beverage
Unidentified sp.		Mokgothwane	Food/vegetable
Grewia flava DC.	Malvaceae	Moretlwa	Food/fruit
Vangueriopsis lanciflora (Hiern.) Robyns ex R.D. Good	Rubiaceae	Mmupudu	Food/fruit

Source: Authors' fieldwork

Discussion

The relationship between natural resources, tourism and community livelihoods is dynamic; at one end tourism is seen as contributing to community livelihoods positively while at the other it brings negative impacts. It is evident that tourism development at both study sites varies in degree and scale of development. At GSENM tourism has advanced if resident participation, diverse tourism activities and secondary attractions are to be used as indicators. Community members own and operate tourism businesses and provide employment opportunities for other members; organise events such as festivals and provide marketing initiatives to promote tourism, something which is missing in the Botswana case. It is important to note that sustainable tourism practice remains a very complex issue and success depends on the level of involvement, diversity of services and empowerment of local communities in sustaining livelihoods and ensuring environmental sustainability. At both study sites, it is evident that the adoption of tourism does not merely endeavour to address economic and environmental conservation concerns only, but also issues of power and equity in society, as advanced by Urry (1990). At CECT a board of trustees has been mandated with spearheading tourism development while at GSENM individuals have embraced tourism by becoming tourism micro-entrepreneurs. The different outcomes further define how tourism benefits trickle down to beneficiaries. At CECT, benefits are distributed to a community CBO that determines how tourism proceeds are distributed to beneficiaries while at GSENM proceeds reach individual community members who directly provide tourism services. Interestingly, at CECT, tourism was/is consumptive in nature as hunting quotas were issued and sold. While the act of hunting wildlife has been challenged, the consumptive nature of tourism through wildlife hunting has increasingly become part of conservation arguments and policy debates (Stone, 2015; Shoo & Songorwa, 2013). However, if it is promoted as a low-impact sustainable use approach, it adds value to natural resources.

Results indicate that natural resource planning and management entails addressing two partly competing and overlapping goals: conserving biodiversity and improving community livelihoods. Resolving potential conflicts between these two goals is particularly challenging at the intersection of natural heritage and economic development (Karanth & DeFries, 2010). It is not disputed that natural resources heritage sites are a popular strategy for managing natural resources conservation. However, their contribution to livelihoods improvement and sustainable development remains contested (Stone & Nyaupane, 2014). Promoting alternative livelihood options within the foundation of available natural resources-based attractions is an obvious management opportunity to reduce pressure on their utilisation and promote co-existence; however, such attempts have mixed results (Kiss, 2004). This is attested by this study's results from CECT, where findings indicate that tourism has reduced pressure on certain wild plants species but has also increased pressure on the use of some that are used to produce handicrafts for the tourism market. Results also allude to increasing human–wildlife conflicts which have implications on community livelihoods. While tourism can be a conservation tool, it can also be blamed for bringing conservation challenges – leading to a "double-edged sword" debate. At GSENM the same argument stands; tourism brings benefits to local residents in many ways, but it also brings challenges through littering, vandalism of the magnificent canyon landscape and off-track driving. It also causes conflicts by not allowing certain activities like mining to be carried out and not hearing all in the community. To bring mitigation measures to reconcile the two tourism outcomes is always a challenge, not only to communities themselves but also to policy makers, community/resource planners and conservation scientists.

Conclusion

As the wave of accommodating tourism through natural resources at protected sites moves forward, it is not only important to evaluate its advancement, but also how such heritage sites and communities incorporate change. As gleaned from this study, a comparison of sites situated in developing and developed countries does not necessarily show major differences in terms of relationships among natural resources, tourism and community livelihoods linkages. Indications are that when tourism is used to strengthen the conservation of natural resources, it becomes an essential component of the processes needed to meet social development priorities and conservation goals.

References

Balmford, A., Beresford, J., Green, J., Naidoo, R., Walpole, M., & Manica, A. (2009). A global perspective on trends in nature-based tourism. *PLoS Biology*, 7(6), 1–6.

Balmford, A., Green, J., Anderson, M., Beresford, J., Huang, C., Naidoo, R., & Manica, A. (2015). Walk on the wild side: estimating the global magnitude of visits to protected areas. *PLoS Biology*, *13*(2), e1002074.

Balmford, A., & Whitten, T. (2003). Who should pay for tropical conservation, and how could the costs be met? *Oryx*, *37*(2), 238–250.

Botswana Tourism Organization (BTO). (2012). Chobe National Park, Northern Botswana. http://www.botswanatourism.co.bw/destination/chobe-national-park [accessed 2 August 2019].

Brand USA. (2015). *Market profiles*. Retrieved from www.the brandusa.com/system/files/programs/documents /FY15BP_MarketProfiles_01.pdf [accessed 21 March 2019].

Bureau of Land Management. (1999). *Grand staircase-Escalante national monument management plan*. Cedar City, UT: Bureau of Land Management.

Bureau of Land Management (BLM). (2015). *Grand staircase-Escalante national monument livestock grazing MMP-A/EIS: Socioeconomic baseline study*. United States Department of the Interior, Bureau of Land Management National conservation land, Utah.

Choi, H. C., & Sirakaya, E. (2006). Sustainability indicators for managing community tourism. *Tourism management*, *27*(6), 1274-1289.

Cooperrider, D., & Whitney, D. (2005). *Appreciative Inquiry a positive revolution in change*. San Francisco: Barrett.

Cornell, J., Page, S.J. & Bentley, T. (2009). Towards sustainable tourism planning in New Zealand: Monitoring local government planning under the Resource Management Act, *Tourism Management*, *30*, 867–877.

The Futures Company and American Express. (2015). *American express travel – future travel trends* (PowerPoint presentation), July 9, 2015.

Government of Botswana. (2011). *Botswana population & housing census*. Gaborone: Government Printers.

Hall, C. (2007). Editorial: Pro-poor tourism; Do tourism exchanges benefits primarily the countries of the South. In C. M. Hall (Ed.), *Pro-poor tourism: Who benefits? Perspectives on tourism and poverty reduction* (pp. 1–8). Clevedon: Channel View.

Jones, B. (2002). Chobe Enclave, Botswana–Lessons learnt from a CBNRM project 1993–2002. Gaborone: IUCN/SNV CBNRM Support Programme.

Karanth, K., & DeFries, R. (2010). Nature-based tourism in Indian protected areas: New challenges for park management. *Conservation Letters, 11*, 1–13.

Kiss, A. (2004). Is community-based ecotourism a good use of biodiversity conservation funds? *Trends in Ecology and Evolution, 19*, 232–237.

Matiku, P. (2008). Poverty and mortality indicators: Data for the poverty conservation debate. *Oryx, 42*(1), 1.

Nyaupane, G., & Poudel, S. (2011). Linkages among biodiversity, livelihood, and tourism. *Annals of Tourism Research, 38*(4), 1344–1366.

Nyaupane, G., & Poudel, S. (2012). Application of appreciative inquiry in tourism research in rural communities. *Tourism Management, 33*(4), 978–987.

Nyaupane, G, P., & Timothy, D. (2013). *Linking communities and public lands through tourism: A Pilot Project Technical Report.* School of Community Resources and Development, Phoenix, Arizona State University, USA.

Ollenburg, C., & Buckley, R. (2007). Stated economic and social motivations of farm tourism operators. *Journal of Travel Research, 45*(4), 444–452.

Reason, P., & Bradbury, H. (2008). *Action research: Participative inquiry and practice*. London, Sage Publications.

Sanderson, S. (2005). Poverty and conservation: The new century's 'Peasant Question?' *World Development, 33*(2), 323–332.

Scheyvens, R. (1999). Ecotourism and the empowerment of local communities. *Tourism Management, 20*, 245–249.

Scheyvens, R. (2007). Exploring the tourism-poverty nexus. In C. M. Hall (Ed.), *Pro-poor tourism: Who benefits? Perspectives on tourism and poverty reduction* (pp. 121–144). Clevedon: Channel View.

Shoo, R., & Songorwa, A. (2013). Contribution of ecotourism to nature conservation and improvement of livelihoods around Amani nature reserve, Tanzania. *Journal of Ecotourism, 12*(2), 75–89.

Stone, M.T. (2015). Community-based ecotourism: A collaborative partnerships perspective, *Journal of Ecotourism, 14*(23) 166–184.

Stone, M.T., & Nyaupane, G. P. (2014). Rethinking community in community-based natural resource management. *Community Development, 45*(1), 17–31.

Stone, M.T., & Nyaupane, G. P. (2016). Protected areas, tourism and community livelihoods linkages: A comprehensive analysis approach, *Journal of Sustainable Tourism, 24*(5), 673–693.

Stone, M. T., & Rogerson, C. (2011). Community-based natural resource management and tourism: Nata Bird Sanctuary, Botswana. *Tourism Review International, 15*, 159169.

Tosun, C., (2001). Challenges of sustainable tourism development in the developing world: the case of Turkey, *Tourism Management, 22*, 289–303.

Timothy, D. J. (2011). *Cultural heritage and tourism: An introduction*. Bristol: Channel View.

Urry, J. (1990). *Tourist gaze: Leisure and travel in contemporary societies*. London: Sage.

U.S. Travel Association. (2012). Retrieved from www.ustravel.org/.

WCED. (1987). World commission on environment and development. Our Common Future, Oxford University Press: Oxford, UK.

WTO. (1993). *Indicators for the Sustainable Management of Tourism*. Report of the International Working Group on Indicators of Sustainable Tourism to the Environment Committee, International Institute for Sustainable Development, Winnipeg, Canada.

17 Economic assessment of tourism-based livelihoods for sustainable development

A case of handicrafts in Southern and Eastern Africa

Gagoitseope Mmopelwa and Lesego Mackenzie

Introduction

Globally, handicrafts is recognised as an informal sector, despite its significant contribution to employment creation, especially in developing countries (USAID, 2006). The sector employs about 200 million people globally (Smith, 2014), resulting in promotion of cultural and economic development (Mahgoub & Alsoud, 2015). Although handicrafts originated from traditional practices, worldwide their designs are currently influenced by fashion (Mahoney, 2012; USAID, 2006). The increased globalisation has led to their commoditisation, with increased competition from producers all-over including China and India (Mahgoub & Alsoud, 2015).

In Africa, the handicraft sector is an important activity linked to tourism (Mbaiwa, 2004) and supporting rural livelihoods. Many countries in Africa have diversified their economies and list tourism as one of the major contributors to Gross Domestic Product (GDP), offering opportunities for welfare improvement to communities (Manyara & Jones, 2007). A few countries, such as Kenya, list the tourism sector as their leading foreign exchange earner, contributing 18% to total foreign exchange earnings (Manyara & Jones, 2007).

As the handicraft sector uses natural and other resources (Egoh, 2012; Le Billon, 2001), its sustainability and contribution to livelihoods will depend on a number of factors, including sustainable use of these resources, transfer of skills from generation to generation (DeMotts, 2017; Mbaiwa, 2004; Ogollah et al., 2011) and training in marketing skills (Harris, 2014). Historically, the handicraft sector has not been regarded as an important economic industry in development, despite the high participation of rural people, especially women, in the sector. This chapter makes an economic assessment of handicraft contribution to livelihoods in the Eastern and Southern African countries of Kenya, Tanzania, South Africa, Botswana and Namibia. The study provides information on the role of players, the market, resource use and economic sustainability of the handicraft sector.

Overview of the craft industry, methods and conceptual framework

Overview of the craft industry development

Southern and Eastern African regions are endowed with abundant natural resources that directly and indirectly contribute to the welfare of their citizens. This is particularly the case for the countries of South Africa, Botswana, Namibia, Kenya and Tanzania. The handicraft sector, considered to be largely an informal sector, is dependent on some of these resources. The development of tourism in southern and eastern African countries has to a large extent promoted the growth of the handicraft sector (Cultural Strategy Group, 1998; Terry & Cunningham, 1993). This is evidenced by the existence of craft markets in most tourism destinations in both regions (Saarinen, 2016; Terry & Cunningham, 1993). Such markets provide employment opportunities for local people who may not have sustainable sources of income (Cultural Strategy Group, 1998).

In Namibia, the development of the handicraft sector started as a small-scale activity linked to certain tribes or ethnic groups where indigenous communities benefited very little from the sale of their products, especially to tourists (Novelli & Gebhardt, 2007). Such patterns of production clearly reflected both local and touristic cultural heritage (Saarinen, 2016). With the passage of time, the quality of the products and the market outlets developed, especially where local organisations offered support to producers who had come together. Since the country attained independence in 1990, the number of stakeholders in the handicraft sector increased significantly (Ashley, 2000). Despite this development, Davel (2016) is of the view that the importance and role of arts and crafts as a source of income generation, an essential part of Namibian heritage, indigenous knowledge and national intellectual capital, is not yet understood because the handicraft sector is not among the top national development goals for the government.

In South Africa, the potential of the handicraft sector was not known until about 1980 due to limited research carried out in the sector (Rogerson, 2000). During the apartheid era, the handicraft sector was marginalised due to its perceived low contribution to the economy (Nyawo & Mubangizi, 2015). Growth and further development of this sector increased after 1994 when South Africa proved to be one of the greatest tourist destinations in Africa and beyond (Eeden, 1996). According to Makhitha (2015), craft producers increased by 40% between 2005 and 2009, a growth estimated at 8% per year, while the sector itself contributes an estimated 0.14% of GDP.

In Botswana, crafts were produced primarily for traditional or household use such as providing storage for agricultural produce before the 1970s (Bishop & Scoones, 1994; Terry & Cunningham, 1993). According to Mbaiwa (2004), tourism in the Okavango Delta has had positive and negative impacts. The positive impacts include elevating community wealth and preserving and promoting cultural tourism, while the negative impacts include loss of traditional structures and pressure to the environment due to harvesting of raw materials. Nevertheless, the growth and expansion of tourism-related activities and the establishment of Botswana Craft, a major craft market in Botswana, has led to expansion of the handicraft sector (Bishop & Scoones, 1994; Terry, 1999). The sector continues to sustain many of the low-income citizens in many parts of the country.

In Kenya the growth of the handicraft sector was estimated to be 70% between 2001 and 2007, which was worth over US$12 million (Waweru as cited in Harris, 2014). It has also been observed that over the years the status of the handicraft sector has gradually shifted from being a cultural to a commercial activity, providing a source of employment, income and poverty reduction (Nyambura, 2014; Ogollah et al., 2011). Not only has the sector become more commercialised, but there has also been a general movement of production locations from rural to urban areas because of the realized potential of the sector to support an export market (Harris, 2014). However, the commercialisation of the sector has not led to loss of cultural value as the cultural appeal of the products has been the main driving factor for the sustained markets (Nyambura, 2014). Kenya's comparative advantage for handicraft production lies on cheap labour, diversity of products stemming from unique ethnical and cultural skills, and the well-known high quality and value of the products (Nyambura, 2014; Mukami, 2012).

In Tanzania the potential of the handicraft sector stems from a wide variety of products made and the existence of an export market in the USA which developed under the African Growth and Opportunity Act (AGOA) of 2000, which was extended in 2015 (Government of Tanzania, 2016). Over time, this potential attracted a large number of immigrants from other parts of Africa, leading to increased output of the sector (Benson, 2014). Nevertheless, the sector faces a number of marketing constraints, such as lack of financial capital, pricing and cost information, institutional support, local and trade networks (Benson, 2014).

Conceptual framework

One of the means of achieving the sustainability of handicraft-based livelihoods is access to assets or capital (Sallu, Twyman, & Stringer, 2010). Figure 17.1 shows the different forms of livelihood assets or capitals and their relationship to the vulnerability context in the production of handicrafts. The vulnerability context refers to factors that can affect the contribution of each capital to the sustainability of the handicraft enterprises and the wellbeing of the crafters. In the context of this study, such factors include climate change, government policies and the availability of local and regional markets. The wellbeing of crafters is a function of livelihood outcomes (Krantz, 2001). For instance, increased income from the sale of handicrafts often leads to increased wellbeing as the income is used to meet various family needs.

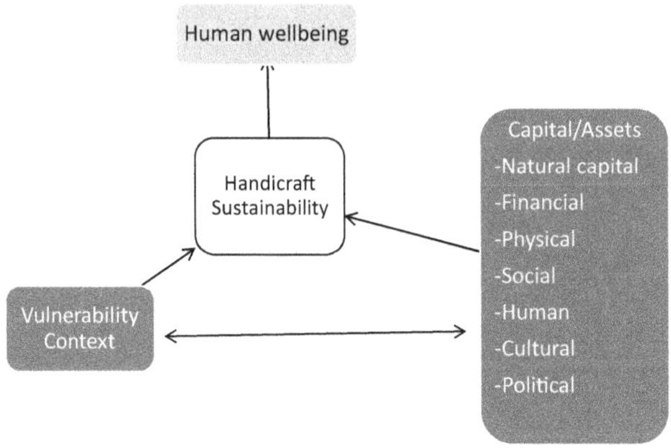

Figure 17.1 Relationship between vulnerability context, livelihood capitals, handicraft sustainability and human wellbeing

Source: Authors (2019)

Natural capital: Most of the crafters need access to natural resources such as plant-based materials in the production process (Mutinda, 2014; Terry & Cunningham, 1993). Sustainable use of these products is expected to enhance the contribution of handicraft enterprises to human wellbeing. Several factors, including climate change and variability, may adversely affect the sustainability of craft-based enterprises. For instance, certain plant habitats may be altered due to reduction in rainfall, leading to changes in abundances of those plants.

Financial capital: Most crafters are poor as they do not have reliable or sustainable sources of income to start or expand their handicraft businesses (Benson, 2014). While start-up capital for the handicraft enterprise is often thought of as very minimal (Tshuma & Jari, 2013), some crafter makers may not have even the minimum financial capital required. For instance, crafters in Kolkol and Lockichar in Turkana district in Kenya do not have adequate financial capital to purchase reeds, dye or specific tools such as a knife and pricking stick to engage in craft production (Watson & van Binsbergen, 2008).

Social capital: Craft making in some countries is a social activity that brings together crafters to share skills (Singh, 1999; Malema & Naidoo, 2017) which are also transferred from generation to generation. According to Benson (2014), handicraft traders in Tanzania obtain their inputs through organised networks as well as organising themselves into supplier groups to reach bigger and well-developed markets. In Kigali, Rwanda, Kappus (2012) found that participation of crafters in cooperatives improved their resilience to shocks as they were able to access economic and non-economic assets.

Cultural capital: The pursuance of different livelihood strategies does not only depend on tangible capital such as natural capital, but also on non-tangible capital such as culture, which essentially gives the identity, pride, self-esteem and resource ownership to the crafters (Daskon, 2015). The transfer of special skills in the handicraft production from one generation to the other forms a significant aspect of cultural heritage among different social ethnic groups.

Physical capital: Physical capital includes infrastructure, tools used to make handicrafts, and storage facilities. The availability of physical capital such as roads accords crafters an opportunity to reach distant markets at much-reduced costs (Tshuma & Jari, 2013). High transport costs incurred during travel on poor roads can adversely affect the sustainability of the handicraft enterprise.

Human capital: The literature indicates that the majority of crafters are women who have no formal education (DeMotts, 2017; Mutinda & Kiumbuku, 2014). Lack of formal education may translate into lack of business or entrepreneurial skills among crafters, despite the existence of traditional skills employed in making various products. The lack of entrepreneurial skills, which includes information about how markets function, may negatively affect the economic viability of the handicraft enterprise. In Zanzibar, Tanzania, Fröcklin, Jiddawi, and de la Torre-Castro (2018) found that while women in the sea shell collection business wanted to expand their handicraft enterprise, they lacked knowledge on market demand and did not have contacts with potential buyers of their products.

Political capital: Among the factors that influence livelihood strategies are the policies, the formal and informal institutions, and the processes that affect people's everyday lives (Baumann & Sinha, 2001; Bennett & Dearden, 2014). For instance, well-established community organisations in the management of craft resources such as craft associations can influence access to local and international markets for crafts as their voices could reach those in authority or decision-makers in government. According to Baumann and Sinha (2001), it is critical that political capital or power is built in such a way that it can effectively address the interests and the realisation of better outcomes for the crafters.

Methods

Data on the economic contribution of handicrafts to households was sourced from studies undertaken in respective countries. This data was expressed in gross value (i.e. inclusive of costs), net values, net present values (NPV) or benefit-cost ratio (BCR) and other measurements such as dollar/m^3 as in the case of wood carving. Since the quantitative data in these studies were collected and analyzed for different years, they were converted to 2016 prices using consumer price indices (CPIs) for each country. The historical CPIs were obtained from World Bank internet sources (https://data.worldbank.org/indicator), with a common base year of 2010. These prices were then converted to US dollars using annual exchange rates for the year 2016, obtained from UNCTADSTAT tables (Table 17.1). Except where stated, the values in parenthesis are the US dollar pricing in 2016.

Table 17.1 Annual exchange rates between US$ and Eastern and Southern Africa countries (2016)

Country and currency	*Annual exchange rates (US$) for 2016*
Botswana (BWP)	0.09173
Kenya (KES)	0.009852
Namibia (NAD)	0.06799
South Africa (ZAR)	0.06798
Tanzania (TZS)	0.0004593

Source: Adapted from United Nations Conference on Trade and the Environment (n.d)

Results and discussion

Definition and description of the role players in the handicraft sector

According to USAID (2006) the definition of craft is centred on three aspects, namely, how the product is made (handmade vs machine made), significance (artistic qualities vs simple craft) and the motivation for production (economic vs. cultural reasons). Specialised skills are employed in the production of these products which may be produced for innovative, decorative, aesthetic, cultural and religious values (Davel, 2016). In general, handicrafts were also defined as customary economic activities commonly found in rural areas that communities carry out to sustain their livelihoods (Allal & Chuta, 1982). For several decades handicrafts have been taken as rural home-based enterprises, however, in recent years they are found in the urban areas (DeMotts, 2017). While there is no universal definition of handicrafts, this study adopts the definitions of Cultural Strategy Group (1998) and that of Rogerson (2010). The Cultural Strategy Group (1998) defines a craft as "the production of a broad range of utilitarian and decorative items manufactured on a small scale with hand processes being part of the value-added content. The production of goods utilizes a range of synthetic and natural materials." Rogerson (2010) defines a handicraft as "a product that is at least 80% handmade from different materials which can include clay, natural fibre, beads, recyclables and textiles."

The craft industry entails production of a range of traditional and contemporary hand products made from various raw materials such as wood, clays, fibre, plant material, stones and paper (Jänis, 2011). A study by Nyambura (2014) in Kenya has denoted that handicrafts are largely dominated by women; there is also the engagement of youth of both genders, and only a few craftsmen can still be found as their numbers have diminished considerably. Similarly, this has been found to be the case in most Southern and Eastern African countries under review. In addition the handcrafters are mostly assisted by their family members and usually work on crafts on a part-time basis, leading to handicraft market potentials untapped (Mbaiwa, 2004). Handicrafts are further influenced by governments as they determine the legal frameworks that guide the production and markets and also the policy regulation and implementations (Ogollah et al., 2011).

Marketing

Generally, the marketing of tourism attraction areas is positively linked to the tourism cultural production and vice-versa (Mbaiwa, 2004). The informality of the handicraft enterprise and illiteracy and lack of local basket institutional structures has led to poor pricing and hence unsustainable livelihoods (Mbaiwa, 2004; Mutua, Massimo, & Mburu, 2004). For many years, basket making was a traditional way of life for people all around Botswana that was used by households for agricultural practices, but has now grown and become a commercialised activity (Mbaiwa, 2004). The demand for handicrafts is believed to be in excess of production as handicrafts are often imported from Kenya, Zimbabwe, Zambia and other countries (Mutua et al., 2004). According to Mbaiwa (2004), Botswana Craft and Botswana Christian Council (BCC) have been the main buyers of locally produced baskets and crafts since 1990. Other important buyers are local, regional and international tourists, retailers and middlemen (Mutua et al., 2004). Conservation International (CI) – Botswana, Trust for the Okavango Cultural and Development Initiatives (TOCADI) and private investors buy handicrafts in order to sell them to tourists in hotels, lodges, airports, urban areas and even export to European and North American markets (Mbaiwa, 2004).

In Kenya, Mahoney (2012) discusses how handcrafters have shifted from selling products as genuine cultural commodities to marketing them as "fair trade" products. Fair trade is a market approach where producers, retailers and consumers have come together to form an association that implements social justice in the economy (DeCarlo, 2011). Such could include advocating for a fair price of crafts. Compared to Botswana, this political capital in Kenya appears to be stronger. The products include wood and soapstone sculptures, sisal baskets, beaded jewelry, pottery, mat weaving, smithing, sewing, basketry and gourd decorations (Könnecke, 2014; Mutinda & Kiumbuku, 2014). The value chain includes artisans who manufacture the products and traders who sell the products in the local markets and export them to external markets in the USA and Europe (Mukami, 2012). According to Nyambura (2014), most of the artisans in the most common market of Kariokor, near Nairobi, face many marketing challenges including lack of skills to promote their products, lack of financial capital, high packaging and shipping costs, high tariffs and high-quality standards set by importing countries. At a general level, handcrafters also lack information on global markets and political capital in the form of weak institutional support (Mukami, 2012).

In Namibia, crafts are marketed at various sites, including craft outlets (curio shops), direct sale to tourists at most destinations, cultural villages as well as in designated places where government or other organisations have a vested interest in the promotion of crafts (Saarinen, 2016). In many cases there are intermediaries or middlemen in the marketing chain who buy directly from producers and sell to other clientele in the chain. In the Caprivi region, Namibia, the craft centres work closely with community trusts to promote production and marketing of craft products. Jänis (2006) found that the Omba Trust sells crafts to a wide variety of clientele, who then export these products to various countries on a fair-trade principle, a situation similar to Kariokor in Kenya.

In South Africa the production of crafts is spread throughout the nine provinces of the country. The Cultural Strategy Group (1998) identified 707 retailers of crafts in all of the provinces, the majority (43%) of which was located in the Kwazulu Natal province. Crafts are marketed through electronic marketing, tourism offices and lifestyle magazines (Cultural Strategy Group, 1998). Craft producers also take advantage of events such as festivals and shows to sell their products. According to Makhitha (2015), craft products are sold in both informal and formal markets.

In Tanzania craft work includes a number of products such as beadwork, basketry, mats, ceramics, pottery, leather bags and wood crafts (Government of Tanzania, 2016). Because of the small nature of the business, individual traders often fail to meet the market demand, especially during the tourism peak period. According to George (2003), Dar es Salaam, the largest city in Tanzania has more than 15 informal handicraft markets, such as Mwenge handicraft centre, Nyumba ya Sanaa (House of Art), Village Museum, and Kariakoo market on Tandamuti Street, amongst others. All are completely different from one another, and they offer different views of African lifestyles. Being one of the most visited tourist destinations in Tanzania, Dar es Salaam city attracts not only locals but also foreign entrepreneurs.

According to Tambwe (2017), the crafts market includes selling to local or domestic craft traders and tourists. The domestic craft traders constitute the Tanzanian Handicrafts who, essentially, are intermediaries selling to the larger local market that exports products directly to the USA, Canada, Australia, South Africa and Japan. Tambwe's study found that the share of the domestic market was 64% (14% to locals and 50% to tourists), while direct export share was estimated at 36%. The study also revealed that the Tanzanian Handicrafts use trade fairs, point-of-sale displays, brochures and the internet to promote the sale of their product. The major challenge in the marketing of handicrafts in Tanzania is the stiff global competition.

Resource use sustainability

Handicrafts are traditional features that are mostly produced on a small scale; however, they essentially are dependent on natural resources which are exposed to increased environmental pressure as more craftsmen and women exploit them (Zargham, 2007). The pressure emanates from the rapidly deteriorating natural resources and their very low regenerative capability (Wambaa, 2011). In Kenya, Wambaa (2011) found that 64% of the carvers reported that depletion of trees affected their enterprises while 36% stated that they could source raw materials from neighboring communities though this increased their production costs due to the distance from their work stations. Nyawo and Mubangizi (2015), interviewed crafters in South Africa and found that about 70% highlighted that there are enough raw materials to use for the production of handicrafts. The main reason for resource sustainability mentioned by crafters was the use of harvesting permits from the KZN Nature Conservation Authority. In addition, Mavundla et al. (cited by Nyawo and Mubangizi (2015) point out an initiative by "Ezemvelo KZN Wildlife and iSimangaliso Wetland Park to protect these resources and promote sustainable harvesting practices." From the interviews, participants from iSimangaliso Wetland Park said they had a craft programme that was aimed at educating crafters about the importance of conserving natural resources.

In order to reduce the pressure by the craftsmen and women on the environment, the concept of community-based natural resource management (CBNRM) became one of the primary concepts in African conservation in the late 1980s. CBNRM is a concept used to explain "the ways in which communities organize themselves to sustainably manage their natural resources" (DWNP, 1999). CBNRM is, therefore, an incentive-based conservation philosophy that links conservation of natural resources with rural development (Mbaiwa & Stronza 2010). CBNRM has been implemented in most Eastern and Southern African countries, including Zimbabwe, Namibia, Mozambique, Zambia, Tanzania and Botswana; there is evidence suggesting that communities involved in CBNRM have gained a greater awareness of the importance of using natural resources in a sustainable way (Mbaiwa, 2015).

Since various areas have diverse cultures and are endowed with distinctive natural resources, this determines decisions not only on a national trade but also on international trade (Mbaiwa, 2015). Economic and ecological sustainability need to be balanced; however, with poverty so eminent although carvers are aware of the environmental impacts, they have showed that they sometimes do not have a viable option (Chingombe & Pedzisai, 2006). In the villages of Shorobe and Etsha in northwest Botswana, a group of women have engaged in growing palm trees on plantations and operating a palm tree nursery, respectively, because of the realization of the need to have a sustainable basket-weaving enterprise (DeMotts, 2017).

Economic sustainability of the handicraft sector

Economic sustainability is presently the main emphasis of many economies worldwide, as it gives a desirable level of welfare security in line with environmental management of natural resources (Markulev & Long, 2013). The interrelationship of the economics, social scopes and the environment are the main aspects that contribute to sustainability of natural resources. Elkington (1998) refers to it as the "triple bottom line." Özkaynak, Devine, and Rigby (2004) state that sustainable development is categorized by economic transformation, subject to constancy of natural resources available while there is growth in the economy to meet social goals.

There is generally a lack of quantitative data to demonstrate the feasibility of craft-based projects, making the sector prone to little or no support from governments or other development agencies (Richard, 2007), and consequently the lack of realisation of the potential contribution to human wellbeing. According to the Cultural Strategy Group (1998), lack of data in the handicraft sector prevents proper assessment of its contribution to the national economy. The reasons for this lack of data include lack of indicators which are usually employed in evaluating the performance of formal sectors, the seasonality of tourism which causes fluctuations in the numbers of participants in the sector, and the high levels of competition from neighboring countries.

South Africa

In South Africa, the handicraft craft sector is a source of employment for many people in rural and urban areas (Rogers, 2000; Nyawo & Mubangizi, 2015; Makhitha, 2015). The sector remains informal in nature with high participation of women, mostly without formal education (Nyawo & Mubangizi, 2015; Makhitha, 2015; Cultural Strategy Group, 1998). According to Cultural Strategy Group (1998), the high participation of black women in the handicraft business is due to the lack of entry barriers when compared to other sectors, creation of new opportunities for women who were marginalised during the apartheid era, building around traditional women's skills and the ease of being integrated with other household activities. Eeden (2011) found in four metropolitan cities of Johannesburg, Pretoria, Cape Town and Durban that the average gross daily turnover was estimated at between R900 (US$206.80) and R1200 (US$275.80) for arts and craft traders and between R500 (US$114.92) and R1600 (US$367.74) for all traders in the informal sectors.

In Kwazulu Natal, a study of women empowerment through economy of affection revealed that the net income derived from the sale of ilala palm (*Hyphaene petersiana*) for nine out of ten women was significantly lower than the monthly minimum of R1400 (US$327.8) required by an urban household in Durban to cater for food, clothing and transport expenses (Singh, 1999). The monthly average profit was R545 (US$91.64). Based on these figures, basket making does not appear to be a profitable enterprise or a livelihood strategy in this area. Furthermore, the Cultural Strategy Group (1998) indicates that in the south coast, ilala palm has been depleted, threatening the sustainability of basket production.

In the Pondoland region of the Wild Coast, Eastern Cape Province in South Africa, where *Cyperus textilis* is the most used plant species for craftwork, Makhado and Kepe (2006) found that 70% of the respondents were engaged in craft work as they needed to supplement their income, 26% indicated that they just needed another income source to depend on, while another 26% stated that they were forced by poverty to engage in this activity. Makhado and Kepe (2006) estimated the mean monthly income from crafts at R86.7 (US$10.68), with a range of R30- R300 (US$3.68–369.47). The contribution of monthly craft income to monthly household income was estimated at 9.5%.

Namibia

In Namibia, Murphy, and Suich (2003) found that purchase and sales data for Mashi Craft Market for individual crafters in Caprivi from 1999 to 2001 amounted to almost N$140,000 (US$19,835.57) with over 4,000 craft items traded. Jänis (2011) found that the income derived from the sale of crafts, which include baskets, hats and bags and products made from palm leaves, mats, and jewelry, was generally much higher than that from other livelihoods and enabled households to pay for basic needs such as education, health, building houses, and food. This is a clear indication of the positive contribution of handicrafts to human wellbeing. On average, Jänis (2011) estimates that a craft producer in Namibia earns between N$200 (US$17.81) and N$500 (US$44.52) on a monthly basis. The variations in the earnings are due to a number of factors, chief among them the operations of the craft shops. The lower limit of this range is higher that the estimated monthly craft work value of US$10.68 reported by Makhado and Kepe (2006) in Eastern Cape Province in South Africa. However, crafts' financial or economic viability in this study were not evaluated. For the Chobe craft market and the Linyanti Craft centre Jänis (2011) shows that the monthly earnings/per person amounted to N$1000 (US$89.04) and N$150 (US$13.36), respectively. These earnings are still higher compared to those reported in South Africa. NACSO (2010) found that crafts generated the second highest income of N$1,956,784 (US$18,299.36) after thatching grass N$2,269,3810 (US$2,122,639.26) among the other sources of incomes outside conservancies which are small tourism enterprises such as campsites, cultural villages and guiding (N$ 1,795,658 or US$167,954.29), Joint Venture tourism contributions (N$ 791,951 or US$74,074.22), veld products (N$ 501,623 or US$4,733.74) and community forests (N$ 143,759 or US$13,446.33). Of the total income from crafts, 77% (N$1499291 or US$140,234) was paid to 1616 producers.

In central Windhoek, Namibia, in a small craft market of less than 20 stands, and controlled by the council, Saarinen (2016) estimated the economic value of craft trade looking at the products purchased by tourists. The results of the interviewed tourists and observations made by the authors led to an estimate of a daily group turnover of N$8,929 (US$607.08), which translates to a daily turnover of US$43.36/person). This is comparable to the general estimate given by (Jänis, 2011). The daily turnover of US$43.36 is however inclusive of the cost of production, hence making it difficult to assess whether this is economically or financially viable. It is also important to note that these values were obtained during a slow or quiet tourism season.

Botswana

In Botswana, craft production forms one of the many diversified livelihoods among rural people. As in South Africa and Namibia, a high participation of women in handicraft production, and with no formal education, is the norm

(DeMotts, 2017; Terry, 1999). Tourism also has had an impact in the shift from traditional use of crafts to commercial use, providing the much-needed cash to pay for school fees, purchase food necessities and meet other family needs such as health care (Lenao & Chanda, 2010; DeMotts, 2017). The income from crafts is also used to sustain families during the lean period when crop yields are significantly low due to the impacts of drought. According to Lenao and Chanda (2010), commercialization of pieces of crafts has subsequently led to a change in both the shape and size of the products being sold, with small-sized craft being in higher demand because of the ease of being transported. Ngamiland District, one of the poorest districts in the country, has the largest number of crafts with a rich resource base for this enterprise (DeMotts, 2017; Terry, 1999). Thus, crafts contribute to both employment and income generation for the rural poor.

In the villages of Gumare and Etsha 5 in northwest Botswana, Bishop and Scoones (1994) used data on collection time, processing time and weaving time to estimate returns to labour (gross margin/labour inputs) for basket making. The labour inputs for making N5 baskets in Gumare and Etsha 5 were 23.6 and 23.9, respectively. The gross annual income obtained from the gross margin and the number of N5 baskets made by a weaver per year (20 in each of the areas) were estimated at P153 (US$69.07) in Gumare and 191 (US$87.32) in Etsha 5. The authors also compared the returns to labour for basket making to other livelihoods (agriculture, beer making and drought relief) and found that, generally, they were lower than those from these livelihoods activities as shown in Table 17.2. The gross margin values for basket making in Gumare are 1.3 times less than those for basket-making in Kwazulu Natal where women were also involved in basket-making (Singh, 1999). For basket-making in Etsha 5, the values are comparable.

Terry (1999) used benefit-cost analysis to evaluate the financial and economic performance of ten handicraft enterprises (basketry, beadwork, skin work, carving, leather, weaving, textile, pottery, jewelry and miscellaneous enterprises (e.g. gourds, corn husks or cow horn) and non-craft enterprises (beer-brewing, grass/reed collection and sale, hut building, wild products collection and sale, and hunting, sewing, bakery, block making, metalwork and milling) in northern Botswana. Analysis revealed that seven of the handicraft enterprises (basketry, beadwork, skin work, carving, leather, weaving, textile) were financially viable

Table 17.2 Returns to labour for basket and non-craft enterprises in Botswana (in 2016 prices)

Livelihood	*Gross margin US$/day)*	*Gross margin (US$/hour)*
Basket making:	US$1.18	US$0.15
Etsha 5 Gumare	US$1.48	US$0.18
Beer brewing	US$4.43	US$0.56
Agriculture	US$4.35	US$0.54
Drought relief	US$2.06	US$0.34

Source: Adapted from Bishop and Scoones (1994)

and worth investing in, while pottery, jewelry and miscellaneous enterprises were not. Economically, all the 10 enterprises were viable, with leather work being the best enterprise (Table 17.3). While the net values for basket making in Botswana are smaller in magnitude as compared to those for baskets in Kwazulu Natal (Singh, 1999), their net present value and the benefit ratio indicate that they are both financially and economically viable at 6% discount rate and 5 year time horizon for both the dynamic and economic model. Thus, these enterprises contribute positively to household livelihoods and wellbeing.

Kenya

In Kenya, the handicraft sector, as in other Southern African countries, is dominated by low income brackets, the majority of which are women (Mukami, 2012). Craft making falls under the small and medium enterprise (SME) and has thrived due to a number of factors, including low or minimal start-up capital, flexible working hours (where women at home can simultaneously participate in other livelihood activities), and freedom to manage own business (Makokha, 2014; Nyambura, 2014).

In many areas in Kenya the sector provides employment opportunities and contributes to rural development (Makokha, 2014; Mukami, 2012), although it faces a number of challenges including political, legal and economic. Among the Maasai of Isinya Location, Kajiado District, Kenya, Sitati (2011) observed that while the income from the sale of handicrafts is considered minimal, it was used to meet welfare needs of the family such as financing children's education, health care, improving housing as well as securing food for the family.

Table 17.3 Summary of financial and economic viability of ten craft enterprises in Ngamiland, Botswana (in 2016 prices)

Enterprises	*Static financial model (Annual net cash income*	*Static economic model (Net value added*	*Dynamic financial model (5 years @ 6% discount rate)*			*Dynamic economic model (5 years @ 6% discount rate)*		
			NPV	*FRT (%)*	*BCR*	*NPV*	*FRT (%)*	*BCR*
Basketry	5.84	26.18	514.14	19.60	1.01	81.55	349	1.24
Beadwork	5.2	69.16	3.12	13.27	1.01	251.41	891.58	2.80
Skin work	19.48	115.76	17.61	20.84	1.01	3.89	805.55	1.39
Carving	42.68	283.09	57.0	27.54	1.03	980.12	584.74	1.62
Leather	22190.53	41183.44	59081.37	35.5	1.16	88367.72	47.8	1.24
Weaving	5513.98	17595.39	1466.33	6.97	1.01	19497.65	18.08	1.11
Textile	7054.59	21959.23	3619.92	8.2	1.01	30142.44	23.34	1.13
Pottery	-5174.46	-2282.41	-20463.8	-100	0.55	-15664.3	-100	0.64
Jewelry	-532.33	1141.66	-419.76	-100	0.83	990.69	23.69	1.05
Miscellaneous crafts	-861.55	-412.28	-4021.58	-100	0.61	-28.98	-100	0.70

Source: Adapted from Terry (1999)

In a study of the wood carving industry in Kenya, Choge (2001) estimated the export value of wood carvings at US$20 million annually and a self-employment opportunity of 60,000 carvers and 350,000 dependents. Using information on costs (low prices: Ksh 1000 to 4000/m^3), medium prices (Ksh 4000 to 8000/m^3) and high prices (above Ksh 8000/m^3) and revenues, the profits were estimated to range between Ksh 30 626 (US$1194/$m^3$) and Ksh 67 234 (US$2623/M^3).

In Wamunyu, Machakos County, also in Kenya, Mothini (2017) found that wood carving has become an important craft activity because livestock and crop farming were no longer viable due to the impact of drought. According to this study, 93% of wood carvers indicated that they were dependent on wood carving as their main livelihood. The gross monthly income for the majority (46.4%) was estimated at 5000–9999 Ksh. While these studies show the contribution of wood carving to livelihoods, the values cannot be easily compared to those in other countries due to the differences in measurement units of the values. For instance, in Bushbuckridge District, South Africa, Shackleton and Shackleton (2004) estimated the annual net values of wood carving and furniture made from *Pterocarpus angolensis* at R3603 (US$566.28) and R7065 (US$1,109.64) where the products made include bowls, spoons and walking sticks, while furniture products include tables, chairs and benches. According to the study, commercialization is threatening continued availability of this important tree species.

Tanzania

In Tanzania, as in Southern Africa, women and young people who are often economically disadvantaged in terms of low income, low level of education and low levels of skills, are the main participants in the handicraft sector (Benson, 2014). A study of handicraft micro and small enterprises in Moshi, revealed a positive correlation between increased handicraft income and household food availability, the state of household education and the state of household health (Kazungu & Njau, 2018). Furthermore, this study found a significant positive relationship between Micro and Small Enterprises (MSEs)'s handicraft income and livelihood outcomes. Thus, handicraft income is critical in sustaining livelihoods.

On the island of Zanzibar, in eastern Tanzania, Fröcklin et al. (2018) found that the shell-handicraft project (Sustainable Coastal Communities and Ecosystems – SUCCESS) funded by USAID to empower and reduce poverty among eight villages, yielded significant benefits in terms of access to financial, physical, human and social resources. According to the study, women derived significant income from the sale of shell products that include earnings, bracelets, and necklaces sold in local shops and to tourists on planned trips. The study estimated a daily income increase of TSZ 420 (US$1.9) to TSZ 6800 (US$3.1) during the project period. With this income most of the women were able to open a savings account, build a house, and purchase a refrigerator (valued at

US$110), mobile phone (valued at US$16) and electricity. The authors also observed that the average daily income for shell crafters was higher than that received by a woman engaged in fishing/gleaning (US$1.34/day), agriculture/forestry (US$1/day) and seaweed farming (US$1.3) in coastal areas, as found by Torre-Castor et al. (2007). The other benefits to women include improved ability in decision-making, particularly at the household level, and increased ability to make strategic choices. While the economic and financial viability of the shell-handicraft was not evaluated in the study, the project seems to have great potential based on the high-valued assets that women were able to purchase. Notwithstanding these positive results, the sustainability of the shell handicraft was reported to be faced with the challenges of lack of market, high maintenance costs of tools or equipment used to make the products, high time demands and possible competition by future scaling up of the project (Fröcklin et al., 2018).

Conclusion and recommendations

The literature review has revealed that the handicraft sector is largely informal in nature in both Eastern and Southern African countries and yet contributes significantly to human wellbeing, especially in rural areas. The sector is dominated by women with a low level of education. The high participation of women is explained by factors such as lower initial capital required, absence of entry barriers and its flexibility to be pursued alongside other livelihoods. The contribution of the sector to the economies of these countries remains unrealized due to its informal nature. Despite this status quo, the sector has expanded alongside the growth and development of tourism over the years. The income from the sale of handicrafts supplements income from other livelihood sources such as agriculture, especially during difficult economic times, such as drought. In some instances, this income exceeds that realized from other livelihood activities and has been used to meet basic family needs such as financing children's education, health care and improving food security. In Botswana, the gross margin/per day for basket-making was found to be 3.5 times less than that for other livelihood activities such as traditional beer brewing and agriculture, but showed to be financially and economically viable. For Tanzania the results show that support for the shell-handicraft project by the USAID yielded much higher incomes than agriculture, forestry or seaweed farming. While a net benefit analysis was not undertaken, the project seems to have great potential based on the high-valued assets that women were able to purchase. The value of crafts enhancement is of cultural significance.

The study also found that the handicraft sector in both regions faces numerous challenges such as lack of access to certain capitals like financial capital. This status quo should lead to a deliberate move by governments of Eastern and Southern African countries and the NGO community to provide a conducive environment for the handicraft sector to thrive as it has been demonstrated to be a source of income and employment generation in poverty reduction. The government should provide access to certain capitals such as political, financial and physical.

The lack of recognition of the sector could imply that studies on economic analysis of the handicraft activities are generally lacking. Therefore, there is a need to direct economic research to investigate the potential for the handicraft sector to address issues of rising unemployment. It is also crucial to assess whether or not support for handicraft projects either by governments or independent agencies can lead to increased economic benefits.

References

Allal, M., & Chuta, E. (1982). *Cottage industries and handicrafts; Some guidelines for employment promotion*. International Labour Office.

Ashley, C. (2000). The impacts of tourism on rural livelihoods: Namibia's experience. Working Paper 128. Overseas Development Institute.

Baumann, P., & Sinha, S. (2001). *Linking development with democratic processes in India: Political capital and sustainable livelihoods analysis*. London: Overseas Development Institute.

Bennett, N. J., & Dearden, P. (2014). Why local people do not support conservation: Community perceptions of marine protected area livelihood impacts, governance and management in Thailand. *Marine Policy*, *44*, 107–116.

Benson, W. (2014). *The benefits of tourism handicraft sales at Mwenge handicrafts centre in Dares Salaam, Tanzania*. Bachelor's Thesis Degree in Tourism, Tampere University of Applied Sciences, Finland.

Bishop, J., & Scoones, I. (1994). Beer and Baskets: The Economics of Women's Livelihoods in Ngamiland. *Botswana, IIED Research Series*, *3*.

Chingombe, W., & Pedzisai, E. (2006). Craft industry impacts on the environmental and a community's social welfare: Masvingo-Ngundu highway in Zimbabwe. *Journal of Sustainable Development in Africa*, *8*(1).

Choge, S. M. (2001). *Study of economic aspects of the woodcarving industry in Kenya: Implications for policy development to make the industry more sustainable*. (MSc Thesis) University of Natal, Durban, South Africa.

Cultural Strategy Group. (1998). *The South African craft industry report*. Unpublished report prepared for the Cultural Industries Growth Strategy, Department of Arts, Culture, Science and Technology, Pretoria.

Daskon, C. D. (2015). Significance of cultural values in securing asset portfolios of rural communities. *Sri Lanka Journal of Social Sciences*, *38*(1), 31–50.

Davel, M. (2016). *A scoping study conducted on the arts, crafts and design domain in Namibia*. Namibia: John Muafangejo Art Centre (JMAC).

DeCarlo, J. (2011). *Fair – Trade: A Beginner's Guide*. One World Publication, Oxford, England.

DeMotts, R. B. (2017). Weaving a living: Gender, craft, and sustainable resource use in Botswana. *Journal of Political Ecology*, *24*(1), 368–385.

Department of Wildlife and National Parks (DWNP). (1999). *Joint venture guidelines. Department of Wildlife and National Parks*, Gaborone, Botswana.

Egoh, B. N., O'Farrell, P. J., Charef, A., Gurney, L. J., Koellner, T., Abi, H. N., & Willemen, L. (2012). An African account of ecosystem service provision: Use, threats and policy options for sustainable livelihoods. *Ecosystem Services*, *2*, 71–81.

Elkington, J. (1998). Accounting for the triple bottom line. *Measuring Business Excellence*, *2*(3), 18–22.

Fröcklin, S., Jiddawi, N., & de la Torre-Castro, M. (2018). Small-scale innovations in coastal communities: Shell-handicraft as a way to empower women and decrease poverty. *Ecology and Society, 23*(2).

George, R. (2003). Tourist's perceptions of safety and security while visiting Cape Town. *Tourism Management, 24*(5), 575–585.

Government of Tanzania. (2016). *National AGOA strategy for the United Republic of Tanzania.* Prepared for the United States Agency for International Development, Ministry of Industry, Trade and Investment.

Harris, J. (2014). Meeting the challenges of the handicraft industry in Africa: Evidence from Nairobi. *Development in practice, 24*(1), 105–117.

Jänis, J. (2011). *The Tourism–Development Nexus in Namibia- A Study on National Tourism Policy and Local Tourism Enterprises' Policy Knowledge.* PhD Dissertation, Department of Political and Economic Studies (Development studies) Faculty of Social Sciences. University of Helsinki.

Kappus, M. B. (2012). *Understanding the impact of handicraft cooperative participation on livelihood strategy (asset accumulation and resiliency) among women: A qualitative case study from Kigali, Rwanda.* Master of International Development and Management, University of Lund.

Kazungu, I., & Njau, J. (2018). The artisans: Livelihood sustainability in the context of handicraft micro and small enterprises in Moshi Tanzania. *International Journal of Research and Innovation in Social Science, 2*(12), 357–364.

Könnecke, C. (2014). *Social capital in handicraft clusters for business startup and growth.* The Case of the Luweero Handicraft Cluster in Uganda. MSc Business Language and Culture Business and Developing Studies, Copenhagen Business School.

Krantz, L. (2001). The sustainable livelihood approach to poverty reduction. *SIDA. Division for Policy and Socio-Economic Analysis, 44.*

Le Billon, P. (2001). The political ecology of war: Natural resources and armed conflicts. *Political geography, 20*(5), 561–584.

Lenao, M., Mbaiwa, J.E. and Raban, R. (2015). Impacts of cultural tourism on the domestic utility of traditional baskets: Case of Gumare and Etsha 6 villages in the Okavango Delta, *Nordic Journal of African Studies, 24*(3 & 4): 301–318.

Mahgoub, Y. M., & Alsoud, K. M. (2015). The impact of handicrafts on the promotion of cultural and economic development for students of art education in higher education. *Journal of Literature and Art Studies, 5*(6), 471–479.

Mahoney, D. (2012). Changing strategies in marketing Kenya's tourist art: From ethnic brands to fair trade labels. *African Studies Review, 55*(1), 161–190.

Makhado, Z., & Kepe, T. (2006). Crafting a livelihood: Local-level trade in mats and baskets in Pondoland, South Africa. *Development Southern Africa, 23*(4), 497–509.

Makhitha, K. M. (2015). Supply chain practices and challenges in the craft industry in Gauteng, South Africa. *Journal of Applied Business Research, 31*(6), 2197–2212.

Makokha, P. P. M. (2014). The place of home-based industries in rural Kenya's socio- economic progress: A case study of Kakamega County. *Journal of Economics and Sustainable Development, 5*(28), 154–173.

Malema, D. R., & Naidoo, S. (2017). Spaces for the empowerment of women: Rural Arts and Crafts Projects. *African Journal of Hospitality, Tourism and Leisure, 6*(2), 1–18.

Manyara, G., & Jones, E. (2007). Community-based tourism enterprises development in Kenya: An exploration of their potential as avenues of poverty reduction. *Journal of sustainable tourism, 15*(6), 628–644.

Markulev, A., & Long, A. (2013). *On sustainability: An economic approach.* Productivity Commission.

Mavros, K. (2015). *Fair trade: An empowering vehicle for the Latin American female artisan.*

Mbaiwa, J. E. (2004). Prospects of basket production in promoting sustainable rural livelihoods in the Okavango Delta, Botswana. *International Journal of Tourism Research, 6*(4), 221–235.

Mbaiwa, J. E. (2015). Community-Based Natural Resource Management in Botswana. In van der Duim, R., Lamers, M., van Wijk, J. (Eds.) *Institutional Arrangements for Conservation, Development and Tourism in Eastern and Southern Africa* (pp. 59–80). Springer, Dordrecht.

Mbaiwa, J. E., & Stronza, A. L. (2010). The effects of tourism development on rural livelihoods in the Okavango Delta, Botswana. *Journal of Sustainable Tourism, 18*(5), 635–656.

Mothini, S. (2017). *Small scale wood carving enterprises and their contribution to rural livelihoods in Wamunyu, Machakos County, Kenya.* A Research Project Submitted in Partial Fulfillment for the Degree of Master of Environmental Planning and Management in the School of Environmental Studies of Kenyatta University.

Mukami, N. S. (2012). *Response strategies adopted by handicraft traders in Kenya to challenges of exporting.* A Research Project Submitted in Partial Fulfillment of the Requirements for the Award of a Master of Business Administration Degree, School of Business, University of Nairobi, Kenya.

Murphy, C. A., & Suich, H. (2003). *Mashi craft market – crafts and livelihoods in Caprivi* (No. 57). Windhoek: Directorate of Environmental Affairs, Ministry of Environment and Tourism.

Mutinda, J., & Kiumbuku, S. (2014). Gender and educational dimensions of handicraft industry in Mwala District, Kenya. *International Journal of Humanities and Social Science, 4*(11), 252–259.

Mutua, K., Massimo, S. K., & Mburu, P. T. (2004). An empirical study of the Botswana handicraft market. *Journal of African Business, 5*(2), 93–112.

NACSO. (2010). *Namibia's communal conservancies: A review of progress and challenges in 2009.* NACSO, Windhoek.

Novelli, M., & Gebhardt, K. (2007). Community based tourism in Namibia: 'Reality show' or 'window dressing'? *Current Issues in Tourism, 10*(5), 443–479.

Nyambura, N. C. (2014). *Challenges facing Kenyan handicraft traders in accessing the global market: A case study of shop owners at Kariokor market.* A Research Project Submitted in Partial fulfilment of the Requirements for the Award of the Degree of Master of Business Administration, School of Business, University of Nairobi.

Nyawo, J., & Mubangizi, B. C. (2015). Art and craft in local economic development: Tourism possibilities in Mtubatuba local municipality. *African Journal of hospitality, Tourism and leisure, 4*(2), 1–15.

Ogollah, K., Bolo, A. Z., & Muchemi, A. W. (2011). Determinants of strategic forces that shape competition in handicraft industry in Kenya. *Prime Journal of Business Administration and Management, 1*(12), 58–67.

Özkaynak, B., Devine, P., & Rigby, D. (2004). Operationalizing strong sustainability: definitions, methodologies and outcomes. *Environmental Values*, 279–303.

Richard, N. (2007). *Handicrafts and employment generation for the poorest youth and women.* UNESCO.

Rogerson, C. M. (2000). Rural handicraft production in the developing world: Policy issues for South Africa. *Agrekon, 39*(2), 193–217.

Rogerson, C. M. (2010). The enterprise of craft: Constraints and policy challenges in South Africa. *Acta Academica, 42*(3), 115–144.

Saarinen, J. (2016). Cultural tourism and the role of crafts in Southern Africa: The case of craft markets in Windhoek, Namibia. *Turizam: međunarodni znanstveno-stručni časopis, 64*(4), 409–418.

Sallu, S. M., Twyman, C., & Stringer, L. C. (2010). Resilient or vulnerable livelihoods? Assessing livelihood dynamics and trajectories in rural Botswana. *Ecology and Society: A Journal of Integrative Science for Resilience and Sustainability, 15*(4).

Shackleton, S. E., & Shackleton, C. M. (2004). The Pterocarpus angolensis DC. based woodcraft industry in the Bushbuckridge district, South Africa. *Forest Products, Livelihoods and Conservation: Case Studies of Non-Timber Forest Product Systems, 2*, 203–228.

Singh, A. (1999). Women and empowerment through the 'economy of affection' in KwaZulu-Natal: Its significance for sustainable development. *Development Southern Africa, 16*(3), 467–488.

Sitati, H. (2011). Commercial handicrafts as a livelihood option for the Maasai of Kajiado District, Kenya, VDM Verlag Dr. Mülle.

Smith, S. L. (2014). *Tourism analysis: A handbook*. London: Routledge.

Tambwe, M. A. (2017). Assessing the efficiency of marketing communication tools used by Tanzanian Handicrafts for enhancing export performance. *Business Education Journal, 1*(3), 1–19.

Terry, M. E. (1999). *The economic and social significance of the handicraft industry in Botswana*. A PhD thesis Submitted to The School of Oriental and African Studies, University of London for the degree of Doctor of Philosophy.

Terry, M. E., & Cunningham, A. B. (1993). The impact of commercial marketing on the basketry of Southern Africa. *Journal of Museum Ethnography*, (4), 25–48.

Torell, E., Crawford, B., Kotowitcz, D., Herrera, M.D. and Tobey, J. (2010). Moderating our expectations on livelihoods in ICM: experiences from Thailand, Nicaragua, and Tanzania. *Coastal Management, 38*: 216–237.

Tshuma, M. C., & Jari, B. (2013). The informal sector as a source of household income: The case of Alice town in the Eastern Cape Province of South Africa. *Journal of African Studies and Development, 5*(8), 250.

United Nations Conference on Trade and the Environment (n.d). Retrieved from https://unctadstat.unctad.org/wds/TableViewer/tableView.aspx?ReportId=117 [accessed 10 July 2019].

USAID. (2006). Global market assessment for handicrafts, Vol. 1. Final draft (8 pages).

Van Eeden, A. (2011). The geography of informal arts and crafts traders in South Africa's four main city centres. *Town and Regional Planning, 59*, 34–40.

Van Eeden, J. (1996). The influence of arts and crafts ideology in South Africa. *South African Journal of Cultural History, 10*(2), 72–87.

Wambaa, J. (2011). *Environmental factors influencing the growth of exports of handicraft products in Kenya*. A Research Project Submitted in Partial fulfillment of the Requirements for the Award of the Degree of Master of Business Administration, School of Business, Doctoral Dissertation, University of Nairobi.

Watson, D. J. and Van Binsbergem, J. (2008). *Livelihood diversification opportunities for pastoralists in Turkana, Kenya*. ILRI Research Report 5. ILRI (International Livestock Research Institute), Nairobi. Kenya, 43.

Zargham, H. (2007). Sustainable tourism development and handicrafts in the developing world. *WIT Transactions on Ecology and the Environment, 102*.

18 Comparative evaluation of visitors' perceptions of park characteristics in Southern Africa

Brijesh Thapa

Introduction

Parks and protected areas in the Southern African region are major wildlife-based attractions that drive domestic and international tourists. But all parks are not equal as some are iconic (e.g. Kruger National Park, South Africa) while others aspire to achieve a certain status and level of visitation flow (e.g. Kafue National Park, Zambia). However, to be competitive and attract visitors, parks need a differentiation strategy along with services that exceed expectations and optimize experiences. Essentially, park managers need to be cognizant of their visitors' preferences, expectations, motivations, experiences and satisfaction as that influences repeat visitation (Akama, & Kieti, 2003; Beh & Bruyere, 2007; Engelbrecht, Kruger, & Saayman, 2014; Kruger, Viljoen, & Saayman, 2017; Morrison-Saunders, Hughes, Pope, Douglas, & Wessels, 2019; Mulholland & Eagles, 2002; Thapa & Lee, 2017), and recommendations either via word-of-mouth or social media (Kim, Thapa, & Kim, 2018). Additionally, knowledge about internal and external factors that impact destination choices (e.g. to visit parks) are also important facets to consider (Castro, Souza, & Thapa, 2015; Formica & Uysal, 2006; Neuvonen, Pouta, Puustinen, & Sievänen, 2010; Souza & Thapa, 2018).

Park-based tourism has been extensively promoted in Southern Africa. However, there is a paucity of research with respect to the examination of visitor experience (Kruger et al., 2017; Okello & Yerian, 2009; Thapa, 2012, 2013; Thapa & Lee, 2017; Wade & Eagles, 2003). Given the level and nature of competition over visitors in Southern Africa, it is vital for park managers to benchmark the quality of services and experiences (Engelbrecht et al., 2014; Thapa, Parent, Mupeta, & Child, 2011; World Bank, 2007). Furthermore, this issue is significant for emerging park-based destinations to demonstrate competitiveness and sustain financial revenues via tourism (Banerjee et al., 2018; Okello, 2014; Thapa et al., 2011; World Bank, 2007). The economic incentive is important as protected area systems in the Southern Africa region are largely underfunded. Hence, increased tourism flows to protected areas are necessary to generate revenues for the country's entire system network (Lindsey et al., 2014).

In Zambia, park-based tourism is a major sector to promote conservation and local economic development as the protected areas' network is extensive and comprises 20 national parks and other designated areas (i.e. Game Management Areas, and Wildlife Sanctuaries). In comparison to other park-based destinations in the region, Zambia receives one of the lowest numbers of visitors (Lindsey et al., 2014). Additionally, the flow of visitors within the country is skewed to a few parks, namely, Mosi-oa-Tunya (23,083 annual visitors in 2015), and South Luangwa (43,653 annual visitors in 2015) (Ministry of Tourism and Arts, 2016), largely due to their iconic status, wildlife concentration and tourism infrastructure (Thapa, 2013; Thapa & Lee, 2017). Hence, the need to invest in park infrastructure, integrated marketing and wildlife concentration in other areas is required to demonstrate competitiveness in the region, as well as dispersal of visitors within the country (World Bank, 2007).

Among the protected areas in Zambia, Kafue National Park (KNP) is the largest spatially (22,480 sq. km), has an abundance of natural and wildlife resources, and opportunities to improve quality of life in the adjacent local communities via tourism and other development initiatives (Thapa, 2013). However, local people have expressed a perceived exclusion from the park and its accrued limited economic benefits, mostly at the community level (Namukonde & Kachali, 2015). The lack of tangible benefits is likely due to the small volume of tourists, as 12,960 annual overnight visitors were recorded in 2015 (Ministry of Tourism and Arts, 2016). The rationale for under-utilization among tourists is related to several factors. International visitors who traveled to Zambia but not KNP perceived fear of personal safety, lack of quality lodges and camps, and not enough information about the park as factors that prevented them to visit. Likewise, domestic visitors noted financial cost and poor road conditions as constraints to park visitation (Thapa, 2012).

KNP is vast and lacks major tourism infrastructure (i.e. inadequate number and price range of accommodations, lack of visitor facilities and rest areas, limited park rangers and visible patrols and so on) along with extensive road networks in comparison to other popular national parks in the Southern Africa region. KNP has various site attributes which make it unique in comparison to other parks in the region. Visitors have highly rated the park for its accessibility to wildlife, such as the opportunity to experience close observation of wildlife, an abundance of biodiversity and wildlife and the ability to view and interact with unspoiled wilderness. Conversely, visitors have also offered low grades for infrastructure (i.e. condition of roads and facilities) and information (i.e. nature and cultural tourism, access to park information) (Thapa & Lee, 2017). Such essential visitor feedback offers the park management authority an opportunity to improve and enhance services.

Given KNP's size and proximity to the capital city (Lusaka) as well as Victoria Falls, the park has the potential to further expand the flow of domestic and international visitors. However, challenges are evident (i.e. advertising resources, lack of brand identity, consumer interest, and so on) and can be mitigated with an emphasis on diversification of tourists. KNP needs to first strategically assess, differentiate and accordingly compete for visitors among park-based destinations in the Southern Africa region. Within this context, the concept of destination competitiveness is most applicable as the guiding conceptual framework (Ritchie & Crouch, 2010). Destination competitiveness is essentially the ability of a destination to deliver superior goods and services that are evaluated to be important by tourists (Dwyer & Kim, 2003). Research related to destination competitiveness has evolved in the tourism discipline and examined from multiple perspectives such as a competitive position for specific

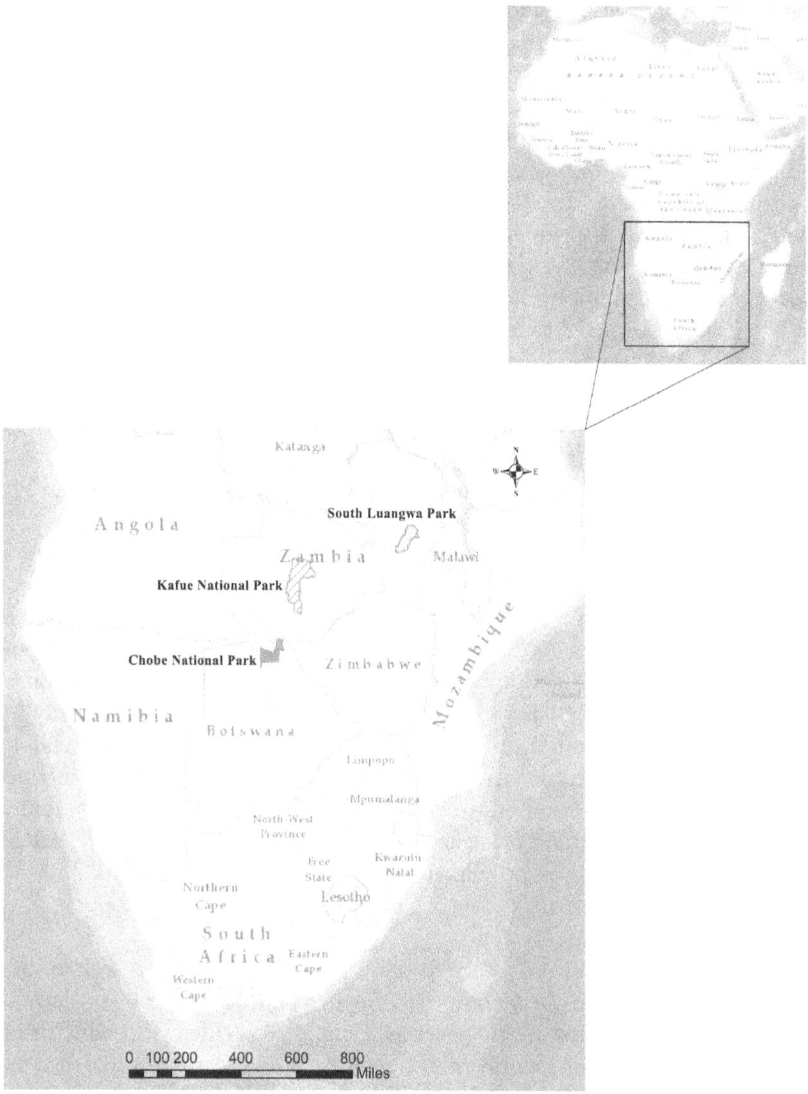

Figure 18.1 Three national parks in Southern Africa

tourism activities, settings and performance attributes, along with determinants and indicators (Manrai, Manrai, & Friedeborn, 2018). With respect to park management, the competitive aspect is indexed against the performance of numerous functions (Engelbrecht, 2015). Moreover, the identification of a park's comparative factors (i.e. resources – cultural, economic, physical) and competitive advantage factors (i.e. effectiveness and efficiency of resource use) are collectively required and subsequently needs to be enhanced to effectively compete (Engelbrecht, 2015; Ritchie & Crouch, 2010; Ritchie & Crouch, 2003). Overall, the competitive level influences the destination's attractiveness among visitors (Dwyer & Kim, 2003).

In the Southern African region, park-based tourism is dependent on wildlife, scenery and tourist infrastructure, hence the need to assess competitiveness. Within this line of inquiry, the purpose of this case study is to build on the baseline evaluation of visitor experience at KNP, with a comparative assessment of visitor experience in competitor parks (i.e. SLNP and CNP). This assessment provides implications to effectively manage, compete and deliver quality visitor experiences to influence satisfaction, repeat visitation, positive word of mouth and social media communications. Further, such visitors' evaluative comparisons offer a benchmark for KNP with regards to competitiveness and market positioning within the Southern Africa's park-based tourism product. Figure 18.1 shows the study areas of the three national parks in Southern Africa.

Methods

Data was collected at various designated sites within KNP, departure terminals in Lusaka and Livingstone International Airports, and several lodging establishments in the city of Livingstone (home of Victoria Falls). Sites were identified based on purposive sample as they were chosen because of the opportunity to interview large samples of visitors. Data collection via visitor intercepts (stratified random selection method) was conducted during the morning, afternoon, and evening from August – October[1]. Respondents were approached by trained interviewers and requested to participate in the study. Upon consent, participants completed the survey questionnaire which usually took less than 30 minutes. Besides solo travelers, in the event of a group, only the identified leader completed the survey. In total, a final sample resulted in 2,395 respondents with a response rate of 61.8%. However, the overall total sample was based on a larger study about KNP and Zambia. For this examination, the service quality dimensions were selected and analyzed. More specifically, for comparative purposes, only responses from participants who had visited KNP and one or both competitor parks (South Luangwa National Park, Zambia, and Chobe National Park, Botswana) within the last three years were included.

Service quality was measured with 15 items and largely grounded within the dimensions related to service delivery, environment and outcome (Brady & Cronin, 2001). The items were formulated to adapt for a park-based context in Southern Africa focused on wildlife, facilities, and service functions.

Respondents were asked to evaluate the quality of various park characteristics which ranged from environmental and wildlife resources to current infrastructure, and visitor experience. A list of 15 characteristics were listed, and respondents rated each item on a scale of 1 to 5 (1=Poor, 2=Fair, 3=Good, 4=Very Good, 5=Excellent). Respondents were also asked to report key wildlife observed since they are the featured attraction which draws visitors to the parks in Southern Africa.

Results

Respondents were requested to evaluate the quality of various characteristics which ranged from environmental and wildlife resources to infrastructure and visitor experience at three parks: KNP, CNP and SLNP. First, a general assessment was conducted for all three parks based on the overall mean values for each characteristic, regardless of respondents' direct comparisons, as some parks were more frequented than others[2] Based on the overall mean scores, respondents generally rated SLNP higher in almost all categories, as 10 out of 14 items reflected mean values of above 4.00. Also, SLNP had the highest mean values for 12 items when compared with KNP and CNP. However, KNP was evaluated the highest among the three parks with respect to "Opportunity to recreate without interference from other visitors." Similarly, CNP was rated the highest among the three parks with regards to the "Condition of road and facilities," and "Accessible roads to the park and/or airstrip." Nevertheless, the mean values across all three parks for these two items were generally low, especially for KNP.

Second, a head-to-head comparative analysis of the two parks (KNP and SLNP) in Zambia was conducted. Since SLNP is the leader (albeit smaller in size – 9,050 sq. miles) with respect to visitor numbers and market share in Zambia, it was prudent to compare and benchmark KNP. While this type of analysis offers respondents' direct evaluation of both parks, the total sample size was reduced to reflect the comparisons. Basically, only evaluations by those respondents who had visited *both parks* were compared. Based on a paired t-test, only 1 out of the 15 items was not statistically significant [see Table 18.1].

Respondents rated SLNP over KNP in almost all categories (12 out of 15 items), except for three items – "Opportunity to experience close observation of wildlife," "Opportunity to recreate without feeling crowded," and "Ability to view and interact with unspoiled wilderness." Conversely, two items that had the lowest mean values (under 3.00) were evaluated for KNP: "Condition of roads and facilities," and "Information about nature and cultural tourism opportunities." On comparative direct evaluation by visitors, it was apparent that SLNP was rated higher in various categories, but KNP still had a comparative advantage with respect to offering a natural wilderness experience.

Third, a head-to-head comparative analysis of three parks (CNP, KNP and SLNP) were conducted. Similarly, only evaluations by those respondents who had visited all three parks were compared [see Table 18.2]. Again, the sample

Table 18.1 Quality of park characteristics at two national parks in Zambia

Quality of park characteristics	*Kafue National Park*		*South Luangwa National Park*			
	Mean	*St. Dev.*	*Mean*	*St. Dev.*	*N*	*t-value*
Opportunity to recreate without feeling crowded	4.14	1.02	3.79	1.09	190	3.78***
Opportunity to experience close observation of wildlife	3.62	1.08	4.45	0.82	191	9.53***
Opportunity to recreate without interference from other visitors	4.23	0.92	3.81	1.10	190	4.72***
Abundance of biodiversity and wildlife	3.49	1.11	4.27	0.89	188	−8.35***
Ability to view and interact with unspoiled wilderness	4.13	1.03	4.11	0.94	189	.306 ns
Condition of roads and facilities	2.62	1.10	3.32	1.10	189	−8.26***
Comfortable and quality lodges/camps	3.64	1.06	4.10	0.92	189	−6.49***
Accessible roads to the park and/or airstrip	3.12	1.19	3.33	1.19	186	−2.32*
Information about nature and cultural tourism opportunities	2.88	1.19	3.39	1.67	183	−7.39***
Ease of obtaining current and accurate park information	3.01	1.16	3.40	1.10	183	−5.61***
Availability of staff to answer questions	3.56	1.20	3.91	1.10	183	−5.32***
Courteous and friendly park staff	4.02	0.99	4.13	0.96	178	−2.47*
Service quality worth the money paid	3.52	1.17	3.78	1.14	185	−4.10***
Value for park entry fee paid	3.38	1.29	3.70	1.21	184	−5.12***
Safety and security within the park boundaries	4.08	0.98	4.18	0.95	190	−2.30*

Note: Items anchored by 1=Poor (SD), 2=Fair (D), 3=Good (N), 4=Very Good (A), 5= Excellent (SA).

St. Dev.=Standard Deviation; N=Frequency.

*p<.05, **p<.01, ***p<.001, ns = not significant

size was dramatically reduced, and the results should be considered within this context. However, the ratings do offer a good basis for direct comparisons by visitors, respectively. Based on a series of one-way ANOVA tests, only two (2) out of 15 items failed to achieve statistical significance (i.e. Courteous and friendly park staff; Safety and security within the park boundaries).

Similarly, results were skewed towards CNP as it was rated higher (10 out of the 15 items) than KNP and SLNP. However, the mean values for some of the respective items were below 4.00 and were relatively close in range with SLNP,

Table 18.2 Quality of park characteristics at three national parks

Quality of park characteristics	*Chobe National Park (Botswana)*		*Kafue National Park (Zambia)*		*South Luangwa National Park (Zambia)*		*N*	*F-value*
	Mean	*St. Dev.*	*Mean*	*St. Dev.*	*Mean*	*St. Dev.*		
Opportunity to recreate without feeling crowded	3.29	1.31	4.12	1.11	3.87	1.11	58	9.22***
Opportunity to experience close observation of wildlife	4.47	0.76	3.39	0.87	4.43	0.87	58	32.63***
Opportunity to recreate without interference from other visitors	3.37	1.27	4.34	1.12	3.97	1.12	58	16.47***
Abundance of biodiversity and wildlife	4.19	1.05	3.28	0.95	4.30	0.95	59	25.37***
Ability to view and interact with unspoiled wilderness	3.79	1.14	4.09	0.95	4.24	0.95	58	5.25**
Condition of roads and facilities	3.82	0.78	2.82	1.02	3.45	1.02	59	29.60***
Comfortable and quality lodges/camps	4.25	0.77	3.62	0.99	4.17	0.99	57	10.95***
Accessible roads to the park and/or airstrip	4.14	0.97	3.26	1.15	3.37	1.15	58	19.39***
Information about nature and cultural tourism opportunities	3.71	1.07	2.89	1.13	3.46	1.13	57	14.80***
Ease of obtaining current and accurate park information	3.79	1.05	3.05	1.27	3.36	1.27	59	11.75***
Availability of staff to answer questions	4.03	1.06	3.61	1.21	3.97	1.21	58	7.15**
Courteous and friendly park staff	4.12	0.86	4.11	1.00	4.23	1.00	58	1.87ns
Service quality worth the money paid	3.94	1.09	3.47	1.21	3.85	1.21	57	6.93**
Value for park entry fee paid	3.69	1.24	3.34	1.26	3.69	1.26	59	4.59*
Safety and security within the park boundaries	4.35	0.85	4.23	0.87	4.33	0.87	59	.849 ns

Note: Items anchored by 1=Poor (SD), 2=Fair (D), 3=Good (N), 4=Very Good (A), 5= Excellent (SA).

St. Dev.=Standard Deviation; N=Frequency.

*$p<.05$, **$p<.01$, ***$p<.001$, ns = not significant

and a few with KNP. In addition, KNP was ranked the highest with respect to two items: "Opportunity to recreate without feeling crowded," and "Opportunity to recreate without interference from other visitors." Nevertheless, two items with the lowest mean values (under 3.00) were again evaluated for KNP with respect to: "Condition of road and facilities," and "Information about

Table 18.3 Type of wildlife observed at three national parks

Type of wildlife	*Chobe National Park (Botswana)*		*Kafue National Park (Zambia)*		*South Luangwa National Park (Zambia)*	
	Yes (%)	*No (%)*	*Yes (%)*	*No (%)*	*Yes (%)*	*No (%)*
Lion	59.4	40.6	64.2	35.8	82.7	17.3
Leopard	31.8	68.3	35.0	65.0	66.8	33.2
Buffalo	84.6	15.4	49.7	50.3	87.4	12.6
Elephant	93.3	6.7	80.0	20.0	95.4	4.6
Rhino	15.7	84.3	8.4	91.6	10.3	89.7
Cheetah	11.9	88.1	20.3	79.7	9.2	90.8
Hippos	88.1	11.9	78.9	21.1	90.2	9.8
Wild Dog	22.9	77.1	23.7	76.3	30.1	69.9
Sable	48.5	51.5	38.8	61.2	25.8	74.2
Roan	27.5	72.6	36.8	63.3	21.0	79.0
Hartebeest	26.8	73.2	40.0	60.0	25.9	74.1
Tsessebe	20.4	79.6	13.7	86.3	10.3	89.7
Red Lechwe	46.6	53.4	32.4	67.6	17.9	82.1

Note: Response could include multiple categories and hence valid percent does not equal 100.

nature and cultural tourism opportunities." Overall, based on the evaluative comparisons, KNP's identity reflects a place to achieve a wilderness experience with minimal infrastructure and services. While KNP's product has an appeal for a niche market segment, it would need to invest in facilities, road networks and visitor services to expand for additional tourist flows.

Since the featured attraction for visitors to parks in Southern Africa is wildlife, a direct comparison was conducted for the three parks. For KNP, the majority of the sightings comprised elephants (80.0%), hippos (78.9%) and lions (64.2%), which are all key animals that attract visitors. However, leopard and cheetah were not as noticeably observed, which was expected given the discreet nature of these animals. For all three parks,[3] sightings of elephants followed by hippos were both rated very high, while cheetahs were not observed by the majority of the respondents. Based on the reported observations, respondents generally rated SLNP higher in most all types of wildlife observed, with KNP and CNP relatively closely. Some of the notable wildlife such as lion, leopard, buffalo, hippo and elephant were reported in high numbers at SLNP [see Table 18.3].

Discussion

The increased tourist demand for nature-based attractions globally has created additional pressure on parks and protected areas (Balmford et al., 2009; Stone & Nyaupane, 2018). This has resulted in increased tourism flows but only in select parks in Southern Africa. Park-based tourism is crucial for national and local economies but is extremely competitive in the region (Lindsey et al., 2014). In addition, tourism provides needed opportunities for employment and entrepreneurial activities for rural and local communities which are in close proximity (Bruyere, Beh, & Lelengula, 2009; Moswete & Thapa, 2015; Mvula, 2001; Namukonde & Kachali, 2015; Ormsby & Mannie, 2006; Saarinen, 2011; Snyman, 2014). Hence, parks are challenged to be more competitive and generate revenues and economic impacts in the region with the goal to integrate nature-based tourism, sustainability and community linkages (Engelbrecht, 2015; Mulholland & Eagles, 2002; Okello, 2014; Saayman & Saayman, 2006). This aspect is relevant to Zambia with respect to the assessment of competitiveness of its established (e.g. SLNP) and emerging parks and protected areas (e.g. KNP).

Based on the conceptual underpinnings of destination competitiveness (Ritchie & Crouch, 2010), in this study, visitors' evaluative judgement of KNP's attributes were compared with competitor parks (i.e. SLNP and CNP) that are more popular in the Southern Africa region. A key aspect to this assessment was related to a direct head-to-head comparative analysis. Visitors were asked to first evaluate the attributes of KNP, subsequently followed to rate the same characteristics at SLNP in Zambia and CNP in Botswana, respectively. The unique contribution of this research was that only visitors who had visited all three parks offered their evaluations. This approach was conducted to ensure parallel assessment by visitors to ascertain service quality and experience across all three parks.

KNP has its comparative advantage with respect to the unspoiled wilderness but also hosts a series of factors that effectively hinder growth in visitor arrivals and length of stay. It is apparent where financial resources need to be mobilized to enhance and strengthen the nature-based tourism product. More specifically, facility and information management, as well as visitor services need to be improved substantially. Also, access and ease of park information along with the conditions of the roads need to be afforded immediate priority status. The issue of roads is notable especially with respect to time and optimizing the wildlife viewing experience (Thapa, 2012), as well as the ability for management to protect the park's resources (Karanja, 2012; Stone, 2013). Furthermore, on comparison with the leading national park in Zambia (SLNP), it was evident that visitors rated it lower in almost all categories, except for close accessibility to wildlife, lack of crowds, and view and interact with unspoiled wilderness. While SLNP rose above KNP in visitors' evaluations, there were certain aspects such as natural wilderness experience which could be promoted.

In comparison to the known parks such as CNP and SLNP, which are major tourist attractions and have a built-in reputation and brand name, KNP was relatively competitive in certain categories, especially with respect to opportunity to recreate without interference from other visitors. These were likely attributed to the vast size of KNP where it is possible to be alone or with a group, and not meet others. Conversely, the size also dictates wildlife sightings as animals have a larger terrain to hide from humans. This could possibly explain the fact that visitors to generally rated KNP much lower than the other two parks with respect to the abundance of biodiversity and wildlife. Additionally, KNP was again ranked the lowest with respect to accessibility of information and the condition of road and facilities. CNP was rated higher in almost all categories among the three parks. However, SLNP was rated the highest with regards to some characteristics such as the abundance of biodiversity and wildlife; ability to view and interact with unspoilt wilderness; and courteous and friendly staff.

Since wildlife is the key attraction that draws visitors to the parks (Kruger et al., 2017), KNP was competitive among certain types of wildlife with the majority of the sightings comprised of elephants, hippos, and lions. Given the smaller size of SLNP and wildlife density, it was rated higher in most types of wildlife observed, with CNP and KNP rated closely. Overall, the type of wildlife sighting at KNP appears to be generally good, but densities of certain species could be increased as the frequency of sighting is an important indicator to engage visitors and encourage repeat visitation. This aspect has been supported via a contingent valuation study whereby both current and non-visitors demonstrated the most support for their willingness to accept and pay a higher park entry fee for improvements of natural resources and amenities (e.g. increase wildlife numbers of the Big Five – lion, leopard, elephant, buffalo and rhino, and other animals) (Thapa & Parent, 2018).

Generally, in comparison, KNP has its strengths (comparative advantage) but needs to focus on certain key areas to better refine its tourism product to enhance its competitive advantage. Moreover, besides government infusion of financial resources for infrastructure and associated upgrades, the Zambia Tourism Board along with the international tour operators need to play an active role for product development, bundling, packaging and marketing for different tourist segments. For example, with the popularity of Victoria Falls, the opportunity exists to leverage on the brand equity and package KNP due to its proximity. There is also much work to be conducted with respect to generating awareness and creating a marketing campaign for KNP. Among those visitors who had not visited KNP, about half of them had never even heard about the park. Also, those who had never ever visited KNP noted two major reasons – not enough information available about the park, and lack of time to visit (Thapa, 2012).

Conclusion

KNP reflects a new frontier with abundant opportunities for a true wilderness experience, but it is also evident that infrastructure, facilities, information and visitor services management are all needed to strengthen KNP as a major competitive nature-based tourism destination in the Southern Africa region. Notably, KNP needs to demonstrate the ability to utilize its comparative factors to ensure a competitive advantage for destination sustainability with respect to resource protection and economic viability (Engelbrecht, 2015; Ritchie & Crouch, 2010; Ritchie & Crouch, 2003). In all, challenges remain and need to be addressed; however, there are some aspects of the KNP product that have unique appeal for visitors. This study provided a baseline assessment for KNP to strengthen, enhance, and improve visitor experience and thus remain competitive in the region.

Notes

1 Since visitation occurs year-round, this cross-sectional sample may not be representative of all tourists.
2 Due to space restrictions, this table is not provided.
3 It should be noted that some type of wildlife (e.g. rhinos in KNP) are not found in all three parks and were likely not accurately reported by respondents.

References

Akama, J., & Kieti, D. (2003). Measuring tourists' satisfaction with Kenya's wildlife safari: A case study of Tsavo West National Park. *Tourism Management, 24*(1), 73–81.

Alavalapati, J. (2018). Financing the sustainable management of Rwanda's protected areas. *Journal of Sustainable Tourism, 26*(8), 1381–1397.

Balmford, A., Beresford, J., Green, J., Naidoo, R., Walpole, M., & Manica, A. (2009). A global perspective on trends in nature-based tourism. *PLoS Biol*, 7(6), e1000144. doi:10.1371/journal.pbio.1000144

Beh, A., & Bruyere, B. (2007). Segmentation by visitor motivation in three Kenyan National Reserves. *Tourism Management, 28*, 1464–1471.

Banerjee, O., Cicowiez, M., Ochuodho, T., Masozera, M., Wolde, B., Lal, P., Dudek, S., & R. R. Alavalapati, J. (2018). Financing the sustainable management of Rwanda's protected areas. *Journal of Sustainable Tourism, 26*(8), 1381–1397.

Brady, M., & Cronin, J. (2001). Some new thoughts on conceptualizing perceived service quality: A hierarchical approach. *Journal of Marketing, 65*(3), 34–49.

Bruyere, B., Beh, A., & Lelengula, W. (2009). Differences in perceptions of communication, tourism benefits, and management issues in a protected area of rural Kenya. *Environmental Management, 43*(1), 49–59.

Castro, E., Souza, T., & Thapa, B. (2015). Determinants of destination attractiveness in the National Parks of Brazil. *Parks: The International Journal of Protected Areas and Conservation, 21*(2), 51–62.

Dwyer, L., & Kim, C. (2003). Destination competitiveness: Determinants and indicators. *Current Issues in Tourism, 6*(5), 369–414.

Engelbrecht, W. (2015). *Developing a competitiveness model for South African National Parks.* Unpublished doctoral dissertation, North-West University, Potchefstroom, South Africa.

Engelbrecht, W., Kruger, M., & Saayman, M. (2014). An analysis of critical success factors in managing the tourist experience at Kruger National Park. *Tourism Review International, 17*(4), 237–251.

Formica, S., & Uysal, M. (2006). Destination attractiveness based on supply and demand evaluations: An analytical framework. *Journal of Travel Research, 44*(4), 418–430.

Karanja, D. (2012). The role of the Kenya wildlife service in protecting Kenya's wildlife. *The George Wright Forum, 29*(1), 74–80.

Kim, M., Thapa, B., & Kim, H. (2018). International tourists' perceived sustainability of Jeju Island, South Korea. *Sustainability, 10*(1), 1–13.

Kruger, M., Viljoen, A., & Saayman, M. (2017). Who visits the Kruger National Park, and why? identifying target markets. *Journal of Travel & Tourism Marketing, 34*(3), 312–340.

Lindsey, P., Nyirenda, V., Barnes, J., Becker, M., McRobb, R., Tambling, C., . . . t'Sas-Rolfes, M. (2014). Underperformance of African protected area networks and the case for new conservation models: Insights from Zambia. *PLoS One, 9*(5), 94–109.

Manrai, L., Manrai, A., & Friedeborn, S. (2018). Environmental determinants of destination competitiveness and its tourism attractions-basics-context, A-B-C, indicators. A review, conceptual model and propositions. *Journal of Economics, Finance and Administrative Science.* doi.org/10.1108/JEFAS-01–2018–0010.

Ministry of Tourism and Arts. (2016). *Zambia tourism statistics digest 2015.* Lusaka: Zambia.

Morrison-Saunders, A., Hughes, M., Pope, J., Douglas, A., & Wessels, J. (2019). Understanding visitor expectations for responsible tourism in an iconic national park: Differences between local and international visitors. *Journal of Ecotourism.* doi:10.1080/14724049.2019.1567740

Moswete, N., & Thapa, B. (2015). Factors that influence residents' support for community-based ecotourism development in Kgalagadi Transfrontier Park, Botswana. *Journal of Ecotourism, 12*(2–3), 243–263.

Mulholland, G., & Eagles, P. (2002). African parks: Combining fiscal and ecological sustainability. *Parks: The International Journal of Protected Areas and Conservation, 12*(1), 42–49.

Mvula, C. (2001). Fair trade in tourism to protected areas: A micro case study of wildlife tourism to South Luangwa National Park, Zambia. *International Journal of Tourism Research, 3*(5), 392–405.

Namukonde, N., & Kachali, R. (2015). Perceptions and attitudes of local communities towards Kafue National Park, Zambia. *Parks: The International Journal of Protected Areas and Conservation, 21*(2), 25–36.

Neuvonen, M., Pouta, E., Puustinen, J., & Sievänen, T. (2010). Visits to national parks: Effects of park characteristics and spatial demand. *Journal for Nature Conservation, 18*(3), 224–229.

Okello, M. (2014). Economic contribution, challenges and way forward for wildlife-based tourism industry in Eastern African countries. *Journal of Tourism and Hospitality, 3*(1), 3–12.

Okello, M., & Yerian, S. (2009). Tourist satisfaction in relation to attractions and implications for conservation in the protected areas of the northern circuit, Tanzania. *Journal of Sustainable Tourism, 17*(5), 605–625.

Ormsby, A., & Mannie, K. (2006). Ecotourism benefits and the role of local guides at Masoala National Park, Madagascar. *Journal of Sustainable Tourism, 14*(3), 271–287.Ritchie, J., & Crouch, G. (2003). *The competitive destination: A sustainable tourism perspective.* Wallingford: CABI.

Ritchie, J., & Crouch, G. (2010). A model of destination competitiveness and sustainability. In Y. Wang & A. Pizam (Eds.), *Destination marketing and management: Theories and applications* (pp. 326–339). North America, MA: CABI.

Saarinen, J. (2011). Tourism development and local communities: The direct benefits of tourism to Ovahimba communities in the Kaokoland, northwest Namibia. *Tourism Review International, 15*(1–2), 149–157.

Saayman, M., & Saayman, A. (2006). Estimating the economic contribution of visitor spending in the Kruger National Park to the regional economy. *Journal of Sustainable Tourism, 14*(1), 67–80.

Snyman, S. (2014). Assessment of the main factors impacting community members' attitudes towards tourism and protected areas in six Southern African countries. *Koedoe, 56*(2), 1–12.

Souza, T., & Thapa, B. (2018). Tourism demand analysis of federal protected areas of Brazil. *Journal of Park and Recreation Administration, 36*(3), 1–21.

Stone, M. T. (2013). *Protected areas, tourism and rural community livelihoods in Botswana.* Arizona State University.

Stone, M. T., & Nyaupane, G. P. (2018). Protected areas, wildlife-based community tourism and community livelihoods dynamics: Spiraling up and down of community capitals. *Journal of Sustainable Tourism, 26*(2), 307–324.

Thapa, B. (2013). Visitor segments and attitudes towards sustainable tourism in protected areas: A case study in Zambia. *Journal of Park and Recreation Administration, 31*(2), 50–60.

Thapa, B. (2012). Why did they not visit? Examining structural constraints to visit Kafue National Park, Zambia. *Journal of Ecotourism, 11*(1), 74–83.

Thapa, B., & Lee, J. (2017). Visitor experience in Kafue National Park, Zambia. *Journal of Ecotourism, 16*(2), 112–130.

Thapa, B., & Parent, G. (2018). Tourists' willingness to accept/pay increased entry fees for park improvement projects. *Current Issues in Tourism*. doi:10.1080/13683500.2018.1533928

Thapa, B., Parent, G., Mupeta, P., & Child, B. (2011). Zambia Tourism Demand Assessment. In *Prepared for U.S. millennium challenge corporation and U.S. department of agriculture-foreignagricultural service* (179 pp.). Washington, DC. U.S. Millennium Challenge Corporation.

Wade, D., & Eagles, P. (2003). The use of importance – performance analysis and market segmentation for tourism management in parks and protected areas: An application to Tanzania's national parks. *Journal of Ecotourism, 2*(3), 196–212.

World Bank. (2007). *Zambia economic and poverty impact of nature-based tourism*. Report# 43373-ZM. Washington, DC: World Bank.

Part V

Conclusion

19 Conclusion

Natural resources, tourism and community livelihoods relationships: contemplating the future

Monkgogi Lenao, Naomi Moswete and Moren T. Stone

Introduction

Almost all of the chapters that made this edition indicate that in Southern Africa problems of natural resources use, community development and unsustainable tourism land-use practices are interdependent and directly linked to institutional deficits in land tenure and property rights systems as highlighted by Van der Duim, Lamers, and Van Wijk (2015). These interconnected complications create a platform of the world's greatest challenges of the 21st century in achieving sustainable development, henceforth questioning the ever-changing resources use regimes between natural resources, sustainable tourism and community livelihoods dynamics. The general indication is that the understanding of the relationship between natural resources, sustainable tourism and community livelihoods is inconclusive and is still open to debate. Therefore, there is, generally, no dispute regarding the question of whether or not the natural resource base needs to be conserved for posterity. However, differences are often notable with regards to the modalities through which this goal is to be achieved. As a result of these differences, there exists an array of approaches ranging from pure protectionism (fortress strategies) (Adams & Hulme, 2001) to people-centred conservation (Bell, 1987) ideas, with several other permutations in between. Protectionism, generally considered the founding ideology of modern-day conservation (Jones, 2006) has not completely disappeared. In fact, the concept of protected areas (e.g. National Parks and Game Reserves) is premised on this ideology, albeit slightly modified. However, there is a general movement towards more people-centred approaches in view of the need to promote societal beneficiation and, subsequently, buy-in. In the wake of this, research concerned with the relationship between natural resources conservation and community livelihoods development has gained currency in recent times. A notable addition to this debate/discourse is that of tourism development, which represents an important user and socioeconomic element of change in wilderness areas (Saarinen, 2015). Therefore, questions concerning the extent to which this relationship is, or can be, sustainable continue to intrigue many scholars around the world (Godwin & Roe, 2001).

These questions are even more pertinent within the developing world where rural livelihoods are predominantly dependent upon the natural resource base (Okello, Seno, & Nthiga, 2009; Stone & Stone, 2011). Within the Southern and Eastern African regions, this concern has been underlined, in part, by a proliferation of doctoral level studies on the topic. Among others, these include; Sandbrook, 2006; Mbaiwa, 2008; Ahebwa, 2012; Stone, 2013; Bedelian, 2014. The common concern among these researchers regarding this topic is by no means accidental. It is common knowledge that these regions are characterised by extreme rural poverty whose consequence often put excessive pressure on the resource base as well as community vulnerabilities that may result from any external shocks experienced by the same resource base. Factors such as human wildlife conflict as well as climate change vulnerabilities and adaptation strategies (or a lack thereof) (Rogerson, 2016; Hambira & Saarinen, 2015) have a potential to compound this relationship. In this volume, the conservation, livelihoods and tourism nexus are explored through a multiplicity of lenses and approaches with the view to draw some key lessons for the southern region and possibly the wider developing world.

Recently, Saarinen (2016) engaged with this topic from an interesting angle where he asked the question: wilderness use, conservation and tourism: what do we protect and for and from whom? In this work, one of the key arguments he drives is that resources conservation should be understood within the framework of coexistence between nature and humans. He asserts, "Thus, in developing countries' situations, especially, we need to challenge the core idea of the 50-year-old act and acknowledge that wilderness is not a place where humans are visitors (i.e. tourists) alone or that tourists are the only viable users of these areas. Instead of aiming to maintain and promote the separation of people and nature for the next 50 years, the focus should be on the full spectrum of sustainability, covering not only ecological integrity but also social equity and inclusions and economic empowerment with responsible limits of wilderness use" (Saarinen, 2016, p. 5).

In this concluding chapter, the future of natural resources, tourism and community livelihoods relationships discourse is contemplated with optimism if adaptive natural resources management strategies are conceptualized or informed by social-ecological adaptive management systems that do not separate human beings from the natural environment. The preceding chapters have demonstrated that natural resources through nature-based tourism have the potential to contribute to community livelihoods improvement and Southern Africa economies. Though the damage has already been done and is irreversible in some parts of the Southern Africa region, for example, the eviction of indigenous communities from their ancestral land to pave way for protection on natural resources through the regime of protected areas, all is not lost. There is still some potential for nature-based tourism to improve local communities' livelihoods; as long as tourists keep visiting Southern Africa, respect for the rule of law, embrace for democracy, political stability and countering terrorism exist.

History and evolution of conservation and livelihoods nexus in Southern Africa: where are we today?

The introduction of different conservation ideologies at various historical epochs in Southern Africa has seen the relationship between natural resources and local communities intermittently shifting in accordance with the latter's prevailing notion of how beneficial it was to co-exist with such resources. In pre-colonial times, local norms and institutions defined and regulated this relationship, with the result that there was practically no reasonable need for external intervention (Gumbo). However, the advent of Western-style conservation efforts from colonial times, which served to distance the locals from the resource base, inadvertently tempered with this balance. These were essentially affirmed by most post-colonial governments in the sub-region resulting in continued conflict and distress to the resource base (Njerekai). A general theme running through this volume is that, perhaps as Saarinen (2016) argues, considerable efforts have to be targeted towards restoring the balance between conservation and livelihoods upliftment through tourism despite the litany of challenges that arise from such dispensations (Chilembwe, Njerekai, Hambira, Hebink Kiaka and Lubilo). Such issues as poor or lack of consultation and engagement among relevant stakeholders regarding the implementation and enforcement of conservation and tourism practices (Chilembwe, Hoogendoorn and Fitchett) are not new, yet they continue to characterise modern-day conservation efforts in the region.

In recent years more seemingly progressive ideas have been piloted and implemented in various settings, including various permutation of community-based conservation through tourism (e.g. CBNRM, Conservancy and CAMPFIRE), inclusive and cross-border collaborative conservation (i.e. transfrontier parks) as well as corporate social responsibility (Hambira). The rationale behind this current wave of efforts has been to address the plight of local communities by somehow including them in the conservation undertaking or demonstrating responsiveness towards their conditions. Despite these efforts, however, some troubling issues remain unresolved. For instance, besides challenges arising from conflicting interests in the community-based conservation through tourism landscape of the Southern African region (Hebink Kiaka and Lubilo), questions are being asked about long-term sustainability of CSR efforts being made (Hambira).

Not least, Hoogendoorn and Fitchett in their review of climate change literature, some 14 years after Preston-Whyte and Watson (2005)'s seminal publication on the same topic, make a very compelling remark concerning how tourism operators go about prioritising resource allocations aimed at improving climate change adaptation strategies in Southern Africa. Their argument is both apt and instructive; there is a certain likelihood of climate change acutely affecting conservation and sustainability of the tourism sector in both the short and long run. However, operators in the region are still at a stage where they need some cajoling in order for them to embrace the development of adaptation strategies that account for local community needs through engagements with the latter.

Matters of policy and practice in present-day conservation

Interestingly, a considerable amount of issues raised in the preceding section mirror the structural challenges relating the regional efforts aimed to reconcile sustenance of community livelihoods and conservation through tourism. After reading through the chapters in this section, it stands to reason that many countries in the sub-region still have a long way to go in ensuring fruitful partnerships and collaborative efforts between different role players in the conservation, livelihoods and tourism nexus. This is, not least, apparent with relation to such issues as land tenure and land management. Using the case of the Okavango Delta region in Botswana, Mogomotsi, Saayman and Saayman identify a strenuous relationship between contemporary (e.g. tourism) and traditional (e.g. agriculture) land uses. The conflict is primarily related to problem animals causing problems to agricultural installations and other investments resulting in negative attitudes towards conservation among locals. Accordingly, the matter is not helped by the apparent lack of capacity in the institutions (e.g. Department of Wildlife and National Parks) tasked with managing the conflict through investigations and government compensations. In the end Mogomotsi et al. note a pendulum swing situation between the good policy intents of promoting co-existence and the harsh reality of conflict resulting from community disgruntlement. The arising conflict situation prevents backwards economic linkages between the tourism industry and the agriculture sector, creating a missed opportunity for tourism to complement agriculture in terms of community livelihoods improvement.

Neatly tied to the foregoing is the realisation that the relations between those responsible for ensuring the realisation of the conservation-based community benefits as well as the intended beneficiaries and other players such as investors. Muzirambi, Musavengane and Mearns note that while the contemporary policy dispensations in the sub-region (e.g. CBNRM) have well intended to decentralise resources management, the challenges of centralised decision-making remain evident. In cases where some kind of decentralisation is being affected, as in their selected case of Zimbabwe, the practical creation of more open, responsive and effective local government for an enhanced and more representative system of community-level decision-making is yet to be achieved. The apparent lack of real engagement with local communities in the conservation processes is seen as one of the factors leading to negative attitudes towards the natural resource base. This point is buttressed by Nyirenda, Milimo and Namukonde in their discussion of the Shantumbu case in Zambia.

In addition to policies emphasising benefits through co-management, other arrangements call for companies doing business in localities to demonstrate commitment towards the promotion of improved community wellbeing and livelihoods. Two contributions in this volume deal with this subject matter.

Hambira notes that a number of companies, from her sample in Botswana, have embraced some form of social upliftment practices ranging from partnerships with communities, investment in community projects and investment in employee and community social welfare matters. Within the same ambit, she observes that these practices do not demonstrate a clear commitment towards addressing local climate change vulnerabilities as well as income inequalities. To this end, Kimaro and Saarinen propose a CSR framework whose primary aim is to promote and improve efficiency of the CSR policy delivery mechanisms adopted by companies in the tourism industry. In recognition of what they perceive as the chaotic nature of CSR initiatives in the Global South, they hope the framework would serve to avert situations where more than one CSR initiative is in conflict with another and thus facilitate improved delivery on the intended goals of livelihoods upliftment among employees and local communities.

Debates

Realisation of policy goals and practices related to the tourism, livelihood and conservation nexus is by no means without debates and controversy. In this volume, Child reminds us that, while we are looking at the efficacy of contemporary livelihoods and conservation policies that aim to achieve inclusive governance, there is a pressing need to acknowledge the role of local communities in shaping the outcomes of these processes. He observes an emergence of agency on their part wherein these communities are increasingly demanding more from their governments by, for instance, challenging the globalisation of their rightful resources through such mechanisms as CITIES. The argument here is that contemporary debates on conservation, livelihoods and tourism development should begin to accommodate local communities not as passive recipients and beneficiaries only but also as real players who can thoroughly influence co-management issues and decisions. In the end, Child emphasises the need to strengthen community ownership of resources as the most prudent way to achieve conservation goals.

An alternative view is proffered by Fairer-Wessels who argues that while we appreciate the agency of local communities in shaping their own destiny, there is a need to consistently remember the effects of structural disadvantages from which they operate. For instance, Fairer-Wessels maintains that even those local communities that are aware of their plight may not necessarily be able to break out of the poverty cycle. With reference to the case of Mangaliso in South Africa, she notes that local communities often fail to attain empowerment owing to their own internal inadequacies such as power struggles and differential awareness and access to benefits. In Mauritius, Seetanah and Sannassee observe that improper institutional arrangement and role designation hamper development of the ecotourism sector. They argue for a holistic approach where efforts are made to ensure that ministries overseeing the Tourism and the Environment sectors co-ordinate their actions and combine their expertise, while clearly delineating each ministry's role to avoid any duplication of initiatives and to ensure policy coherence.

Southern Africa in the world: cases and lessons

The Southern African region presents a very interesting case in the study of conservation and livelihoods nexus. The region provides a wide spectrum of issues: natural resource abundance, economically vulnerable and resource-dependent communities, human wildlife conflict and implementation of various conservation initiatives as well as different appraisals on their performance by researchers, academics and practitioners. Overall, there is a general agreement that tourism development has great potential to promote a healthy relationship between resources conservation and improvement of local community livelihoods. This argument is largely premised on the idea that, if properly planned and implemented, tourism can accommodate local participation as well as benefit, which would in turn promote ownership and enhanced conservation practices (Ashley, 2000).

While the foregoing remains a highly persuasive argument, some serious structural challenges make a successful realisation of the twin objectives of conservation and livelihoods upliftment a distant dream. For example, one of the Mmopelwa and Mackenzie decry the informal nature of craft businesses in the region, making it difficult to account for their success and contribution to the local economy. Mmopelwa and Mackenzie attribute this to the apparent lack of well-articulated strategies geared towards supporting craft-based enterprises. This is particularly instructive, given that the craft industry is one of the oldest and most enduring businesses in the region. The region's inability to leverage on the potential of an industry with historical connections to the local cultures is worrisome. This is even more so considering that local craft production largely relies on indigenous technical knowledge whose contribution towards resource conservation has never been in doubt.

Notably, Thapa cites another example from Zambia demonstrating the fact that despite the internal resource richness of some areas in the region, there is an existence of an apparent inability to capitalise on such comparative advantages. When comparing Kafue National Park to South Luangwa and Chobe National Parks, Thapa observes a dire need for infrastructural development as well as the provision of visitor information with the hope to strengthen competitiveness for Kafue National Park. Importantly, Kafue National Park in Zambia is being compared to other nature-based parks within this same region. The level of development of nature-based tourism in the area is already significant, meaning that for Kafue communities to start reaping benefits, it would only take a little political will and action.

In the same light, Stone et al. compare tourism development between Chobe Enclave Conservation Trust in Botswana and the Grand Staircase-Escalante in the USA. They discover an interesting dichotomy between the manner in which tourism benefits are realised in Chobe through the Chobe Enclave

Conservation Trust (CECT) and in the Grand Staircase-Escalante area. While in the former case benefits are realised through a collective means, in the latter case, at the community level, benefits accrue from an individual locus and begin to spread outwards. Therefore, the case of CECT calls for devising and implementing some benefit-sharing mechanisms that will ensure that no section of society is inadvertently left out.

Overall, the many case studies presented in this book indicate that the relationships between natural resources and tourism; conservation and community livelihoods; natural resources and livelihoods; tourism and conservation are dynamic, complex and locally specific such that there is no single theory or conceptual framework underpinnings to examine the complex relationships, posing a challenge to attain sustainable development or sustainable development goals. It is evident throughout the book chapters that the relationships between natural resources, tourism and community livelihoods are complex. At times, natural resources conservation, tourism and community livelihoods appear to be complementary, and at other times they appear to be directly competitive, and at other, in conflicts. The purpose of this book, therefore, was to explore how tourism, particularly nature-based tourism, can help link biodiversity conservation and destination community livelihood improvements. The overall findings of the contributing chapters to this volume have some implications for community development planners, conservation architects and policy makers. Appropriate intervention measures through the guidance of government policies and institutional arrangement and/or realignment are key to establish and maintain direct and positive linkages among natural resources conservation, tourism and community livelihood improvements. Failure to realign community development planners and conservation architects understandings and government policies with these complexities of relationships pose a challenge for Southern African countries to attain sustainable development, hence the contribution of this book through the general theme of natural resources, tourism and community livelihoods in Southern Africa: challenges of sustainable development.

References

Adams, W. M., & Hulme, D. (2001). Conservation and communities: Changing narratives, policies and practices in African conservation. In D. Hulme & M. Murphree (Eds.), *African wildlife and livelihoods: The promise and performance of community conservation*, (pp. 9–23). Oxford: James Currey and New Hampshire: Heinemann.

Ahebwa, W. M. (2012). *Tourism, livelihoods and biodiversity conservation; An assessment of tourism related policy interventions at Bwindi Impenetrable National Park (BINP), Uganda*. PhD thesis, Wageningen University, Wageningen, the Netherlands.

Ashley, C. (2000). *The impacts of tourism on rural livelihoods: Namibia's experience*. London: Overseas Development Institute.

Bedelian, C. (2014). *Conservation, tourism and pastoral livelihoods: Wildlife conservancies in the Maasai Mara, Kenya*. PhD Thesis, University College London, London.

Bell, R. (1987). Conservation with a human face: conflict and reconciliation in African land use planning. In D. Anderson & R. Grove (Eds.), *Conservation in Africa: People, policies and practice* (pp. 79–101), Cambridge, Cambridge University Press.

Godwin, H., & Roe, D. (2001). Tourism, livelihoods and protected areas: Opportunities for fair trade tourism in and around national parks. *International Journal of Tourism Research, 3*, 377–391.

Hambira, W. L., & Saarinen, J. (2015). Policy-makers' perceptions of the tourism – climate change nexus: Policy needs and constraints in Botswana. *Development Southern Africa, 32*(3), 350–362.

Jones, S. (2006). A political ecology of wildlife conservation in Africa. *Review of African Political Economy, 33*(109), 483–495.

Mbaiwa, J. E. (2008). *Tourism development, rural livelihoods, and conservation in the Okavango Delta, Botswana*. PhD thesis, Texas A&M University, USA.

Nyaupane, G. P., & Poudel, S. (2011). Linkages among biodiversity, livelihood, and tourism. *Annals of Tourism Research, 38*(4), 1344–1366.

Okello, M. M., Seno, S. K. O., & Nthiga, R. W. (2009). Reconciling people's livelihoods and environmental conservation in the rural landscapes in Kenya: Opportunities and challenges in the Amboseli landscapes. *Natural Resources Forum, 33*, 123–133.

Preston-Whyte, R. A., & Watson, H. K. (2005). Nature tourism and climatic change in Southern Africa. In C. M. Hall & J. Higham (Eds.), *Tourism, recreation and climate change* (pp. 130–142), Bristol: Channelview Publications.

Rogerson, C. M. (2016). Climate change, tourism and local economic development in South Africa. *Local Economy, 31*(1–2), 322–331.

Saarinen, J. (2016). Wilderness use, conservation and tourism: What do we protect and for and from whom? *Tourism Geographies, 18*(1), 1–8.

Sandbrook, C. G. (2006). *Tourism, conservation and livelihoods: The impacts of gorilla tracking at Bwindi Impenetrable National Park, Uganda*. PhD thesis, University College London, University of London, London.

Stone, L. S., & Stone, T. M. (2011). Community-based tourism enterprises: Challenges and prospects for community participation; Khama Rhino Sanctuary Trust, Botswana. *Journal of Sustainable Tourism, 19*(1), 97–114.

Stone, M. T. (2013). *Protected areas, tourism and rural community livelihoods in Botswana*. PhD thesis, Arizona State University, Arizona, USA.

Van der Duim, R., Lamers, M., & Van Wijk, J. (Eds.). (2015). Novel institutional arrangements for tourism, conservation and development in Eastern and Southern Africa. In R. Van der Duim, M. Lamers, & J. Van Wijk (Eds.), *Institutional arrangements for conservation, development and tourism in eastern and Southern Africa: A dynamic perspective*. New York and London: Springer.

Index

acacia savannah 209
active *vs.* passive community tourism participation 229–230
Adair, R. 143
Adams, W. M. 269
adopt-a-school program 58
adventure tourism 83–84
Africa, handicrafts sector in *see* handicrafts
African Growth and Opportunity Act (AGOA) 237
African Ministerial Conference on Environment (AMCEN) 56
African Wildlife Economy Summit 165, 171
Africa's national parks, tourism in 38–49; conference resolutions 48–49; discussion 47–48; findings 41–47; literature review 39–40; methodology 40–41; *see also specific* parks
Agenda for Sustainable Development, 2030 205
Ahebwa, W. M. 270
Ahmed, R. R. 26
Albion Beach 197
anti-poaching unit (APU) 46–47
anti-poverty tourism (APT) 129
Antiquities Act 224
appreciative inquiry (AI) approach 226–227, 228
arable farming 94, 96, 98–102, 178, 226
aridity 82
Arusha Declaration 47
assets, defined 178
Association des Hôteliers et Restaurateurs de l'île Maurice (AHRIM) 195
attitudes: locals' perceptions and 108–109; negative 27, 103; Shantumbu (Zambia) residents' 116–117
awareness campaigns/sensitization programs 29, 200–201
Baird, T. 84
Bali 60
Balint, P. J. 152
Bangladesh 185
Barnes, G. 145
Barrow, E. 162
Basariya, S. R. 26
Basubiya 179, 226
Baumann, P. 239
beach recession 84, 198
beach tourism 81, 84, 191, 199
Bedelian, C. 270
Bell, R. 269
Bello, F. G. 28, 34
Binga case study (Zimbabwe) 149–150
birdwatching 58, 84, 211
Bishop, J. 246
Blantyre's Michiri Tourist Conservation Area 29
Blue Bay, Mauritius 195–196
Blue Flag status 191, 197
Botswana: forest reliance in 177; hunting ban in 163, 164; impact of climate change on tourism 82; trophy hunting in 13, 43; wildlife policies in 166
Botswana, community social upliftment practices 52–60; findings 58; methods 57–58; politics of 'sustainable' tourism and community impact 54–55; resilience to climate change 55–56; results 59–60; sustainable tourism development framework 56–57
Botswana, historical evolution of conservation and tourism 11–22; conservation debates 13–14; conservation in the colonial era 15–17; conservation landscape in the pre-colonial period 14–15; framework 12–13; historical development of

tourism 20–22; hunting ban 18–20, 163–165; methodology 13; post-colonial conservation trajectories 17–18
Botswana, nature-based tourism and community livelihoods in 221–233; active *vs.* passive community tourism participation 229–230; AI approach 226–227, 228; background 222–223; CECT *see* Chobe Enclave Conservation Trust (CECT); CNP 222, 224, 225–226, 227, 257, 258–263; data collection and analysis 226–227; discussion 232–233; dynamics 227, 229; handicraft sector 236–237, 240, 241, 243, 245–247; hunting elephants in 165; positive and negative outlooks 230–231; study sites and methods 224–225
Botswana Christian Council (BCC) 241
Botswana National Tourism Policy of 1990 54
Boulder community 224
Bowen, H. 124
Bramwell, B. 123
Branding 201–202
Braun, V. 181
Brejning, J. 127
Brundtland report 206, 223
Bryce Canyon National Park 224
Bryman, A. 31
Buhalis, D. 33
building poles 184
Butcher, J. 26
Butterworth, K. J. 177

Campbell, C. 143, 147
CAMPFIRE programme (Zimbabwe) 146–153, 170; Binga case study 149–150; case study analysis, implications from 153; communities in 171; devolution of power and community participation in CBNRM 148–149; Hwange case study 152–153; Mahenye case study 151–152; political decentralisation and devolution of power to local levels, impact on 148
camps and lodges sector 57–58
Camp Xakanaka *see* Okavango Delta
Cannonville 224
Cape Maclear 28
Cape St Francis 84
Cape Town Declaration on Responsible Tourism in Destinations 123
Cape Vidal 210
capital and livelihood 237–239
Capital Reef National Park 224
Carroll, A. B. 126–128, 130, 134
carving 239, 246, 248
cattle ranching 165, 171
Cavanagh 57
CBNRM (Conservancy and CAMPFIRE) 271, 272
CECT *see* Chobe Enclave Conservation Trust (CECT)
Central Kalahari National Park (Botswana) 38
certification and environmental sustainability 197, 201
Chambers, R. 207
Chichewa (local language) 30
Chigwere Cultural Village 29
Child, B. 5, 145, 273
Chilembwe, J. M. 5, 33, 34, 271
χ^2 test 105
Chi-square (χ^2) test of independence, Pearson 96
Chobe 165; craft market 245; foot and mouth disease in 102–103
Chobe Enclave Conservation Trust (CECT) 274–275; active *vs.* passive community tourism participation 230; AI process 227, 228; data collection process 227; discussion 232–233; natural resources, tourism and community livelihoods dynamics 227, 229; positive and negative outlooks, natural resources and tourism 230–231; study sites and methods 225–226
Chobe enclave residents *see* Chobe Forest Reserve (CFR) (Botswana), conservation of
Chobe Forest Reserve (CFR) (Botswana), conservation of 176–187; building poles 184; ecotourism 186–187; ethical composition 179–180; fencing poles 182, 183–184; firewood 182–183; forest dependency for livelihood needs 185–186; forest products, use of 182; kraal poles 182, 184–185; other resources 185; questionnaires and interviews 180–181; respondents' profile 181–182; Statistical Package for Social Sciences (SPSS) version 23 181; study setting, description 178; sustainable livelihood framework (SLF) 178; thatching grass 183; wild fruits 183
Chobe National Park (CNP), Botswana 222, 224, 225–226, 227, 257, 258–263, 274
Chobe River (Botswana) 178, 226
Choge, S. M. 248
Chongoni Art Rock 29

Chose Hills 28
CITES (Conservation on International Trade in Endangered Species of Wild Fauna and Flora) 42, 169, 171
citizen participation 54
Clarke, V. 181
Cleaver, F. 67
climate change: benefits to tourism 80–81; ecotourism industry and 193; and the environment 213; research in Southern Africa *see* Southern Africa, sustainable tourism in; tourism and 208; tourism-dependent communities' resilience to 55–56
climate-related stressors 80
Clinton, B. 224
coastal communities (South Africa) 212–213
Cohan, D. 152
Colbry, S. 143
community: contributions 125; cultural village activities 29; engagement 49; governance and money flow 171–172; resilience 55–56, 59–60; sustainability, investment in 58; tourism participation, active *vs.* passive 229–230
community-based conservation in Malawi *see* Malawi
community-based natural resource management (CBNRM) 52, 55, 130, 243; in Botswana 170; Chobe Forest Reserve (CFR) (Botswana) 179; in Mozambique 170; in Namibia 64, 71, 74, 75, 170–171; in Zambia 170; in Zimbabwe *see* Zimbabwe
community-based organisations (CBOs) 179, 227, 229, 230
community-based projects in Malawi *see* Malawi
community-based tourism (CBT) in Malawi 26–28; research methods 30–31; results and discussions 31–35; and sustainable tourism 29–30
community livelihoods: in Botswana *see* Botswana; Chobe Enclave Conservation Trust (CECT) 227, 229; forest resources in *see* Forest resources in community livelihoods; iSimangaliso Wetland Park, South Africa 206–208; in Lusaka National Park *see* Zambia, Lusaka National Park; rural livelihoods and forest reliance 176–177; United States *see* United States
community participation in Zimbabwe 143–153; CAMPFIRE programme 146–153; comparative documentary analysis 144–145; governance of nature conservation and ecotourism 145–146
consensus building 207
conservancy and CAMPFIRE (CBNRM) 271, 272
conservation agencies 119, 167
Conservation International (CI) 241
consumer price indices (CPIs) 239
contagious bovine pleuropneumonia (CBPP) 97
content analysis method 57
controlled hunting areas (CHAs) 97; *see also* hunting
Conway, G. 207
coral reefs, damage to 79
Corporate Finance Institute (CFI) (Botswana) 57
corporate social responsibility (CSR), poverty alleviation through 123–139; conceptual framework 126–127, 129–130; involvement of employees 125–126; Namibian case study 130, 134–138; role of CSR in the tourism sector 124–125
corporate social responsibility (CSR) initiatives 53
Cradle of Humankind (South Africa) 85
crop damage/raiding 99–101, 102, 106, 116, 118
Cuganesan, S. 57
Cukier, J. 126
cultural and heritage tourism 84–85
cultural capital 238–239
Cultural Strategy Group 240, 242, 244

Dar es Salaam, Tanzania 242
Darkoh, M. 98
deep-sea fishing 211
deforestation 15, 29, 32–35, 167–168, 194
De Jongh, D. 126
DeLacy, T. 59
de la Torre-Castro, M. 239, 248
desert tourism 81–82
destinations management 214
Devine, P. 243
Ditlev-Simonsen, C. 125–126
Dixie National Forest 224
Dodds, R. 134
Donija Cultural Village 29
Drakensberg and Table Mountain 81
drought 19, 56, 79, 80–83, 81, 82, 186, 193, 209, 212, 246, 248

dryland farming 98–99
Dube, N. 153

ecocentrism 118
eco-Label scheme 191
eco-lodges 199
economic sustainability, of handicraft sector 243–249; Botswana 245–247; defined 243; Kenya 247–248; Namibia 245; South Africa 244; Tanzania 248–249; *see also* handicrafts
ecotourism: Chobe Forest Reserve (CFR) (Botswana), conservation of 186–187; governance of 145–146; income generated from 164; in Lusaka National Park *see* Zambia, Lusaka National Park; Mauritius 195–196
education: educational programs 200–201; employee, and CSR 137–138; environmental tourism 33; influence on locals perception 116, 118
Elinor Ostrom 167
Elkington, J. 243
Ellis, F. 178
employees involvement and CSR 125–126, 134–137
employment: Namibia 72; nature-based tourism and 28; opportunities 21, 72, 130, 164, 206, 208, 212–214, 234, 236–237; self-employment opportunities 238
Enclave community 226
environmental degradation: due to community-based tourism 32–34; due to nature-based tourism 28; due to poverty 41–42
environment and societal wellbeing 56
Escalante 224
Ethiopia, forest income in 177
ethnical composition, Chobe Forest Reserve (CFR) (Botswana) 179–180
Etosha National Park (Namibia),) 38
Ezemvelo KZN Wildlife 212–213, 243

Fairer-Wessels, F. A. 6, 273
fair trade 125, 241
Fair Trade (certification) 201
Farmaki, A. 214
farming 211; arable 94–96, 98–102; in Kenya 248; livestock 102–105; mixed 14; mushroom 29; seaweed 249; subsistence 211; as sustainable livelihood 58, 229; in Tanzania 248–249
fencing poles 182, 183–184
financial capital 238
firewood 182–183, 185–186
fishing harvesting 96, 102–105
Fitchett, J. M. 5, 271
flooding 82, 84
flood recession farming 99
focus group discussions (FGDs) 65, 95–98, 101–102, 210–212
forest dependency 185–186
forest products 177, 181–183, 185
forest resources in community livelihoods 176–187; conceptual framework 178; discussion 185–186; ecotourism 186–187; materials and methods 178–181; results 181–185
Framework for Anti-Poverty Tourism (APT) 127–128
Frey, N. 134
Friends of Queen Elizabeth National Park 45
Fröcklin, S. 239, 248
Futing Liao 31
Fynbos biome 79

Game Scouts 46
game viewing 54, 83–84, 116
Garekae, H. 5
Gartner, C. 126
Gautam, K. H. 177
gender and perception 116, 118
George, R. 134, 242
ggplot2 package 111
Ghana, forest income in 177
Giddy, J. K. 83
Glen Canyon National Recreation Area 224
Global Compact Network Spain (GCNS) 53
Global Reporting Initiative (GRI) 53
Godwin, H. 269
Goeldner, C. 2
Gonarezhou National Park (Zimbabwe) 150
Goodwin, H. 124
Google Scholar (search engine) 145
Gorongosa National Park (Mozambique) 41–42
Grand Staircase-Escalante National Monument (GSENM) 222, 274–275; active *vs.* passive community tourism participation 229–230; AI process 227, 228; data collection process 227; discussion 232–233; natural resources, tourism and community livelihoods dynamics 227, 229; positive and negative outlooks, natural resources and tourism 230–231; study sites and methods 224–225

grass/reed collection 246
Greater St Lucia Wetland Park 208
Great Limpopo Transfrontier Park 82
Green Globe certification 197
greenhouse gas emissions 208
green productivity 198
groundwater depletion 212
GSENM *see* Grand Staircase-Escalante National Monument (GSENM)
Gumbo, B. 5
Guthrie, J. 57

habitat destruction 82, 167
Hall, D. 84
Hambira, W. L. 5, 82, 270, 271, 273
handicrafts, in Southern and Eastern Africa 235–250; Botswana 245–247; conceptual framework 237–239; craft industry development 236–237; data on economic contribution 239–240; defined 240; economic sustainability 243–249; Kenya 247–248; marketing 241–242; Namibia 245; resource use sustainability 242–243; role players 240–241; South Africa 244; sustainability 237–238; Tanzania 248–249
Harrison, E. P. 148
Healy, M. G. 42
Hebinck, P. 5
Hebink Kiaka 271
helicopter patrols 46
Helmsing, A. H. J. 148
Hess, C. 168
hiking 84
hippo and crocodile watching 211
Hluhluwe Dam (South Africa)
Hluhluwe-Umfolozi 80
Hoogendoorn, G. 5, 271
horseback riding 211
Hospitality and Tourism Association of Botswana (HATAB) website 57
Hotel Development Strategy (HDS), Mauritius 191, 194
hotel sector 57–58, 197
Hove, M. 152–153
Hulme, D. 269
human capital 239
human–wildlife conflicts 4, 34–35, 41, 43, 74, 103, 150
hunting 246; ban in Botswana 18–20, 163–165; in Botswana 11, 14–18, 226, 229, 232; rural livelihoods and 211–212; wildlife economy and 162–165, 169; in Zambia 149
Hurwitz, M. 143
Hwange case study (Zimbabwe) 152–153

incentives 126
Induna (Namibia) 73
Industrial Action Plan (DTI) 209
intensive conservation areas (ICA) (Zimbabwe) 167
intensive protection zone (IPZ) 46
Inter-governmental Panel on Climate Change (IPCC) 56
IRDNC (NGO) 68, 72
iSimangaliso Wetland Park, South Africa 205–215, 243; awareness of climate change and the preservation of the wetland park 211–212; case study 208–209; climate change and the environment 213; climate change and tourism 208; primary livelihood activities of the selected rural communities 211; research methods 210–211; skills and employment opportunities 212–213; skills development, employment opportunities and tourism developments 213–214; study sites 210; sustainable tourism development and community livelihoods 206–208
isiZulu 208, 210
Isle Aux Aigrettes, Mauritius 195–196
IUCN 162, 166–167

James, M. 84
Jänis, J. 245
Jiang, M. 59
Jiddawi, N. 239, 248
Jones, S. 269
Jozini Dam (South Africa) 212

Kafue National Park (KNP) (Zambia) 38, 255–263, 274
Kaján, E. 81
Kamphorst, D. 144
Kariakoo market 242
Kasane community (Botswana) 179
Kasane Forest Reserves (Botswana) 186, 226
Kenya: foreign exchange earnings 235; handicraft sector 237, 238, 240, 241, 242–243, 247–248; non-electrified fences in 101; political capital in 241
Kgalagadi Transfrontier Park regions 82
Kgathi, D. L. 97
ǂKhoadi ǁHôas Conservancy 68–69, 74–75
Kiaka, R. D. 68

Kilimanjaro 28, 79
Kilungu, H. 34
Kimaro, E. 5, 82, 273
knowledge, attitudes and practices (KAP) framework 111
knowledge exchange 48–49
Kolkol, crafters in 238
Kosi Bay: coastal communities of 212–213; rural communities of 210–211; women 213
Kosi Bay Lake system 208, 209
kraal poles 182, 184–185
Kruger National Park 38, 162; drought and water shortage in 82–83; eutrophic expansion of 80; revenue generated by 164
Kuehnel, J. 134
Kurebwa, J. 148
Kwazulu Natal, handicraft sector 244, 246
KZN Ezemvelo Wildlife 206, 209, 210–211
KZN Nature Conservation Authority 243

Lake Chilwa 28
Lake Malawi National Park 28, 30
Lake Sibaya 209
Lake St Lucia 209, 211, 212–213
Lamers, M. 269
land constraints, Mauritius 193
land-use conflicts 43; *see also* Okavango Delta
land-use policies 55
land uses, conflict and coexistence 93–106; contemporary land uses and livestock farming, conflict between 102–105; Okavango Delta, land-use conflicts in 98–102; Okavango Delta, land use in 96–98; study site and methods 94–96
Lane, B. 123
leadership 130, 207, 209
leather work 165, 242, 246, 247
Lepetu, J. 5
Lesotho 83
Lewis-Beck, M. S. 31
Likert scale 111, 134
Linyanti Craft centre 245
livelihood capitals 237–239
livestock farming 96, 102–105
livestock grazing 96
livestock predation: in Okavango Delta 103–104; in Shantumbu (Zambia) 108
Lockichar, crafters in 238
London Conventions of 1900 and 1933 166
Long, N. 68
Luangwa Valley 164
Lumbasi, J. A. 144
Lusaka National Park *see* Zambia, Lusaka National Park
Luvushi Manda National Park (Malawi) 38

Mackenzie, L. 6, 274
Madikwe Game Reserve 164
Mahenye case study (Zimbabwe) 151–152
Mahoney, D. 241
Makhitha, K. M. 236, 242
malaria-prone region of Southern Africa 79–80
Malawi 26–35; community-based tourism and sustainable tourism 29–30; nature-based tourism 28; overview 26–28; research methods 30–31; results and discussions 31–35
Mamo, Y. 177
Maphelane nature reserve 208
Maputaland marine reserve 209
marital status and locals' perception 116
marketing and promotion: of handicrafts in Southern and Eastern Africa 241–242; of National Parks in South Africa 43; of National Parks in Uganda 45
Mashi Craft Market 245
Mashinya, J. 152
masimo (land) 105
Mateke, S. M. 177
Mauritius 190–202; current tourism policies 194; ecotourism, challenges 192–194; ecotourism policies and regulations 191–192; financial constraints 199; interviews 195; island's topography 198; lack of stakeholders' collaboration 199; policy framework 195–200; as a resort destination 198–199; supply-side constraints 199–200
Mauritius Ile Durable (MID) concept 191
Mauritius Standards Bureau (MSB) 197–198
Mayeni people 72
Mbaiwa, J. E. 98, 236, 241, 270
Mearns, K. F. 147, 272
Measham, T. G. 144
meat production 164–165
media 165, 181, 213, 228, 254, 257
MET (NGO) 68, 71–72
micro and small enterprises (MSEs) 248
Midttun, A. 125–126
Milimo, C. 5, 272
Mkhuze River papyrus swamps 209

Mmopelwa, G. 6, 274
Mnangagwa, President (Zimbabwe) 165
Mogomotsi, K. 5, 272
molapo arable farming 99
moraka (cattle post) 105
moral management and poverty alleviation 130
Moremi Game Reserve (Botswana) 3
Mosi-oa-Tunya 255
Moswete, N. 82
Mothini, S. 248
Mozambique 166
Mpale Cultural Village 29
Mt. Carmel 224
Mtubatuba 210–213
Mubangizi, B. C. 243
Muchapondwa, E. 147
Muchemwa, C. 152–153
Mudimba, T. 147
Mukul, S. A. 185
Mulanje Mountains 28
Murphree, M. W. 162
Murphy, C. A. 245
Musavengane, R. 149, 152, 272
mushroom farming 29
mussel harvesting 211
Mutana, S. 149
Muzirambi, J. M. 147, 272
Mwenge handicraft centre 242
Mzipah Primary School 47

Nama Karoo 80–81
Namaqualand daisies 79
Namibia 64–76; communities in 171; conservancy politics and everyday life 75–76; handicraft sector in 236, 241–242, 245; institutional complexities 71–75; ǂKhoadi ǁHôas 68–69; theoretical framing 65, 67–68; wildlife ownership and game fencing 167; wildlife policies in 166; Wuparo 69, 71
Namibian Association of CBNRM Support Organisations (NACSO) 245
Namibian case study 130, 134–138
Namukonde, N. 5, 272
national aspirations indictor 58
natural capital 238
Natural Resources Act of 1941 (Zimbabwe) 167
Natural Resources Board (Zimbabwe) 167
nature-based tourism: capital essential for 56–57; impacts of drought on 82–83; impacts of tropical cyclones on 82; in Malawi 28; and national parks 81–83; in Southern Africa 80–81; *see also specific* country
nature conservation 145–146, 243
Nature Tourism and Climatic Change in Southern Africa (Preston-Whyte and Watson) 78
negative reciprocity 124–125
Nepal 186
Ngwerume, E. 152–153
Njerekai, C. 5, 271
Nkasa Lupala Lodge (Namibia) 74–75
Nkasa Ruparo National Park 68
non-timber forest products (NTFPs) 177, 187
North, D. 170
Nthiga, R. W. 270
Ntuli, H. 147
Nyambura, N. C. 240, 241
Nyaupane, G. P. 102
Nyawo, J. 243
Nyika National Park 28, 30
Nyirenda, V. 5, 34, 272
Nyumba ya Sanaa (House of Art) 242
Nzama, A. T. 207, 210, 212

occupation and locals' perception 116
ocean acidification 79
ocean warming 79
Okavango Delta 55, 82, 93–106; arable farming 98–102; farmer's perception on land uses 96–98; livestock farming 102–105; revenue generated by 164; site analysis 94–96; tourism in 236; wildlife resources in 93–94; *see also* Botswana; land uses, conflict and coexistence
Okello, M. M. 270
Omba Trust 242
Orderville 224
Ostrom, E. 168, 172
Ovahimbas (ethnic group) 130
Özkaynak, B. 243

park-based tourism 162–163, 254–264; data collection 257–258; discussion 261–263; overview 254–257; results 258–261
park management model 44
Parks and Wildlife Act of 1975 166–167
Peace (2010)'s observation 47
Pearls of Uganda 45
Petrified Forest (Namibia) 85
photo-tourism, potential of 162–165
physical capital 239

Pilanesberg 82
Pilanesburg park 164
Platteau, J. P. 67
poaching 32, 33, 42, 43, 46, 80
policy reforms 153–154, 161–162; integrating wildlife with economy *see* Southern Africa drylands wildlife economy; *see also* wildlife, policies
political capital 239, 241
political decentralisation 148–149
political leadership 207
poverty alleviation 59, 205, 207; and CSR *see* corporate social responsibility (CSR) initiatives
power devolution to local levels 148–149
precipitation patterns 80
predator-prey interactions in national parks 83
Preston-Whyte, R. 78–81, 271
Pringle, J. 161
Prinsloo, P. 126
Problem Animal Control (PAC) unit 100
Protected Areas Act of 2011 209
Prusak, L. 152
public awareness 108, 200, 212; *see also* education
public-private-partnership (PPP) approach 44–45

raffia-palm forests 209
Ram, Y. 84
Ramawati, P. I. 59
Ramsar sites 94, 206, 209
Rantšo, T. 28, 34
Raosoft sample size calculator 95
Reed Festival (iNcema) 213
Republic of Gabon 48
residency, length of 118–119
resource use sustainability, handicraft sector 242–243
revenue 6, 24, 45, 52, 83, 144, 164, 222; Botswana economy 52; Botswana National Tourism Policy 54; from CAMPFIRE *see* CAMPFIRE programme (Zimbabwe); generated by Victoria Falls 164; generated by Victoria Falls 164; Kruger National Park for 83; and nature-based tourism 28; nature tourism 6; park-based tourism 254, 262; public lands and resources through 229; Solimar International Integrated Development model 44, 45; tourism tax 190, 222; from wildlife 144, 171, 172
rewards 126
Ribot, J. C. 148
Rigby, D. 243
Ritchie, B. 2, 127, 129
rock art 85
Roe, D. 269
Rogerson, C. M. 240, 270
Roosevelt, T., President 166, 168
R statistical software 111
Rubin, H. J. 96
Rubin, I. S. 96
rural communities: empowering 55–56; *see also* community livelihoods; of Kosi Bay 210; primary livelihood activities of 211; *see also specific* country
rural development 207
Rural Development Council (RDC) 144, 148
Rural District Councils (RDCs) 146
rural household economies, forests and 176–177

Saarinen, J. 81, 82, 245, 269, 270, 271, 273
Saayman, A. 5, 272
Saayman, M. 5, 272
Sacchetti, S. 143
safari hunting 19, 20, 55, 147, 150–151
Sahlins, M. 124
Sandbrook, C. G. 270
Sannassee, R.V. 6, 273
San of Khwai residents (Botswana) 3
SANParks' business model 83
savanna ecoregion 80
Save River (Zimbabwe) 150
Scheyvens, R. 53
Scoones, I. 207, 246
Scopus (search engine) 145
scuba diving 79, 211
sea level rise, impact 84
Sebele, L. S. 34
Seetanah, B. 6, 273
Seno, S. K. O. 270
service sector 57–58
Seychelles 42, 48
Shackleton, C. M. 248
Shackleton, S. E. 248
Shale, L. 27–28, 34
Shannon, S. E. 57
Shantumbu (Zambia): animal species in LNP 113; attitudes, residents' 116–117; conservation strategies 119; data collection and analysis from inhabitants of 109–111; perceptions, residents' 113–114, 116, 118; practices and actions,

residents' 117; size and boundary 112–113; socio-demographics 112
Simatele, D. M. 149, 152
Sinha, S. 239
Sitati, H. 101, 247
Sjaastad, E. 177
skills training 205–208, 212–214
skin work 246
ski tourism industry at Afriski 83
small and medium enterprise (SME) 247
Small Island Development State (SIDS) 190
Smith, A. 169
snorkelling 42, 79, 211, 212
Snyman, S. 27
soapstone sculptures 241
social capital 238
social learning 214
social relations, defined 178
social upliftment: defined 53–54; practices *see* Botswana, community social upliftment practices
societal wellbeing 56
socio-cultural factors 11, 190, 206
socio-economic impacts 59
socio-economic segregation 54
socio-economic wellbeing 53
Sodwana Bay 209, 210–213
Solimar International Integrated Development model, The 44–45, 47
Soli-speaking people 110; *see also* Shantumbu (Zambia)
Somerville, K. 170
South Africa: firewood reliance 186; wildlife ownership and game fencing 167; wildlife policies in 166
Southern Africa: forest income in 177; handicrafts in 244–245; *see also* handicrafts
Southern Africa, sustainable tourism in 78–86; adventure tourism 83–84; beach tourism 84; climate change and tourism sustainability, Preston-Whyte and Watson's concerns 78–81; cultural and heritage tourism 84–85; nature-based tourism and national parks 81–83; regional climate threats to tourism sub-sectors 81
Southern Africa drylands wildlife economy 161–173; colonial policies of public ownership 166–167; community-based natural resource management 169–172; methods 162; wildlife as a private good 168–169; wildlife economy 162–165
Southern African Sustainable Use Specialist Group 162
South Luangwa National Park (Zambia) 162–163, 255, 256, 258–263, 274
species domestication 187
Spenceley, A. 124
SSR Botanical Garden, Mauritius 195–196
stakeholder collaboration 49
Stanley and Livingstone private game park (Zimbabwe) 42, 45–46
StatsSA 214
St Francis Bay 84
Stone, L. S. 270
Stone, M. T. 6, 102, 270
Stone, T. M. 270
storm damage 82, 84
subsidiarity 169
Suich, H. 245
Sustainable Coastal Communities and Ecosystems (SUCCESS) 248
sustainable development goals (SDGs) 53, 123, 205
sustainable ecotourism, Mauritius 195–196
sustainable livelihood development 212–213
sustainable livelihood framework (SLF) 178
sustainable tourism: defined 206–207; development 1–3, 109, 212–213; development, Botswana *vs.* USA 223
"Sustainable Tourism Management in African National Parks" conference 39
Swaziland 166

Tambwe, M. A. 242
Tanzania, handicraft sector 237, 238, 242, 248–249
Tao, T. C. H. 207, 214
tattoo rocks 28
Taylor, F. 177
Terry, M. E. 246
Tervo-Kankare, K. 82
Thapa, B. 6, 274–275
thatching grass 183
Tichaawa 147
Tonga ethnic group (Zimbabwe) 149
Tongaland, turtle beaches/coral reefs of 209
tour-guiding activities 29
Tourism, Recreation and Climate Change (Higham) 78
tourism climatic index 82, 84
tourism governance structures 214
Tourism Land Bank (Botswana) 55
tourism multinational companies *see* Botswana, community social upliftment practices

Trans-frontier Conservation Areas (TFCAs)/ 'Peace 'Parks' 38
TripAdvisor reviews 82–83
trophy hunting: in Botswana 13, 43; controversy about 162; Namibia 64, 65, 67, 68, 69, 73–74; in national parks 42–43
tropical cyclone Idai 82
tropical rainforest regions 177
Trust for Okavango Cultural and Development Initiatives (TOCADI) 241
Tsiseb Conservancy in Namibia 82
turtle beaches/coral reefs of Tongaland 209

Uberhuaga, P. 177
Uganda: canopy and guerilla tourism in 48; Solimar International Integrated Development model in 44–46
Umfolozi River 209
UN Climate Change Conference in Durban 2011 213
UN Conference of Environment and Development World Sustainability Forum, 2002 209
UNCTADSTAT 239–240
United Nations Conference on Trade and Development (UNCTAD) 54
United Nations Development Programme (UNDP) 57, 144
United Nations Environment Program (UNEP) 56
United Nations World Tourism Organisation (UNWTO) 39, 49, 53, 193, 206–207
United States, natural resources, tourism and community livelihoods 221–233; active *vs.* passive community tourism participation 229–230; AI approach 226–227, 228; background 222–223; data collection and analysis 226–227; discussion 232–233; dynamics 227, 229; GSENM *see* Grand Staircase-Escalante National Monument (GSENM); overview 221–222; positive and negative outlooks 230–231; study sites and methods 224–225

Valley Ferney, Mauritius 195–196
Van der Duim, R. 269
Van Wijk, J. 269
Vedeld, P. 177
veterinary fences 97, 98, 102
Victoria Falls (Zambia and Zimbabwe) 47, 85, 164, 255
Victoria Falls National Park (Zimbabwe) 42
Vigor 42
Visser, G. 126
vocational skills training 212–213; *see also* skills training
Voices from the Frontline (Botswana) 164
Vredefort Crater (South Africa) 85
vulnerability context, production of handicrafts 237–238

Wall, G. 207, 214
Walpole, M. J. 101
Wambaa, J. 242
Ward, C. 57
waste disposal, Mauritius 193
water shortage 212
Watson, D. J. 78–81
Watson, H. K. 271
Web of Science (search engine) 145
Wetland of International Importance 94
whitewater tourism in southern Africa 83
wild fruits 183, 186
wildlife 161, 167; conservation in Lusaka National Park *see* Zambia, Lusaka National Park; ownership 168–169; policies 166–167, 169; privatization within protected areas 45–47; products 162; resources and nature tourism *see* community-based tourism (CBT) in Malawi; resources in Okavango Delta 93; trade bans 171
wildlife economy 162–165; communities and ecosystem management 164, 165; hunting and 163–165; meat production and 164; park-based tourism 162–163; wildlife products and 165
wildlife management areas (WMAs) 97, 102
wild products collection 246
World Bank internet sources 239
World Bank's "attacking poverty" framework 129
World Business Council for Sustainable Development 125
World Commission on Environment and Development 56
World Heritage Convention Act 208
World Tourism Organization 223

World Tourism & Travel Council (WTTC) 52–53
World Travel and Tourism Council 207
Wuparo Conservancy 69, 71; bureaucracy 74–75; commoners 73–74

Yasarata, M. 212

Zambezi River (Zimbabwe) 149; flow reduction 79; regions surrounding 80
Zambia: handicraft sector 241; park-based tourism 255–263
Zambia, Lusaka National Park 108–119; discussion 117–119; methods and materials 109–111; results 112–117
Zhao, W. 127, 129
Zimbabwe 143–153; finding and discussion 145–153; handicraft sector 241; methodology 144–145; national parks in 38; wildlife ownership, legislation for 168–169; wildlife policies in 166; wildlife ranches and CAMPFIRE areas in 164
Zimbabwe's Parks and Wildlife Authority Act of 2014 169

For Product Safety Concerns and Information please contact our EU representative GPSR@taylorandfrancis.com
Taylor & Francis Verlag GmbH, Kaufingerstraße 24, 80331 München, Germany

www.ingramcontent.com/pod-product-compliance
Lightning Source LLC
LaVergne TN
LVHW011116300926
849160LV00008B/403
9781032085173